Agamemnon Despopoulos · Stefan Silbernagl

Color Atlas
of Physiology

145 Color Plates
by Wolf-Rüdiger Gay and Barbara Gay

1981
Georg Thieme Verlag Stuttgart · New York

Professor Dr. *Agamemnon Despopoulos*
Department Medizin
Ciba Geigy
CH-4004 Basel, Switzerland

Professor Dr. *Stefan Silbernagl*
Physiologisches Institut der Universität
Fritz-Pregl-Str. 3
A-6010 Innsbruck, Austria

Wolf-Rüdiger Gay and *Barbara Gay*, Stuttgart, FRG

Deutsche Bibliothek Cataloguing in Publication Data

Despopoulos, Agamemnon:
Color atlas of physiology / Agamemnon Despopoulos ; Stefan Silbernagl. 145 color
plates by Wolf-Rüdiger Gay and Barbara Gay. – Stuttgart ; New York : Thieme, 1981.
 Dt. Ausg. u.d.T.: Silbernagl, Stefan:
 Taschenatlas der Physiologie
 ISBN 3-13-545001-5

NE: Silbernagl, Stefan:

Important Notice: Medicine is an ever-changing science. As new research and
clinical experience broaden our knowledge, changes in treatment and drug therapy
are required. The editors and the publisher of this work have made every effort to
ensure that the drug dosage schedules herein are accurate and in accord with the
standards accepted at the time of publication. Readers are advised, however, to
check the product information sheet included in the package of each drug they plan
to administer to be certain that changes have not been made in the recommended
dose or in the contraindications for administration. This recommendation is of
particular importance with regard to new or infrequently used drugs.

This book is an authorized translation from the 1st German edition published and
copyrighted 1979 by Georg Thieme Verlag, Stuttgart, Germany. Title of the German
edition: Taschenatlas der Physiologie.

© 1981 Georg Thieme Verlag, Herdweg 63, 7000 Stuttgart 1, FRG

Typesetting by Tutte Druckerei GmbH, Salzweg-Passau, FRG (Monophoto 400)
Printed in West Germany by Aumüller, Regensburg

ISBN 3-13-545001-5   5 4 3 2 1 0

Preface

In the modern world, visual pathways have outdistanced other avenues for informational input. This book takes advantage of the economy of visual representation to indicate the simultaneity and multiplicity of physiological phenomena. Although some subjects lend themselves more readily than others to this treatment, inclusive rather than selective coverage of the key elements of physiology has been attempted.

Clearly, this book of little more than 300 pages, only half of which are textual, cannot be considered as a primary source for the serious student of physiology. Nevertheless, it does contain most of the basic principles and facts taught in a medical school introductory course. Each unit of text and illustration can serve initially as an overview for introduction to the subject and subsequently as a concise review of the material. The contents are as current as the publishing art permits and include both classical information for the beginning student as well as recent details and trends for the advanced student.

A book of this nature is inevitably derivative, but many of the representations are new and, we hope, innovative. A number of people have contributed directly and indirectly to the completion of this volume, but none more than *Sarah Jones**, who gave much more than editorial assistance. Acknowledgement of helpful criticism and advice is due also to Drs. *R. Greger*, *A. Ratner*, *J. Weiss*, and *S. Wood*, and Prof. *H. Seller*. We are grateful to *Joy Wieser* for her help in checking the proofs. *Wolf-Rüdiger* and *Barbara Gay* are especially recognized, not only for their art work, but for their conceptual contributions as well. The publishers, Georg Thieme Verlag and Deutscher Taschenbuch Verlag, contributed valuable assistance based on extensive experience; an author could wish for no better relationship. Finally, special recognition to Dr. *Walter Kumpmann* for inspiring the project and for his unquestioning confidence in the authors.

Basel and Innsbruck, Summer 1979
*Agamemnon Despopoulos**
Stefan Silbernagl

* On November 2nd, 1979, *Agamemnon Despopoulos* and his wife *Sarah Jones-Despopoulos* started from Bizerta/Tunisia in their 30 foot-sail boat intending to cross the Atlantic. Since then they have been missing. There is but little hope of seeing them again.

December 1980
Stefan Silbernagl

Table of Contents

2   Table of Contents

4   Table of Contents

"... if we break up a living organism by isolating its different parts it is only for the sake of ease in analysis and by no means in order to conceive them separately. Indeed when we wish to ascribe to a physiological quality its value and true significance we must always refer it to this whole and draw our final conclusions only in relation to its effects in the whole."

Claude Bernard

The science of **physiology** originated in the study of function in normal animals. Later, the development of invasive techniques permitted the examination of normal functions of smaller and smaller subunits: organ systems, organs, cells, and subcellular organelles. Two auxiliary sciences developed from physiology, namely, **biochemistry** which deals with chemical reactions of substances originating within the organism, and **pharmacology**, which deals with chemical reactions of substances originating in the environment (xenobiotics).

SI Units

The understanding of any process depends upon the precision with which it can be measured; however, expression of a process in quantitative terms requires agreement among investigators on the magnitude and meaning of fundamental units. The variety of units used in the past contributed to confusion and misunderstanding.

The International System of Units was adopted in 1960 in an attempt at standardization. The SI is founded on **seven basic units**:

	Unit	Symbol
Length	meter	m
Mass	kilogram	kg
Time	second	s
Electrical current	ampere	A
Temperature	Kelvin	K
Luminous intensity	candela	cd
Amount of substance	mole	mol

Multiples and submultiples of the basic units are described by the following prefixes. Multiples of 3 are preferred; the prefixes deca-, hecto-, deci-, and centi- are discouraged.

Multiples			Submultiples		
Prefix	Symbol	Value	Prefix	Symbol	Value
(deca-	da	10^1)	(deci-	d	10^{-1})
(hecto-	h	10^2)	(centi-	c	10^{-2})
kilo-	k	10^3	milli-	m	10^{-3}
mega-	M	10^6	micro-	μ	10^{-6}
giga-	G	10^9	nano-	n	10^{-9}
tera-	T	10^{12}	pico-	p	10^{-12}
peta-	P	10^{15}	femto-	f	10^{-15}
exa-	E	10^{18}	atto-	a	10^{-18}

There is only one unit for each physical quantity. Those quantities not included in the basic units may be described by **derived units**. These are formed by multiplication or division of the basic units but without use of numeric factors. Denominators are expressed preferably as negative exponents rather than as fractions. Amount of substance (mol) is preferred to mass (kg), especially in expressing concentrations; $mol \times l^{-1}$ is preferred to $kg \times l^{-1}$. Certain terms not included in the basic units have been retained provisionally (liter, bar). Liter is now defined as 1 cubic decimeter (dm^3) instead of as the volume of 1 kg of water at 4 °C.

Length, Area, Volume

The basic unit for length is the meter (m).

Conversions of other units are:

Ångström = 10^{-10} m = 0.1 nm
Micron   = 10^{-6} m = 1 µm
Inch   = 25.4 mm
Foot   = 0.3048 m
Yard   = 0.9144 m
Mile   = 1.609 km

Area $(m \times m = m^2)$ and   volume $(m \times m \times m = m^3)$ are derived units.

For fluids and gases, the special unit, liter (l), is used for volumes

1 l   = 10^{-3} m^3 = 1 dm^3
1 ml = 1 cm^3

Conversions of other units are:

1 fluid ounce (USA) = 29.57 ml
1 fluid ounce (GB) = 28.47 ml
1 gallon (USA)   = 3.785 l
1 gallon (GB)   = 4.54 l

Frequency, Velocity, Acceleration

Frequency refers to how often a periodic event occurs and is given as a time unit in Hertz (Hz) as s^{-1}. Expressed in minutes, min^{-1} = 1/60 Hz ≈ 0.0167 Hz.

Velocity of a mass (such as an automobile) is distance per time $(m \times s^{-1})$. Movement of fluids is described both by linear velocity as $m \times s^{-1}$ and by "volume velocity" or rate of flow as $l \times s^{-1}$ or $m^3 \times s^{-1}$.

Acceleration refers to rate of change of velocity or $m \times s^{-1} \times s^{-1} = m \times s^{-2}$. A negative value refers to deceleration.

Force and Pressure

Force is acceleration times mass, or $kg \times m \times s^{-2}$; the unit of force is the newton (N). Conversions of other units are:

1 dyne = 10^{-5} N = 10 µN
1 pond = 9.8 mN

Pressure is force exerted on an area or $N \times m^{-2}$; the unit of pressure is the pascal (Pa). Conversions of other units are:

1 mmHg = 1 Torr = 133.3 Pa = 0.133 kPa
1 cm H$_2$O = 98 Pa
1 technical atmosphere (at) = 98.067 kPa
1 physical atmosphere (atm) = 101.324 kPa
1 dyne/cm^2 = 0.1 Pa
1 bar = 100 kPa

Work, Energy, Heat, Power

Work is force times distance, $(N \times m)$, or pressure × volume $(N \times m^{-2} \times m^3 = N \times m)$; the unit is the joule (J).

Energy and **heat** have the same unit, the joule. Conversion of other units are:

1 erg   = 10^{-7} J = 0.1 µJ
1 calorie   ≈ 4.185 J
1 kcalorie   ≈ 4.185 kJ
1 Watt second = 1 J
1 kilowatt hour = 3600 kJ = 3.6 MJ

Power is work per time, or $J \times s^{-1}$ = watt (W). Conversions of other units are:

1 erg/s   = 10^{-7} W = 0.1 µW
1 calorie/h   = 1.163 mW

Mass, Quantity, Concentration

The unit of **mass** is the kilogram (kg).

Quantity of a substance is the molecular weight expressed in grams, the gram molecular weight or **mole**. Quantity of atoms or ions is also expressed in moles, though ions may be expressed as equivalents, or mole × ionic charge.

Example: 1 mol Ca^{2+} = 2 equivalents Ca^{2+}.

Concentration refers to various relationships:

Mass per volume   = $g \times l^{-1}$
Amount per volume = $mol \times l^{-1}$
Mass per mass   = $g \times g^{-1}$ = 1
Volume per volume = $l \times l^{-1}$ = 1

The last two relationships are relative or *fractional concentrations*.

Examples of fractional concentrations:

1 % = 1 part in 100 parts
1 ‰ = 1 part in 1.000 parts
1 vol % = 1 volume in 100 volumes
1 ppm = 1 part per million parts

Temperature

The SI unit for temperature is the kelvin (K). On the kelvin scale, 0 K is absolute zero. Conversions to other units are:

Celsius: $°C = K - 273.15$
Conversion of °F to °C and vice versa are:
Fahrenheit: $°F = (9/5 °C) + 32$
Celsius: $°C = 5/9 (°F - 32)$

Electrical Units

Current is the movement of electrons or ions; current strength is determined by the number of electrons moving past a point per unit time. The unit is the ampere (A). Current can flow from one point to another only when a **potential difference** exists across the two points. The unit is the volt $(V = W \times A^{-1})$. A potential difference occurs when there is an inequality of distribution of ionic charge, as in a battery or across a cell membrane. **Resistance** to flow of current is measured in ohms (Ω). The reciprocal of resistance is **conductivity**, measured in Siemens $(S = 1/\Omega)$.

The conductivity of ions is a measure of their ability to permeate a membrane.

Electrical work or electrical energy is described in joules (J) or watt seconds (Ws). **Power** is expressed in watts (W).

Direct current (DC) has a fixed polarity, but alternating current (AC) changes its polarity at a fixed frequency, which is described in Hertz (Hz)

Solutions

Solutions are composed of **solutes** dissolved in a solvent, most commonly water.

Avogadro's law states that equal volumes of substances in gaseous or in liquid form at standard conditions of temperature and pressure contain an equal number of molecules.

Avogadro's number. One mole of a substance has a volume of 22.4 l at standard conditions of temperature and pressure and comprises $6.02 \times \times 10^{23}$ molecules. One mole of a substance is the molecular weight of this substance expressed in grams. Example: 1 mol NaCl = 58 g.

Solutions can be prepared to contain the same number of **solute** molecules either per unit volume of solution (molar solution) or per unit number of solvent molecules (molal solution). A 1 mol**ar** solution contains 1 mol of solute in 1 l of solution. A 1 mol**al** solution contains 1 mol of solute in 1 kg of solvent.

The **colligative properties** of water depend on the number of solutes in solution. These properties include freezing and boiling points, vapor pressure, and osmotic pressure. A solute dissolved in water affects the colligative properties according to its concentration rather than to its chemical characteristics; thus, a solution of $1 \, mol \times l^{-1}$ urea produces the same effects as a solution of $1 \, mol \times l^{-1}$ glucose. Solutions may be described by the unit osmolarity $(Osm \times l^{-1})$ or milliosmolarity $(mOsm \times l^{-1})$. Osmolarity and osmolality, like molarity and molality, are determined by the concentration of osmoles per liter of solution and per kg of solution, respectively.

At 37 °C a solution of 1 Osm × l⁻¹ exerts an osmotic pressure of 2.58 MPa (c. 1 atm) and has a freezing point of -1.86 °C and a boiling point of 100.51 °C. The osmolar concentration of body fluids is c. 300 mOsm × l⁻¹ and is maintained relatively constant by intricate physiologic mechanisms. Osmolarity in the extracellular (ECF) and intracellular fluids (ICF) is the same even though the chemical compositions of these fluid compartments differ.

Osmolarity is readily determined by measuring the freezing point of a solution; other methods are more cumbersome. A solution of 1 Osm × l⁻¹ freezes at −1,86 °C; a 2 osmolar solution freezes at −3.72 °C (2 × 1.86). Solutions of nondissociable solutes have the same osmolarity as molarity. Example: glucose 1 mol × l⁻¹ = 1 Osm × l⁻¹. Solutions of dissociable solutes behave differently because they ionize: each ion acts as a discrete solute particle. If a solution of NaCl could ionize completely, a 1 mol × l⁻¹ solution would equal 2 Osm × l⁻¹ because of the combined action of Na⁺ and Cl⁻ions. In reality, the extent of ionization is dependent upon the concentration and is complete only at infinite dilution. The fraction ionized at any concentration can be calculated from the difference between the theoretical freezing point (if ionization were complete) and the actual freezing point.

According to the Brønsted-Lowry theory, an **acid** is a substance with a tendency to lose a proton, whereas a **base** is a substance with a tendency to gain a proton. This definition suffices for aqueous biologic systems.

In aqueous solutions, a weak acid (HA) is in equilibrium with its dissociated components: $HA \rightleftarrows H^+ + A^-$. The dissociation constant of the acid, $K = [H^+] \times \frac{[A^-]}{[HA]}$. From this relationship, the Henderson-Hasselbalch equation can be derived: $pH = pK + \log \frac{[A^-]}{[HA]}$ where A^- is the salt form and HA the acid form ($\rightarrow$ p. 98). When H⁺ions are added to the salt of a weak acid, the titration curve (**Fig. 1**) initially will show large changes in pH for small additions of H⁺, as the salt form A^- accepts H⁺ to form acid HA. At this early stage, the concentration ratio $[A^-]/[HA]$ is large. As the concentration ratio approaches 1.0, the solution resists a change in pH. Lar-

ger quantities of H⁺ produce smaller changes in pH. At a concentration ratio $[A^-]/[HA] = 1.0$, the pH of the solution equals the pKₐ. As more H⁺ is added, the pH falls below the pKa and the concentration ratio $[A^-]/[HA]$ becomes smaller and smaller. Once again, small quantities of H⁺ produce large changes in pH. The titration curve shows that the greatest resistance to a change in pH occurs near the pK_a (**Fig. 1**).

A solution containing a weak acid and its salt is a **buffer solution**; the weak acid and its salt are a **buffer pair**. A buffer solution offers the greatest resistance to changes in H⁺ when the pH of the solution is closest to the pK_a of the weak acid. At the pK_a, the ratio of $[A^-]/[HA]$ equals 1.0. At ± 1.0 pH unit from the pK_a, the ratio of $[A^-]/[HA]$ is either 1/10 or 10/1 and indicates that the buffer pair has used up 90% of its buffering capacity. At ± 2.0 pH units from the pK_a, the ratios are 1/100 or 100/1, and the buffer resists only very small changes in H⁺.

Properties of Living Systems

A living system is one that actively exchanges substances with the environment. Preservation and perpetuation of life processes require regulation of the rates of exchange. As the organism becomes more complex, sensitivity and specificity of the regulatory mechanisms increase. Claude Bernard, the pioneer French physiologist of the 19th century, wrote of the stability that the organism maintains in its body fluids: "All the vital mechanisms, however varied they may be, have only one object, that of preserving constant the conditions of life in the internal environment."

Integration

Complex organisms are organized to generate an integrated, consistent, and uniform response to a stimulus. The central nervous system (CNS) is the ultimate integrator. It receives and transmits both electric (neural) and chemical (humoral) messages. Neural input to the CNS is predominantly sensory, whereas output is chiefly motoric. Sensory input is coded under several categories ($\rightarrow$ p. 253): (1) **quality** of the stimulus, according to the form of stimulus energy, is detected by sensory receptors: touch, sight, smell, taste and hearing; (2) quantity or **intensity** of the stimulus is recognized by the extent receptors become stimulated and the frequency of stimulation of the nerves originating from the receptor; (3) **localization** of the stimulus, or its direction, is recognized by segmental and symmetric distribution of various receptors.

Control of responses to stimuli is achieved by reflex arcs, biologic rhythms, and feedback loops ($\rightarrow$ p. 218):

$$\text{afferent} \overset{\nearrow \text{ integration } \searrow}{\underset{\longleftarrow \text{ feedback} \longleftarrow}{}} \text{efferent.}$$

Defense

The organism has three major **defensive responses** against changes in the internal environment:

Homeostasis, a term introduced by Walter Cannon, has come to refer both to the constancy (steady state) of body processes and to the mechanisms by which constancy is achieved. A homeostatic mechanism has several characteristics: there is a systemic pool (content) with input (gain) and output (loss); there are, in addition, a means to detect changes in size of the pool (detector), a process to adjust gain or loss (rectification), and a signaling

Concentration ratio of the buffer pair

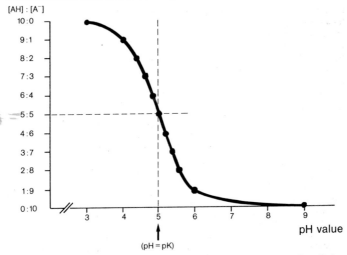

Fig. 1   Titration curve of an acid (HA) and its buffer base (A⁻). The point of inflection of the sigmoid curve indicates a pK_a of 5.0. Optimal buffering or the greatest resistance to change in pH occurs within a range of $pK_a \pm 1$ pH unit. At the limits of this range, the concentration ratio of acid and buffer base is 1:9 or 9:1 indicating that the buffering capacity is 90% utilized.

pathway (feedback) between the detector and the rectifier. When a change occurs, reactions are initiated to restore the original functional level. In some pathologic states, although a homeostatic mechanism may still operate, it may be set at a different (abnormal) functional level (example: fever).

Immune responses resist changes that occur when the organism is invaded by foreign chemicals or organisms (→ p. 60).

Stress response. Stress is a term introduced by Selye c. 1940 to describe characteristic physiologic responses to a variety of stimuli. The term has been so greatly misused that it has lost much of its original significance. In physics, **stress** represents the applied or external force (stimulus) and **strain** the deformation (response). In physiology, stress has been variously used to define both the stimulus and the response. The organism responds to external emergencies with a "fight or flight" response. Stress occurs when the physiologic activity of the organism increases faster than the adaptive responses. For the immediate reaction to a stress, many biologic substances are generated and made available (e.g., epinephrine, glucose). Simultaneously, restorative processes that have a longer latency are activated. These achieve maximal effect after the emergency has ended. They serve to restore supplies of the substances consumed during the immediate response. In "**chronic stress**," the continuing activity of the acute stress response may produce breakdown of control systems with pathologic changes.

Cell Theory

1. All free-living organisms are composed of cells and their products.
2. All cells are basically similar in their chemical construction.
3. New cells are formed from preexisting cells by cell division.

4. The activity of an organism is the sum of activities and interactions of its cells.

The cell is the structural unit of living organisms (→ **A–C**). It is surrounded by a **cell membrane** within which are found the cell **cytoplasm**, or matrix, and the subcellular structures, the cell **organelles**.

Cells may be described as **prokaryotic** or **eukaryotic**. Prokaryotic cells, such as bacteria, have a minimum of internal organization and no membrane-bound intracellular organelles. The genetic material, desoxyribonucleic acid (DNA), is distributed throughout the cytoplasm.

A eukaryotic cell has many characteristic membrane-limited intracellular organelles. The genetic material is confined within the nucleus. Respiration is associated with the mitochondria, and reproduction involves meiosis and mitosis.

The cell membrane is a complex plastic structure that maintains the internal environment of the cell. It has the additional important functions of recognizing extracellular messengers and of transmitting the information they carry to intracellular organelles. The membrane furnishes a lipid barrier to penetration which is irregularly interrupted by a protein mosaic. Penetration of the membrane by extracellular solutes may occur by **nonspecific diffusion** of solutes through the phospholipid bulk of the membrane or by **specific transport** involving receptor protein molecules in the membrane.

Almost half of the cell membrane mass comprises lipids of three classes: **phospholipids**, which are the major class, **neutral lipids**, of which cholesterol is the chief representative, and **glycolipids**. All of the lipids are **amphipaths**, i.e., they have a polar or hydrophilic head and a nonpolar or hydrophobic tail.

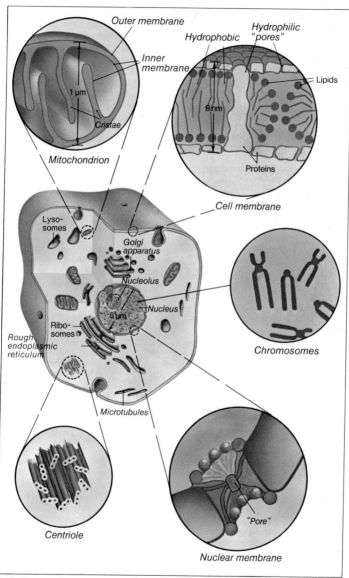

Outer membrane

Inner membrane

1 µm

Cristae

Mitochondrion

Hydrophilic "pores"

Hydrophobic

Lipids

9 nm

Proteins

Cell membrane

Lyso-somes

Golgi apparatus

Nucleolus

Nucleus

5 µm

Ribo-somes

Rough endoplasmic reticulum

Chromosomes

Microtubules

Centriole

"Pore"

Nuclear membrane

A. Cell organelles (schematic)

The phospholipids provide the structural framework of the membrane. In an aqueous environment, phospholipids spontaneously form a **double layer** in which the nonpolar tails are directed inward, toward each other, in a palisade-like formation. This arrangement leaves the polar heads jutting out of the surfaces of the bilayer toward the aqueous environment.

The lipid molecules in each half of the bilayer are free to move laterally, having an exchange rate of c. 10^6 times per second. In contrast, exchange from one monolayer to the other, a **flip-flop** exchange, is unusual. Since the two monolayers of the cell membrane differ in their lipid compositions and are therefore structurally asymmetric, the restriction of flip-flop exchange serves to maintain the asymmetry.

A bilayer composed only of phospholipids will change from a liquid to a crystalline (gel) phase over a broad temperature range. Introduction of cholesterol into the bilayer partly immobilizes those parts of the phospholipid chains closest to the outer membrane surface and hinders the gel transition. Thus, in the presence of cholesterol, fluidity of phospholipids is maintained, but their lateral mobility is restricted. The function of glycolipids in the membrane is obscure but they have been implicated as receptors for bacterial toxins.

More than half of the membrane mass is protein. The protein components of the membrane are responsible for (1) receiving and transducing specific extracellular signals carried by messengers, such as hormones, neurotransmitters ($\rightarrow$ p. 212), growth factors, antigens; (2) forming junctions with adjacent cells; (3) exocytosis or secretion of cellular contents into the environment; (4) transport of solutes across the membrane.

Some proteins (extrinsic) adhere to the membrane surface, while others (intrinsic) are built into the bilayer. Protein composition of the bilayer is asymmetric. Intrinsic proteins extend through the bilayer; some of these are exposed at both surfaces of the bilayer; others are limited to a monolayer but are in contact with adjacent proteins that extend across the layers.

Proteins tend to immobilize a collar of surrounding lipid molecules, the **boundary lipids,** and thereby create specific topographic units. These units form a mosaic of hydrophilic areas that may act as channels to conduct electrolytes across the membranes. Many proteins in the membrane are mobile; they can diffuse laterally in the membrane or can rotate on an axis perpendicular to the membrane.

The concentration of receptor proteins on the membrane can be **modulated** by the availability of messenger substrates. When a membrane protein binds with an extracellular messenger, it may redistribute in any of three ways: (1) **patch formation** is the passive cross-linking aggregation of several macromolecules in the plane of the membrane; (2) **cap formation** is the active segregation of the polar ends from other membrane components; (3) **pinocytosis** (v.i.) results in active removal of the protein complex from the membrane; the process is reversible when the cell synthesizes new protein to replace that which has been removed.

Exocytosis ($\rightarrow$ **B**) is an energy-dependent process whereby intracellular substances are extruded into the extracellular space. The substance is encapsulated in a membrane-lined vesicle. The vesicle migrates to the cell membrane, with which it fuses, opens to the outside, and discharges its contents.

Endocytosis is the reverse process by which bulk material, either solid or in solution, is taken into the cell.

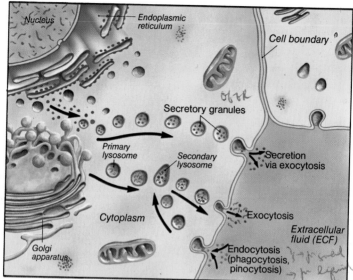

B. Endocytosis and exocytosis

Pinocytosis is a process of endocytosis for liquids: the cell membrane engulfs a droplet of extracellular fluid and discharges it into the cyctoplasm in a membrane-lined vesicle.

Phagocytosis is a similar process for solid matter.

Organelles (→A)

The **ribosomes** are the sites where protein synthesis takes place. They are found either free in the cytoplasm or bound to the endoplasmic reticulum. They consist of two subunits and contain transcripts (ribonucleic acid, RNA) of the nuclear DNA.

The **mitochondria** are the power-house of the cell. They contain enzymes of the citric acid (Krebs) cycle and of the respiratory pathway. They are the principal site for oxidative reactions that generate energy. The energy thus produced is stored in chemical form in the adenosine triphosphate (ATP) molecule. Synthesis of ATP provides almost all of the energy stores of the body; breakdown of ATP by various enzymes (phosphatases, ATPases) liberates energy for utilization in cellular reactions. The mitochondria also contain ribosomes and can synthesize certain proteins.

The **Golgi apparatus** is an extensive membrane system within the cell that includes large and small vesicles. It processes the proteins synthesized within the cell and prepares them for excretion. It also participates in synthesis of glycolipid components of membranes, particularly in nerve, kidney, and spleen.

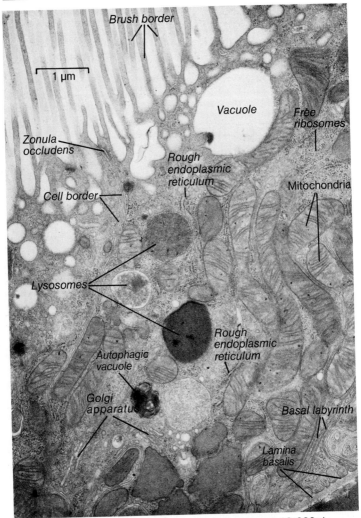

C. Cell structure, electron micrograph (enlargement 13,000 x)
(Proximal tubule cell of rat kidney; micrograph by Dr. W. Pfaller, Innsbruck)

The **endoplasmic reticulum** (ER) is an extensive membrane network within the cell. Two forms may be recognized: the **rough** ER, which is associated with the ribosomes, and the **smooth** ER, which is not. Proteins synthesized on the ribosomes penetrate the membrane of the ER into its lumen. They are then carried by secretion vesicles to the cell borders, where the vesicles fuse with the cell membrane and discharge their contents to the outside.

Lysosomes are dense bodies containing a number of hydrolytic enzymes. Synthesis of lysosomes is at least partly associated with the Golgi apparatus. Their enzymes contribute to the breakdown of damaged cell components and of endocytized material.

Other organelles whose action is less well understood include **microfilaments** and **microtubules**.

The **nucleus** is a complex structure that contains the genetic material of the cell. It is essential for cell reproduction. It contains the **chromosomes**, which are composed of DNA and several specialized proteins organized into **chromatin**. Two main classes of specialized proteins are: (1) **histones**, strongly basic proteins that vary little in structure from one tissue to another and **non-histone proteins**, which are much more heterogeneous; there may be more than 500 species of non-histone proteins in a single nucleus, with considerable variation among tissues.

General Aspects of Genetic Expression

DNA is a double helical strand held by bonds between paired nucleotides. Only four nucleotides are represented: adenosine binds only with thymidine and guanosine only with cytidine. The nucleotide composition in the two strands is therefore complementary; the structure of one strand determines that of the other. RNA is a single strand in which nucleotide units corresponding to thymidine in DNA are replaced by uridine.

The process of protein synthesis is fundamentally a transfer of information, initially encoded in the gene (DNA) as a polynucleotide to form the final protein, a polyamino acid. It has been estimated that a typical cell synthesizes c. 100,000 different proteins during its life cycle.

The first step in protein synthesis, **transcription**, requires formation of RNA in the nucleus according to the informational template on the gene (DNA). Formation of RNA is controlled by a **polymerase**, which is normally prevented from acting by a repressor protein on the DNA and is activated when the repressor is removed (derepression). This precursor of messenger RNA (mRNA) undergoes excision and rejoining of selected segments and modification of its terminals in the second intranuclear step, **post-transcriptional modification**. The mRNA next binds to polyribosomes in the cytoplasm and assembles amino acids (polymerization) carried to it by transfer RNA. The anticodons of this tRNA match corresponding bases in the codons of the mRNA. The rate of polymerization is approximately four to eight amino acids per second. This step, **translation**, ends in the formation of a polypeptide chain. The last step, **post-translational modification**, involves cleavage of bonds within the new protein, modification of selected amino acids in the chain, and folding of the protein into its characteristic configuration. The final protein is then delivered to its site of action, e. g., to the nucleus, to cytoplasmic organelles, or out of the cell into the blood.

Proteins to be retained by the cell are formed on free ribosomes, whereas proteins destined for export from the cell are synthesized on a ribosome matrix, which is attached to the endoplasmic reticulum (rough ER). As these latter proteins are synthesized, the initial segment functions as a leader to carry the molecule across the reticular membrane and into the cisternal space. The protein then moves within these channels to the Golgi complex, where it is incorporated into vesicles that ultimately reach the cell membrane and open to the extracellular space (exocytosis).

Energetics

Free energy is a thermodynamic concept. When a substance changes its state, it is possible to describe a **change** in free energy (ΔE), which is equal to the work done (W) and the heat absorbed ($-Q$). The **first law of thermodynamics** defines this relationship for a given change in energy state: $\Delta E = W - Q$. ΔE provides a measure of the maximum amount of energy that can be obtained as useful work during a change of state; it also describes the ability of a substance to undergo a transformation. A reaction associated with a decrease in free energy ($-\Delta E$) can take place spontaneously. On the other hand, if ΔE is positive, the reaction will take place only if energy is supplied to the system. By definition, **endergonic** reactions have $+ \Delta E$, **exergonic** reactions $-\Delta E$.

The **second law of thermodynamics** defines the conditions under which spontaneous change may occur: $\Delta E = \Delta H - T \times \Delta S$, where ΔH is the change in heat content (**enthalpy**) in a reaction that proceeds at a constant pressure, T is the absolute temperature, and ΔS is an expression for change in disorder or randomness (**entropy**) of a system. ΔS has the dimensions of energy; when $\Delta S = 0$, a system is at equilibrium; when $\Delta S > 0$, the system can undergo spontaneous change.

ATP. In all living cells, adenosine triphosphate (ATP) functions as a common reactant linking endergonic and exergonic processes. ATP contains two high-energy phosphate bonds, so called because the molecule undergoes a large decrease in free energy when the phosphate bonds are hydrolyzed. When one high-energy phosphate bond is hydrolyzed, energy is liberated and ATP is converted to adenosine diphosphate (ADP); ADP is then capable of accepting a phosphate group from exergonic reactions, thus regenerating ATP. This cycle drives many synthetic reactions and is an energy source for movement, work, secretion, absorption, and electric conduction.

Although other high-energy molecules exist, ATP is the most widely distributed and the most frequently encountered. ATP may be split by **orthophosphate cleavage** to yield $ADP + P_i + 30.6$ kJ (7.3 kcal) or by **pyrophosphate cleavage** to yield adenosine monophosphate (AMP) + $PP_i + 36.0$ kJ (8.6 kcal).

The phosphate is liberated by a hydrolytic enzyme, ATPase. There are several ATPases; they vary in location, pH optimum, cofactors, and in the energy-requiring reactions with which they are coupled.

Biologic Transport

Transport in biologic systems refers to the movement of solutes or water across membranes. Movement across a membrane is described as a **flux**; fluxes may be **active or passive**. Movement into a compartment is **influx**, out of a compartment **efflux**. Movement of solute is always from a region of higher to a region of lower electrochemical potential; it is always in accordance with the existing gradient. In passive flux, all of the operant forces can be identified and measured, and movement corresponds to the gradient. In active transport, some of the forces are not known, and movement may occur in opposition to the **apparent** gradient.

Filtration is movement of water across a membrane, the driving force being the hydrostatic pressure ($\rightarrow$ p. 144). Movement is in accord with the hydrostatic pressure gradient. Solvent movement, **bulk flow**, carries solute with it across the membrane by an action called **solvent drag**.

Passive diffusion is a process by which a substance contained in a gas or liquid extends to fill the available volume. The driving force is the thermal energy of the particles, and movement is in accord with the concentration gradient. Diffusion through a membrane is defined by Fick's first law: $Q = P \times A \times \Delta C$, where Q, the rate of diffusion, depends on the permeability constant (P) of the membrane, the membrane area (A), and the difference in concentration (ΔC) between the substances on either side of the membrane. If bulk flow occurs, Fick's law is not strictly applicable and must be modified.

Trapping. When a substance moves across a membrane into a compartment, its physical state may be altered: it may be bound to a protein or it may be metabolized. By these means and others, its concentration in the compartment is kept low so that the original concentration gradient is maintained and its flux can continue.

Non-ionic diffusion is a special form of trapping in which passive diffusion of a weak acid or base occurs between compartments of different pH ($\rightarrow$ p. 98). The undissociated form of a weak acid generally penetrates membranes more readily than its ionic form. When the undissociated form of the acid penetrates into a relatively alkaline compartment, it ionizes. Since the ionic form cannot diffuse as readily through the membrane in the reverse direction, it is "trapped" in the relatively alkaline

compartment. Moreover, since it is ionized, the concentration of undissociated acid in that compartment remains low and permits diffusion to continue.

Active transport processes can take place in opposition to the apparent electrochemical gradient. They are dependent on cellular energy, they are usually oxygen dependent, and they have a high temperature coefficient. A change in temperature of of $\pm 10\,°C$ produces a large change in rate of transport (Q_{10}). The energy for transport is furnished almost exclusively by high-energy phosphate compounds (ATP) in the cell. There is usually a high structural specificity for transport: only substances with closely related molecular structures may participate as substrates of individual transport systems. For example, many cells distinguish between the structure of D- and L-isomers of amino acids, transporting one and not the other. Because of such structural specificity, a reaction between the substrate and a **cellular receptor** is believed to be the first step in transport. The transport process has a limiting or maximal rate and therefore shows saturation kinetics. Active transport can be inhibited competitively by other molecules acting as substrates and noncompetitively by anoxia and by metabolic inhibitors.

Receptors

A **receptor** is a unique molecule that recognizes a circulating messenger molecule (most often a hormone) or substrate for transport. A receptor occurs only in cells that are responsive to the messenger or the transport substrate. The receptor may be localized on the cell membrane, in the cytosol, on the membrane of intracellular organelles, or in the nucleus. The messenger or transport substrate initiates a response by binding with its receptor. The binding process has the following characteristics: it exhibits strict structural and steric specificity, it is saturable because the number of receptors in each cell is finite and limited, it is rapid and reversible, and it initiates a characteristic physiologic response in the cell.

In most receptor reactions, the rate of reaction varies with the substrate concentration according to a hyperbolic curve which is characteristic of Michaelis-Menten kinetics. The reaction velocity eventually reaches a maximum value at higher substrate concentrations and is not increased by further increases in substrate concentration (saturation kinetics). **Allosteric** receptors deviate from Michaelis-Menten kinetics: they show a sigmoid relationship to increasing substrate concentration. The reason offered is that the substrate induces a change in state or structure (**allosterism**), which alters the reactivity or affinity of the receptor for the substrate. Allosterism is also used in a less specific sense to describe changes in activity brought about by structural changes in the receptor.

Electrogenic pumping. In most cells, there is an asymmetric distribution of ions and of electric charge across the cell membrane ($\rightarrow$ p. 22). The extracellular fluid (ECF) has a high concentration chiefly of sodium (Na$^+$) and chloride (Cl$^-$), whereas the intracellular fluid (ICF) has a high concentration chiefly of potassium (K$^+$) ($\rightarrow$ p. 22). The distribution of concentrations of these ions is such that an electric voltage (potential difference or p. d.) can be measured across the membrane, whereby the outer surface is positive relative to the inner. Thus, there are both electric and chemical gradients established that influence movement of cations (+ charges) and anions (− charges) through the membrane according to their permeabilities.

For chloride ion (Cl$^-$), for example, the chemical gradient would establish an influx, whereas the electric gradient would motivate an efflux. The membrane potential at which Cl$^-$ fluxes equilibrate is the **equilibrium potential** and is defined by the Nernst equation. The equilibrium potential E at 37 °C can be calculated from:

$$E = -61.5 \cdot \frac{1}{z} \cdot \log [\text{ion}]_{ECF} / [\text{ion}]_{ICF}$$
(Nernst)

where [ ] is concentration and z the charge per ion. z = −1 for Cl$^-$, z = +1 for K$^+$ or Na$^+$, z = +2 for Ca^{2+}.

The electrochemical gradient for Na$^+$ across the cell membrane is appropriate for influx, but influx is resisted because of the low membrane permeability for Na$^+$. Nevertheless, small quantities of Na$^+$ do manage to leak into the ICF and to reduce the asymmetry of Na$^+$ distribution.

The membrane permeability for K$^+$ is somewhat higher than for Na$^+$. The chemical gradient for K$^+$ can power an efflux that is resisted, however, by the antagonistic electric gradient for K$^+$. Nevertheless, small quantities of K$^+$ do leak out of the ICF and reduce the asymmetry of K$^+$ distribution.

To antagonize these small but continuous leaks, there exists a specific enzyme, an ATPase, which has been found in every cell membrane that separates an asymmetric distribution of Na$^+$ and K$^+$. The ATPase within the membrane is able to alter its position so that it faces either toward the ECF or the ICF. When it faces into the cell, it has a high affinity for Na$^+$. When Na$^+$ binds to ATPase inside the cell, it initiates a stepwise hydrolysis of ATP, the first step of which is phosphorylation of the enzyme. A conformational change then rotates the enzyme so that it faces the ECF, carrying Na$^+$ with it. In this position, affinity for Na$^+$ decreases and affinity for K$^+$ increases. The phosphoenzyme exchanges Na$^+$ for K$^+$ in the ECF and activates the next step, hydrolysis of the phosphoenzyme bond. A second conformational change then rotates the K$^+$

ATPase so that it faces into the cell where K^+ is exchanged for Na^+. The cycle then repeats.

By this means, Na^+ that leaks into the cell is extruded: K^+ that leaks out of the cell is replaced. Energy for the process is derived from the high-energy bonds of ATP. However, a one-for-one exchange of Na^+ for K^+ is exceptional. More often, 1 mol of ATP is used to pump 3 mol Na^+ out of the cell and 2 mol K^+ into the cell. In the process, an outward current is induced that contributes to maintaining the cell interior negative.

Ion Gates

Many cells use an electric charge on the membrane, the cell membrane potential, to regulate their physiologic functions. These functions include ion permeability of nerve and muscle membranes, neurotransmitter release from the presynaptic terminal of nerves, intracellular availability of Ca^{++} for muscle contraction, and adrenal medullary secretion of catecholamines. It is to be expected that more examples will be found as more cells are subjected to electrophysiologic analysis.

Excitable cells, chiefly nerve and muscle, have the property of altering the electric characteristics of the membrane when they are properly stimulated. They produce thereby a wave of excitation called the action potential, which is initiated by movement of positively charged particles (cations) into the negatively charged cell interior. The link between a change in membrane potential and the physiologic response is furnished by a **voltage**

sensor. The sensor is believed to be a polar molecule in the membrane bearing a large charge or dipole moment. The molecule responds to an electric field by changing its conformation, thus opening or closing channels through which ions may penetrate the membrane. Because of this action, the voltage sensor is called a **gate**, and the current it generates when its conformation changes is called a **displacement** or **gating current**.

Sodium (Na^+) penetrates the membrane through fast channels that are influenced by two gates (h and m) in series. Under resting conditions, the h gate is open and the m gate is closed. When the cell is stimulated and its resting charge of -70 to -90 mV is neutralized, the gates change. At the membrane threshold voltage (-60 mV), the m gates swing open quickly (<1 ms). At a similar voltage near the threshold, the h gate moves shut, but it is more sluggish than the m gate, requiring 3 to 4 ms to close. Thus, there is a brief period during which both gates are fully open and Na^+ rushes into the cell along its electrochemical gradient ($\rightarrow$ p. 22).

Membranes of many excitable cells also have slow channels for Ca^{++} and Na^+, which are not the same as the fast channels. They seem to be controlled by a similar double-gating mechanism. The slow channels open at a more positive potential than the fast channels. Metabolic energy is required continuously to maintain the slow channels. The slow channels figure prominently in determining the rate of discharge of the cardiac pacemaker.

Excitable Tissue. Nerves

An excitable cell is able to respond to a stimulus by changing the electric properties of its cell membrane. In primitive animals, neuroeffector cells could both transmit a stimulus and respond to it. In more complex animals, these effector cells have specialized to form (a) nerve cells to transmit an impulse and (b) muscle cells to respond to the stimulus by contracting.

The human nervous system (NS) contains 10^{10} nerve cells or **neurons**. The neuron is both the structural and functional unit of the NS. A typical motor neuron ($\rightarrow$ **A**) has a **soma** or cell body, an **axon**, and **dendrites**. The soma contains the usual intracellular organelles as well as Nissl substance and neurofibrils. The dendrites furnish a large area for contact with other neurons (up to 0.25 mm^2). They are **afferent** or centripetal fibers because they receive signals from other neurons and transmit them *to* the soma. The axon is an **efferent** or centrifugal fiber because it carries signals **from** the soma to neighboring neurons. It terminates in a swelling, the **synaptic knob** or **terminal button**, in which are the **vesicles** that store **neurotransmitter** ($\rightarrow$ **B**). The neurotransmitter may be synthesized throughout the neuron, but the major portion of synthesis takes place in the soma. From here, the transmitter and other compounds are transported along **microtubules** in the axon by **axoplasmic flow** (c. 200 to 400 mm/day) until they reach the terminal button. On a single motor neuron, there may be more than 5500 contact points with terminal buttons of other neurons; these can cover up to 40% of the total surface of a neuron.

The cell membrane of the soma extends along the axon as the **axolemma** ($\rightarrow$ **A**). The axon is surrounded by **Schwann cells**. In many neurons, these cells wrap around the axon in spiral layers and condense to produce myelin ($\rightarrow$ **C, 1**), a lipoprotein sheath outside the axolemma. In such myelinated nerves, the sheath is interrupted by the **nodes of Ranvier** at intervals of up to 1.5 mm. In unmyelinated neurons ($\rightarrow$ **C, 2**), the Schwann cells are present but do not form myelin lipoproteins.

The **synapse** ($\rightarrow$ **A**, **B**) is the area where signals are transmitted from the axon of one neuron to the axon, dendrite, or soma of another. In mammals, there is no actual contact at the synapse (with rare exceptions). The **synaptic cleft** (10 to 40 nm) separates the two neurons and acts as an insulator. For transmission of a signal, the electrical impulse that reaches the **presynaptic membrane** must release a chemical transducer, the **neurotransmitter**, into the synaptic cleft. There are several transmitters: acetylcholine, norepinephrine, dopamine, glycine, etc., but in general, each neuron has only one neurotransmitter. The transmitter diffuses to the **postsynaptic membrane** and initiates a new electric signal. Since there is no neurotransmitter within the postsynaptic membrane, the synapse functions as a one-way valve transmitting only from the presynaptic to the postsynaptic neuron. The transducer may produce either an excitatory or an inhibitory response depending on the nature of the transmitter. In muscle, the counterpart of the synapse is the **neuromuscular junction** ($\rightarrow$ **D** and p. 30), which allows transmission of an impulse from nerve to muscle.

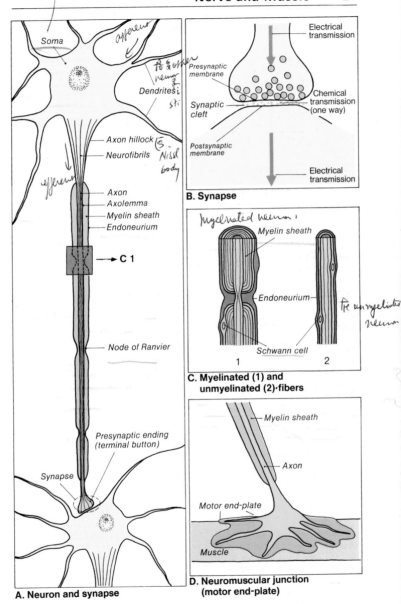

B. Synapse

Electrical transmission

Presynaptic membrane

Chemical transmission (one way)

Synaptic cleft

Postsynaptic membrane

Electrical transmission

A. Neuron and synapse

Soma

Dendrites

Axon hillock

Neurofibrils

Axon

Axolemma

Myelin sheath

Endoneurium

→ C 1

Node of Ranvier

Presynaptic ending (terminal button)

Synapse

C. Myelinated (1) and unmyelinated (2)-fibers

Myelin sheath

Endoneurium

Schwann cell

1   2

D. Neuromuscular junction (motor end-plate)

Myelin sheath

Axon

Motor end-plate

Muscle

Resting Membrane Potential

An electrical tension or potential difference ($\rightarrow$ p. 7) can be recorded across the membrane of living cells. According to cell type this so-called **resting potential** amounts to **60–100 mV** (*cell interior is negative*) in muscle and nerve cells. The resting potential is due to the *uneven distribution of ions* ($\rightarrow$ **B**) between the intracellular fluid (**ICF**) and the extracellular fluid (**ECF**).

The following phenomena are involved:

1. By means of **active transport** ($\rightarrow$ p. 17) Na^+ is continuously "pumped" out of the cell and K^+ "pumped" into it ($\rightarrow$ **A2**), so that the K^+ concentration inside the cell is about 40 times higher and the Na^+ concentration 15 times lower than outside ($\rightarrow$ **B**). The so-called **Na^+-K^+-ATPase** (splits ATP [$\rightarrow$ p. 16]) plays an important role here.

2. Under resting conditions the *permeability of the cell membrane to Na^+ ions is low* (i.e. its Na^+ conductance is low) so that the Na^+ concentration gradient ($\rightarrow$ **A3–5**) cannot be reversed by passive back diffusion ($\rightarrow$ p. 17).

3. Most of the anions inside the cell are the *negatively charged proteins* and *phosphate* ($H_2PO_4^- \rightleftharpoons HPO_4^{2-}$). The *permeability* of the cell membrane to both of these anions is *extremely low* ($\rightarrow$ **A4, 5**).

4. The *permeability* of the cell membrane to K^+ is relatively *high* ($g_K \gg g_{Na}$). On account of the steep concentration gradient (point 1) K^+ ions diffuse from the ICF into ECF ($\rightarrow$ **A3**). Due to the positive charge of the K^+ the diffusion of only a few such ions is sufficient to produce a distortion in the charge (potential) across the membrane, since the greater part of the intracellular anions do not follow and no significant inward diffusion of Na^+ is possible (see point 2). This **diffusion potential** continues to increase until the further efflux of K^+ ions (driven by the concentration gradient) is prevented by the rising potential.

Since the *permeability* of the cell membrane for Cl^- is relatively high (see below) the rising potential drives the Cl^- out of the cell against its chemical gradient ($\rightarrow$ **A4**). The diffusion of K^+ (down a chemical gradient) is increasingly opposed by the rising potential, and that of Cl^- (driven by the potential) by rising chemical gradient. Finally, an **equilibrium potential** is attained **for K^+** (**E_K**). At E_K the driving force for outward K^+ diffusion (the chemical gradient) is exactly equal to the inwardly directed force due to the potential (electrical gradient), i.e. the electrochemical gradient for K^+ is 0.

E_K can be calculated using the **Nernst equation**, which is (for 37°C) $E_K = -61.5 \frac{1}{z} \log [K^+]_i / [K^+]_e$ (mV) where z is the valency of the ion in question (for K^+: z = +1) and $[K^+]$ is the K^+ concentration in the ICF (i) or ECF (e). If, for example, $[K^+]_i / [K^+]_e$ is as 40/1 ($\rightarrow$ **B**), then $E_K = -61.5 \times 1.6 = -99$ mV, since log 40/1 = 1.6. Although the permeability of the resting membrane to **Na^+** is very low, Na^+ ions diffuse continuously into the cell (high electrical and chemical gradients, $\rightarrow$ **A5**). The resting membrane potential is consequently usually lower than E_K.

Due to the relatively high permeability of the cell membrane to Cl^- (lower in nerve cells than for K^+, higher in muscle cells than for K^+) the Cl^- is so distributed between ICF and ECF that the Cl^- equilibrium potential equals the resting membrane potential. The distribution of Cl^- between ICF and ECF is thus numerically approximately equal to that of K^+ except that the concentration gradient for Cl^- (on account of its negative charge) is in the opposite direction ($\rightarrow$ **B**). If, however, the E_{Cl} calculated from the Cl^- distribution (Nernst equation with z = −1) is smaller than the resting membrane potential (as in certain epithelial cells) it can be concluded that Cl^- is being transported against an electrochemical gradient, i.e. active Cl^- transport is taking place ($\rightarrow$ p. 17).

All living cells exhibit a resting membrane potential but the excitable cells (nerve, muscle) possess the additional property of *altering the ion permeability* (conductance) of their membranes, which leads to considerable changes in potential ($\rightarrow$ p. 24).

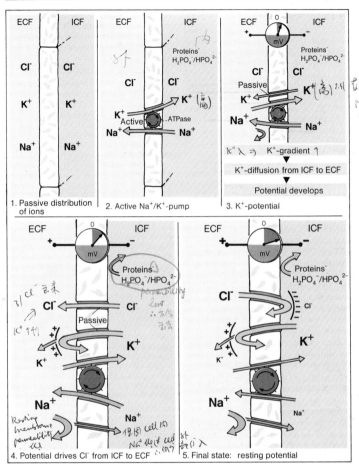

A. Development of the resting potential

	Concentration		Equilibrium	Diameter of
	ECF	ICF	potential	hydrated ion
K⁺	4 mmol/l	160 mmol/l	−98 mV	396 pm (3.96 Å)
Na⁺	145 mmol/l	10 mmol/l	+71 mV	512 pm (5.12 Å)
Cl⁻	114 mmol/l	3 mmol/l	−97 mV	386 pm (3.86 Å)
H⁺	pH 7.4	pH 6.9	−31 mV	150 pm (1.50 Å)

Apparent pore size: 0.8 nm (8.0 Å)

B. Distribution of ions; equilibrium potential

Action Potential (AP)

Maintenance of the resting membrane potential (RMP) is a property of all living cells. **Excitability**, however, is shown only by specialized cells, nerve, and muscle. These cells respond to a stimulus by producing transient changes in their membrane properties that result in generation of an **action potential** (AP). Details differ in nerve and muscle, but the basic processes are similar.

Passive leaks of ions through the membrane will change the RMP. Na^+ influx, for example, produces a depolarizing current that reduces negativity of the RMP. When negativity of the RMP falls to a critical value, the **threshold potential** (TP), there is a sudden increase of the Na^+ conductivity (g_{Na}) and an overwhelming Na^+ influx (→ **A, 2; B**). The membrane depolarizes rapidly, and in the process it opens more Na^+ channels to increase the g_{Na} still more (regenerative depolarization, a positive feedback). At 0 mV, Na^+ influx continues, being driven by the concentration gradient; the cell interior overshoots and becomes (+). (The chemical and electric gradients for Na^+ balance each other at c. +60 to 70 mV) (→ p. 22, and p. 23, C). By now, the g_{Na} has been inactivated, but in the meantime g_K has been increasing. K^+ efflux proceeds to remove (+) charges from the cell and to lower the membrane potential (MP) to its original value (**repolarization**) (→ **A, 3; B**). Before g_K is inactivated, the MP may become more (−) than the original RMP for a few milliseconds (**hyperpolarization**). The actual amount of Na^+ gained and K^+ lost during one AP is quite small; the cell can be repeatedly fired without depleting K^+. In addition, the Na-K-ATPase pump functions continuously to restore the original Na^+ and K^+ gradients (→ **A, 3**). In nerves, the AP starts at the axon hillock where the TP is lowest. The response is always total and maximal regardless of the stimulus (all-

or-none). Early in repolarization, the cell is in its **absolute refractory period** and does not respond to any stimulus; later, it is in its **relative refractory period** when a larger stimulus than normal is necessary to reach the TP.

During AP inward Na^+ current (I_{Na}) depends on the MP *before* the start of AP. I_{Na} is maximal at an MP more negative by c. 40 mV than RPM and is c. 60% of the max. I_{Na} at the RMP. At a MP of c. −30 mV Na^+ influx can no longer be activated (refractory period). Dependence of I_{Na} on MP is influenced by Ca^{2+}.

Cardiac muscle has two types of cells: contractile and automatic. The rate of depolarization of **contractile** cells resembles that of neurons (dV/dt = 500 to 800 V/s), but repolarization is slower (→ pp. 150–152). In **automatic** cells, dV/dt rarely exceeds 1 V/s. In **contractile** cells, the upstroke of the AP (phase 0) has two components: (1) There is an immediate inward Na^+ current through fast channels. These channels require metabolic activity of the cell for maintenance. (2) When the potential reaches c. −40 mV, slow channels become available for an inward current chiefly of Ca^{2+} but also of Na^+. The slow channels are not inactivated at the same rate as the fast channels so that **repolarization** has several **phases**: phase 1 − slow influx of Ca^{2+} and Na^+; inactivation of fast channels; phase 2 − plateau; predominantly Ca^{2+} influx; phase 3 − rapid repolarization; predominantly K^+ efflux; phase 4 − isoelectric period between two consecutive APs; inactivation of K^+ current and Ca^{2+} current. In **automatic** or pacemaker cells of the SA node, the AP can be generated spontaneously. These cells have several distinctive features (→ p. 153, A): depolarization is slower (phase 0), the plateau is shorter (phase 2) and conduction velocity is slow. They have no isoelectric period (phase 4); there is no RMP. Instead, phase 4 is unstable; the MP reaches a **maximum diastolic potential** and then becomes progressively less negative. As in contractile cells, the outward K^+ current is inactivated at the end of phase 3, but although little K^+ leaves the cell, there is a leak of Na^+ and Ca^{2+} into the cell. This inward current reduces the MP until threshold is reached and AP is produced. This process is called **diastolic depolarization** (→ p. 150).

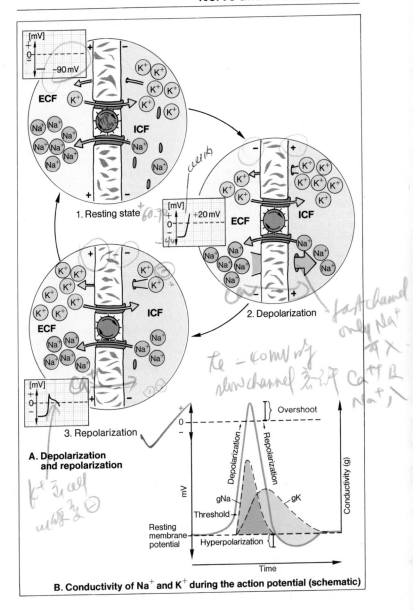

A. Depolarization and repolarization

1. Resting state

2. Depolarization

3. Repolarization

B. Conductivity of Na⁺ and K⁺ during the action potential (schematic)

Propagation of the Action Potential

Nerve and muscle fibers are analogous to an electric cable; they are cylindric conductors of electric current and are surrounded by an insulating sheath, the cell membrane. Unlike a cable, however, there are large electric leaks through the membrane, and energy is dissipated in the core. As a result, a single voltage stimulus applied to a point on a nerve fiber would fade out logarithmically before it traveled 1 mm or so. For the impulse to reach the end of the fiber, there must be a relay system in the membrane that boosts the signal to full strength along the entire fiber length and compensates for the losses.

At rest, the cell membrane is electrically polarized with $(+)$ charges on the outside. When a voltage (PD) is applied to a cell, current will flow from high $(+)$ PD to low $(-)$ PD. The current, by convention, coincides with the movement of $(+)$ ions. Thus, when threshold (TP) ($\rightarrow$ p. 24) is reached, there is a sudden influx of Na^+ at the point of stimulus and also an increase in inward flowing current ($\rightarrow$ **A, 1 a**). When current flows the local area loses its $(+)$ charges and the membrane becomes depolarized ($\rightarrow$ p. 25, A, 2). The local area then acts as an electric sink; $(+)$ charges move toward it from adjacent areas and a circular current flow is established ($\rightarrow$ **A, 1 b**). When enough $(+)$ charges have been withdrawn, adjacent areas of the membrane reach their TP and also produce an AP ($\rightarrow$ p. 24). The process repeats along the fiber length until the end is reached. The AP is therefore a *local* process that propagates by successive serial depolarizations ($\rightarrow$ **A 1, B 2**).

The velocity at which the impulse travels depends on the rate at which the membrane ahead of the impulse is discharged beyond its threshold. This, in turn, is related to fiber diameter and resistance. The larger the diameter, the higher the conduction velocity ($\rightarrow$ **C**). However, in a complex nervous system, there is not enough space for a large number of large diameter fibers. The optic nerve in man, for example, contains more than 10^6 closely packed nerve fibers. An alternate process for increasing velocity while permitting small diameter is seen in myelinated nerve fibers. In such fibers, the myelin sheath improves the insulation and reduces electric leaks through the membrane. Leaks can occur, nevertheless, at the nodes of Ranvier, where the myelin sheath is lacking ($\rightarrow$ p. 20). In these fibers ($\rightarrow$ **A, 2**), an electric impulse sets up the same circular currents as in an unmyelinated fiber, but because of the myelin sheath, the current loops must extend from one node of Ranvier to the next. Propagation of the AP takes place from node to node and the AP appears to jump the gap (saltatory conduction) ($\rightarrow$ **A 2, B 1**). In myelinated nerves, conduction velocity is 20 to 25 times greater than in unmyelinated nerves ($\rightarrow$ **C**).

An axon can, in principle, conduct in any direction. If it is stimulated at its center, an AP can be propagated toward the synapse (**orthodromic conduction**) and toward the soma (**antidromic conduction**). Once under way, the propagation cannot reverse direction because the membrane behind the wave is refractory ($\rightarrow$ p. 24; **A 1**) and cannot respond. When an orthodromic impulse reaches a synapse, it stimulates the next nerve by releasing neurotransmitter. When an antidromic impulse reaches a synapse, it dies out because the valve function of the synapse will not allow transsynaptic conduction (neurotransmitter is present only in the presynaptic neuron) ($\rightarrow$ p. 20).

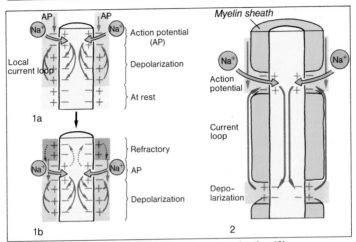

A. Serial depolarization (1a, 1b) and saltatory conduction (2)

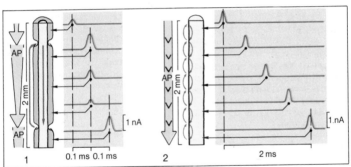

B. Impulse propagation in myelinated (1) and unmyelinated (2) fibers

Fiber	Function	Diameter (µm)	Conduction velocity (m/s)
Aα	Afferents - muscle spindle, tendon organ. Efferents - skeletal muscle	15	70–120
Aβ	Afferents - touch	8	30– 70
Aγ	Efferents - muscle spindle	5	15– 30
Aδ	Afferents - temperature, fast pain	3	12– 30
B	Sympathetic, preganglionic	3	3– 15
C	Sympathetic, postganglionic afferents - slow pain	1 (unmyelinated)	0.5– 2

C. Classification of neurons

(after Erlanger and Gasser)

Electric Events in the Neuron

Characteristics of neural stimulation. When an electric stimulus is applied to a neuronal membrane, current flows out of the cell at the cathode and reduces negativity of the resting membrane potential (RMP): depolarization. When the potential reaches the threshold potential (TP), an action potential (AP) is generated (→ p. 25, B). The stronger the stimulus, the less time is required to fire the AP. A weak stimulus must be applied for a long time before an AP is fired. The *rheobase* (→ **B**) is the weakest signal that can fire an AP provided it is applied for an infinite time. *Chronaxie* (→ **B**) (excitation time) is the time needed to generate an AP if the stimulus is twice the rheobase; it is a measure of **excitability** of a neuron. Rheobase is a current, chronaxie is a time.

After an AP is generated, there occurs an **absolute refractory period** during which the neuron cannot produce another AP no matter how great the stimulus. It is followed by the **relative refractory period** during which a second AP can be generated if a greater than normal stimulus is applied.(p. 24).

The **process of excitation** involves the action of two different kinds of membrane. The first is an **electrically excitable membrane** and is characteristic of the conductile regions of axons. It supports an overshooting AP and propagates the AP without decrement by successive electric excitation of adjacent regions of the same membrane (→ p. 24). The second is an **electrically inexcitable membrane**. It undergoes graded depolarization and decremental conduction in response to chemical and physical stimuli (but not electric stimuli). It cannot support a full regenerative AP. Both classes may occur in the same neuron. The first type produces an **all-or-none-response**, the second type a **graded response**. Graded responses are seen at (1) the postsynaptic membrane, (2) the region of sensory receptors and their afferent neurons (generator potential) (→ p. 254), and (3) neuromuscular junctions (→ p. 30).

Synaptic potentials. When a neuron is stimulated, the AP that it generates travels to the presynaptic membrane (→ p. 21, B) where it releases a quantity of neurotransmitter. The amount of neurotransmitter liberated at a synapse depends upon the *frequency* of firing of a single neuron and on the *number of neurons* that are firing. The postsynaptic membrane responds with either a local depolarization or a hyperpolarization depending on the transmitter. γ-aminobutyric acid (GABA) and glycine are often hyperpolarizing transmitters, for example. These responses are respectively *excitatory* or *inhibitory postsynaptic potentials* (EPSP or IPSP) (→ **D**). Such potentials are graded and have the following characteristics (→ **C**): (1) *synaptic delay* or the lag time between stimulus and the onset of the EPSP; (2) *rise time* or the time required to reach peak amplitude (c. 1 ms for the EPSP); (3) *decay time* – also described by the *half-time* or the *time constant* (time to decay to $1/e$ or $1/2.7$ of the peak). For the EPSP, the time constant is c. 5 ms and is independent of the amplitude (→ **C**).

Although a single EPSP is insufficient to generate an AP (its *maximum* value is 20 mV), it does increase excitability of the postsynaptic membrane. Several EPSP may summate until the TP is reached, at which time an AP will be fired. In contrast to the EPSP, an IPSP reduces excitability of the membrane (maximum value of IPSP is c. 4 mV). The sum of all EPSPs and IPSPs determines whether an AP will be generated at any particular time (→ **D**).

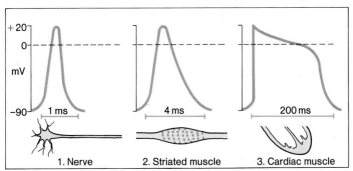

A. Action potential - nerve and muscle (see also p. 43)

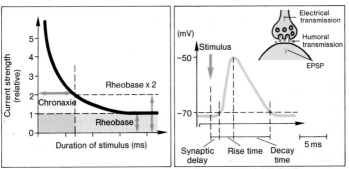

B. Stimulus/response

C. Characteristics of EPSP

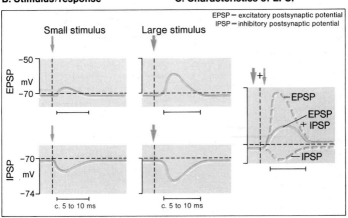

EPSP = excitatory postsynaptic potential
IPSP = inhibitory postsynaptic potential

D. Graded response of EPSP and IPSP

Neuromuscular Junction

The functional unit for motor activity is not the muscle alone but the **motor unit**, which is a single efferent motor neuron, and all of the motor fibers that it innervates. In different muscles, the innervation ratio may range from 100 (M. lumbricalis) to more than 1000 (M. temporalis) muscle fibers/neuron.

The muscle fibers of a single motor unit may be distributed over the whole muscle. Two different types of motor units, **fast-twitch** and **slow-twitch units** can be distinguished. To which type a unit belongs is determined late in fetal or early in postnatal life by properties of the respective motor neuron. The signal responsible for differentiation probably is the discharge frequency. In the fetus all units belong to the slow-twitch class. A certain part of them do not change their main properties in later life. They have a more active oxidative metabolism, are more sensitive to hypoxia, have more capillaries, are more resistant to fatigue, have a lower resting potential and a broader Z line, if compared to the fast-twitch units differentiating around birth. Occurrence of the two types is different in different muscles ("red" and "white" muscles).

Muscular activity is **graded** according to how many motor units become activated. One muscle contains c. 10^2 (M. lumbricalis) to 10^3 (M. temporalis) motor units. **Recruitment** of more motor units increases the tension developed. The **order of recruitment** of slow and fast or big and small units depends on the type of motor activity (phasic or tonic contraction, reflex activity or voluntary effort). After all of the motor units have been recruited, tension can be increased still further by increasing the frequency of stimulation (**summation**) ($\rightarrow$ p. 39, B).

The neuromuscular junction (NMJ) between the terminal motor neuron and the **motor end-plate** ($\rightarrow$ **A**) of the muscle has a transducer function similar to that of the synapse. As in most peripheral synapses, the neurotransmitter is **acetylcholine (AcCh)**. On the muscle fiber ($\rightarrow$ pp. 20, 52), only the motor end-plate is sensitive to AcCh. In the resting muscle, there is always a continuous release of some AcCh granules from the motor nerve ending. Each granule represents a quantum of AcCh that can increase permeability of the motor end-plate to Na$^+$ and K$^+$ for as long as 1 to 2 ms. It thus produces a small endplate potential (EPP). The miniature EPP spreads along the membrane and is quickly attenuated; it is not large enough to activate an AP ($\rightarrow$ p. 24) on the muscle fiber to initiate contraction. However, when the motor neuron is activated by an AP, it releases up to 10^6 quanta of AcCh. These can produce an EPP large enough to propagate along the muscle fiber and to initiate muscle contration. Simultaneously, AcCh-esterase hydrolyzes AcCh and terminates its action ($\rightarrow$ p. 52).

In electron micrographs of the nerve terminals some vesicles are lined up in a double row at the *active zone* of the presynaptic membrane. This presynaptic region is the site where the transmitter is released. The release of AcCh results from the fusion of vesicles with the presynaptic membrane of the active zone ($\rightarrow$ **A**). On the opposite side of the synaptic cleft the postsynaptic membrane is folded: *Postsynaptic folds*. In this region the acetylcholine receptors are found. AcCh-esterase is probably localized in the *basement membrane* ($\rightarrow$ **A**). **Neuromuscular blockade**. Blockade of transmission in the NMJ interferes with stimulation of the muscle and results in muscle weakness or paralysis. **Competitive** blockade develops when compounds compete with AcCh for receptors in the motor end-plate (e.g., curare). In this type of blockade, depolarization of the muscle is prevented. The effect can be reversed by increasing the local concentration of AcCh as with cholinergic drugs or with inhibitors of cholinesterase ($\rightarrow$ p. 52). **Noncompetitive** blockade is produced by maintaining the muscle in a continuously depolarized state so that it cannot respond to a stimulus. Such an effect is achieved by excesses of cholinergic drugs or by cholinesterase inhibitors (nerve gases) that result in paralysis. The blockade is enhanced by K$^+$, which can also produce depolarization.

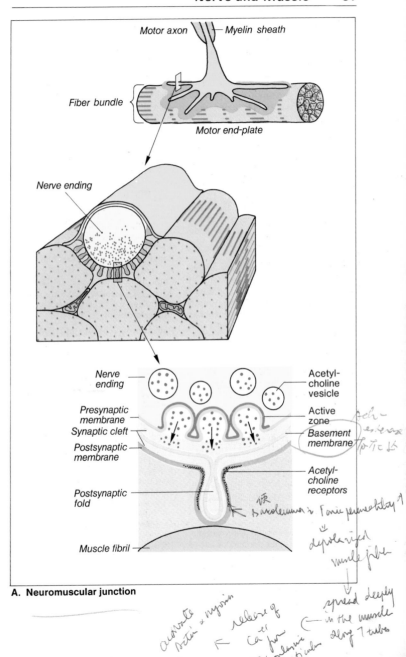

A. Neuromuscular junction

Contractile Machinery

Muscle transforms the chemical energy of ATP ($\rightarrow$ p. 16) into mechanical energy. The components of the muscle cell are therefore simultaneously structural and enzymatic elements. The muscle cell is a **fiber** ($\rightarrow$ **A, 1**). One muscle contains c. 10^4 (M.lumbricalis) to 10^6 (M.temporalis) fibers. Each fiber contains 1,000 to 2,000 **fibrils** ($\rightarrow$ **A, 3**), which are composed of **sarcomers** ($\rightarrow$ **A, 3**; pp. 34, 35), arranged in series (4,000/cm of fibril). The sarcomere is the contractile unit of the muscle and is built up by interdigitating **thick (myosin)** ($\rightarrow$ **A, 6**) and **thin (actin)** ($\rightarrow$ **A, 8**) **filaments**. Contraction of muscle occurs when the filaments ($\rightarrow$ **A, 4**; p. 34) slide past each other and shorten the sarcomere.

Myosin (My) ($\rightarrow$ **A, 5**) is a highly charged molecule with a long tail and two heads at one end. My is composed of **light** and **heavy meromyosin** (LMM and HMM) ($\rightarrow$ **A, 5**). HMM has two components: a double head unit (HMM-S1) and a neck unit (HMM-S2). The S1 contains the ATPase activity of My ($\rightarrow$ pp. 34–37). The thick filaments ($\rightarrow$ **A, 6, 7**) are made up of c. 360 molecules of My, arranged in longitudinal bundles with the heads at each end ($\rightarrow$ **A, 6**); the center segment is composed only of My tails (LMM); at the ends, the S1 heads project in six longitudinal rows ($\rightarrow$ **A, 7**). The ends of S1 and S2 form joints that enable them to swivel for attachment to the actin chain. Each S1 (with ATP) forms one bond with one actin molecule. **Actomyosin** is the complex formed by combination of My with actin.

Actin (Ac) is an asymmetric globular protein; 400 single molecules of G actin associate into a bead-like strand of F actin. Two such strands intertwine to form the thin filament ($\rightarrow$ **A, 8**). The asymmetry of the individual G actin molecules is arranged so that the thin filaments have opposite polarity extending from the center of the strand.

Tropomyosin (*TM*) ($\rightarrow$ **A 8**) resembles My but lacks the heads. It is a helical, two-stranded filament that winds around the Ac strand. At one end, it binds to one molecule of troponin. One molecule of TM extends along seven molecules of Ac. During contraction, TM acts as a switch and shifts in the groove of the thin filament to expose reactive sites for formation of cross-bridges between Ac and My. It does not bind Ca^{2+} but, with troponin, it is essential in order for Ca^{2+} to regulate contraction.

Troponin (*TN*) ($\rightarrow$ **A, 8**; p. 36) is a globular protein with three subunits. TN-C binds with Ca^{2+}; TN-T binds the TN complex to TM; TN-I prevents formation of cross-bridges between Ac and My. One molecule of TN-I turns off four to seven units of Ac. When TN-C is saturated with Ca^{2+}, it inactivates TN-I and allows My cross-bridges to attach to Ac.

Ratios: by weight, Ac : My : TM : TN = 27 : 54 : 12 : 7; by molarity, Ac : My : TM : TN = 28 : 7 : 1 : 1.

Sarcotubular system ($\rightarrow$ **B**). The fibril lies behind a curtain, the sarcoplasmic reticulum (SR), which has an active transport system for uptake of Ca^{2+}. Near the **terminal cisternae** (TC) of the SR, the cell membrane invaginates to form **transverse tubules** (TT). One TT and two adjacent TCs form a **triad** near the junction of A and I bands ($\rightarrow$ p. 34). The TT is continuous with the ECF; it does not communicate with the TCs, which are closed chambers. The triads control flow of Ca^{2+}. They are the site of coupling of electric events (excitation) with mechanical events (contraction) ($\rightarrow$ pp. 34–37).

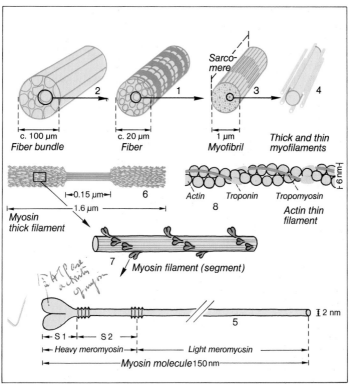

A. Fine structure of striated muscle

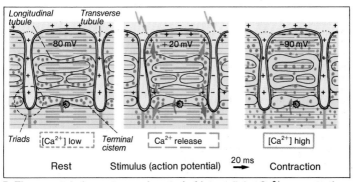

B. The sarcotubular system and control of intracellular Ca²⁺ concentration

Sliding Filament Hypothesis

The **sarcomere** (→ **A**). Under the microscope, a Z line can be distinguished. This is a plate-like protein structure that marks the ends of the sarcomere; one sarcomere lies between two Z lines. Each actin (Ac) filament is fixed at the center of its strand to the Z line; half a strand extends into each sarcomere on either side of the Z line. The two half-strands are oppositely polarized. On either side of the Z line is a region made up exclusively of thin filaments, the I band. The thick myosin (My) filaments comprise the darker A band. Over part of the A band, thick and thin filaments overlap, but at its center, there is a less dense area of only My filaments where there is no overlap, the H zone. An area of increased density, the M line, marks the center of the sarcomere.

The **sliding filament hypothesis** (→ **B**) states that contraction or tension in muscle is achieved by parallel interdigitating filaments of Ac and My. When they are activated, they slide along each other by cyclic attachment and detachment of cross bridges that project from the My filaments and react with the Ac filaments. Activity of the cross-bridges is regulated by Ca^{2+} and ATP within the cell.

As the muscle contracts (→ p. 39, C), the Z lines approach each other and the sarcomere shortens. The ends of the Ac filaments come closer and the degree of overlap with the My filaments increases so that the H zone and the I band become narrower. The limit of contraction is reached when the thick My filaments butt against the Z line. At this stage, the ends of the Ac filaments overlap at the center of the sarcomere. Both sets of filaments slide without changing their length. The sarcomere may shorten, but the length of the A band remains constant.

The driving force for sliding of the fibers comes from the S1 heads (→

p. 33, A) that have ATPase activity. The heads (cross bridges) attach to the thin filament (→ **B**). When the heads swivel to a new angle, the thin filament is pulled past the thick filament. The heads on each end of the My filament swivel in opposite directions so that the Z lines are pulled toward each other. A single cycle shortens one sarcomere by 2×10 nm, or by c. 1 % of its length. In a standard muscle twitch, the fibril may shorten 30 to 50 %. For this contraction the swivel cycle has to be repeated, therfore, many times.

Activation of skeletal muscle contraction (→ p. 44) starts when the action potential reaches into the muscle fiber via the transverse tubules (TT) (→ p. 33, B). The signal releases Ca^{2+} from the terminal cisternae and initiates contraction. Immediately, Ca^{2+} is actively accumulated by another part of the sarcoplasmic reticulum and is recirculated to the terminal cisternae. Activation is independent of extracellular Ca^{2+}.

Activation of myocardial contraction, in contrast to skeletal muscle, is greatly influenced by extracellular Ca^{2+}; in a Ca^{2+}-free medium, there is no contraction. The TT have a fivefold greater diameter than in skeletal muscle and furnish a larger surface for Ca^{2+} inflow. The Ca^{2+} that enters through slow channels during repolarization does not immediately take part in contraction; it is stored in the sarcoplasmic reticulum and is available for release at subsequent contractions. Reuptake of Ca^{2+} from the cytoplasm is chiefly into the sarcoplasmic reticulum but also partly into the mitochondria.

Activation of smooth muscle contraction. The cell surface has many invaginations (caveolae), which may increase surface area up to 70 %. Near the caveolae, there is a sarcoplasmic reticulum that comprises only c. 5 % of cell volume. Ca^{2+} may be stored on the plasma membrane, the reticulum and in the mitochondria (7 % of cell volume). Smooth muscle has two types of electric activity: (1) a slow or pacemaker potential that is metabolically regulated and (2) a spike potential that is caused by flow of Ca^{2+} into the cell.

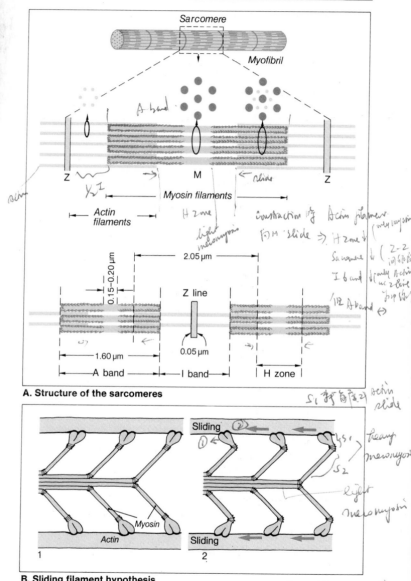

A. Structure of the sarcomeres

B. Sliding filament hypothesis

Molecular Events in Muscle Contraction

Excitation of a muscle cell to produce an action potential takes place on the cell surface; contraction of the muscle takes place inside the cell. Calcium is the universal coupler ($\rightarrow$ p. 34) between the excitation and contraction processes. In the absence of Ca^{2+}, even though an action potential can be elicited, contraction does not occur. There are two targets for Ca^{2+} action. In invertebrates, Ca^{2+} regulation of contraction is linked with myosin (My); in vertebrates, it is linked with actin (Ac) where a subunit of troponin (TN) is the Ca^{2+} receptor.

The make and break cycle ($\rightarrow$ **A**). The S1 heads of My ($\rightarrow$ p. 33, A) bind with ATP to form My-ATP. The affinity of this bonding is so great that as long as ATP is available, every S1 head is occupied.

My-ATP is altered to an intermediate or **charged state**. In this state, the My-ATP complex is stable but has a high affinity for Ac. TN on the Ac chain prevents contact with My, however, and the Ac-My complex cannot be formed. The Ac chain is switched "off" by the tropomyosin (TM) and TN.

Activation of the muscle begins with depolarization of the cell membrane. As the action potential sweeps over the membrane, it is carried inward by the transverse tubules ($\rightarrow$ p. 33, B). The signal is transmitted to the terminal cistern of the sarcoplasmic reticulum (SR). Ca^{2+} is released to the ICF where it binds with TN-C ($\rightarrow$ pp. 32) and switches "on" the Ac chain. This occurs by shifting TM in the grooves of the thin filament, thus exposing active sites on the Ac molecules for binding with My. The SR has a large capacity for Ca^{2+} (30 mmol $\cdot$ l^{-1}) and can vary the intracellular concentration from 0.1 μmol $\times$ l^{-1} at rest to 10 μmol $\times$ l^{-1} for activation of con-

traction. My-ATP binds with Ac to form an **active complex** with a short life span (c. 10 ms) ($\rightarrow$ **A2**). Normally, the ATPase of the S1 heads has low activity (20% of maximal), but when the Ac-My complex is formed, its activity increases and ATP is split.

The energy released by hydrolysis of ATP allows contraction or tension to develop ($\rightarrow$ **A3**). Energy is converted to work by swiveling of the S1 heads according to the sliding filament hypothesis.

When ATP is split, the Ac-My complex stabilizes in a **low energy state** ($\rightarrow$ **A4**). Simultaneously, Ca^{2+} is pumped out of the intracellular space as fast as it enters; one ATP is used by the SR for each two Ca^{2+} pumped.

At low Ca^{2+} concentration, ATP reacts with the Ac-My complex and splits it, forming My-ATP. In the absence of Ca^{2+}, the TM-TN chain shifts to its former position and switches the Ac chain off ($\rightarrow$ **A5**).

In the **resting state** ($\rightarrow$ **A1**), the Ac and My filaments are free to slide past each other, since there are no cross-bridge attachments to Ac. This low resistance to stretch is important for passive diastolic filling of the heart and also for motion since shortening of one muscle often requires simultaneous lengthening of its antagonist muscle. If there is no ATP available, as occurs when ATP is totally consumed after death, the Ac-My complex is not split; it persists and the muscle remains inextensible. This stable Ac-My complex is called a **rigor complex** ($\rightarrow$ **A4**) and is the basis for development of rigor mortis in the postmortem state.

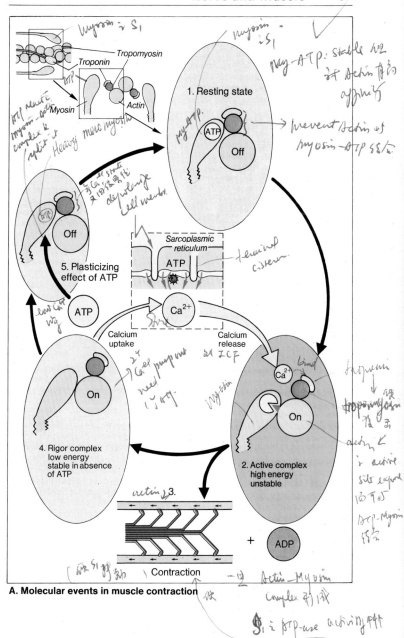

A. Molecular events in muscle contraction

Mechanical Properties of Muscle

In muscle, fibrils make up 50% of tissue volume, mitochondria 30% to 35%, reticulum and tubular systems 5%, and connective tissue most of the remainder. Connective tissue surrounds the fibrils, muscle bundles, and muscle groups (perimysium and epimysium). At the muscle endings, the connective tissue becomes continuous with the tendons. In this way, connective tissue is arranged both in series and in parallel with muscle fibers. When a muscle contracts, three elastic components participate ($\rightarrow$ A). (1) the **myofilaments**, the fundamental contractile elements ($\rightarrow$ p. 32); (2) the **parallel elastic component** (PEC): when the fibril is stretched in its resting state, the filaments slide and the sarcomeres elongate, but they do not pull completely apart because they are surrounded by connective tissue; this tissue is elastic and resists stretch like a rubber band (resting tension) ($\rightarrow$ p. 41, D, E); the PEC contributes to the tension that is generated when a muscle fiber is stretched; (3) the **series elastic component** (SEC): when a muscle tendon is cut, the muscle shortens because even at rest, the muscle is under tension. When a muscle is stimulated, the fibrils contract, but the muscle does not achieve its full shortening because the connective tissue that is in series with the fibrils stretches and counteracts the shortening. The action is comparable to the slack in a train, which must be taken up before all the cars can move. In muscle, such slack points occur to some extent at the cross-bridges between filaments but more significantly at the terminal sarcomeres where they join connective tissue and in the tendons. The total effect of the SEC slack is called **series compliance**. At maximum tension, it amounts to 3% of skeletal muscle length (10% of cardiac muscle length) and thus differs from passive stretch, which can be as much as 65%.

Summation ($\rightarrow$ B). Because of the SEC slack, a **single** stimulus, even if supramaximal, is ended before full or maximal fiber tension can be developed. If a second stimulus is applied before the contraction is completed (but after the refractory period), the tension of the second response adds to that of the first. The shorter the interval between two stimuli, the greater is the summated response. Repeated stimuli produce stepwise summation (**Treppe**) ($\rightarrow$ B) until all the SEC slack is taken up. If the maximum tension is sustained, the condition is **tetanus** ($\rightarrow$ B); it is the equivalent of prolonged summation and represents the maximum tension that the motor unit ($\rightarrow$ p. 30) can develop, which may be as much as four times the tension of a single twitch.

Muscle contraction can be measured unter two conditions ($\rightarrow$ A): (1) **isometric**: muscle length remains constant while tension is generated; (2) **isotonic**: muscle length changes while tension is maintained constant.

Length-tension (L-T) relationships ($\rightarrow$ C, p. 41, E): the total tension generated by a muscle is the sum of the active and resting tensions. **Active tension** is determined by the number of cross-bridges formed between actin and myosin and varies with the initial fiber or sarcomere length. The maximal active isometric tension (P_0) is developed at the **resting length** (L_{max}) at which the greatest number of cross-bridges will be formed. As the sarcomere shortens, the thin filaments begin to overlap and tension falls off. At still shorter lengths (70% of L_{max}), the thick filaments buckle against the plates of the Z line and tension falls off still more. When the sarcomere is stretched to lengths greater than L_{max}, the degree of overlap of the filaments decreases, fewer cross-bridges can be

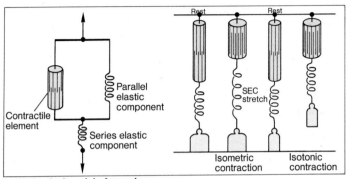

A. Mechanical model of muscle

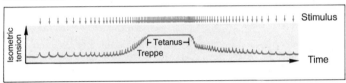

B. Stimulus frequency and muscle tension

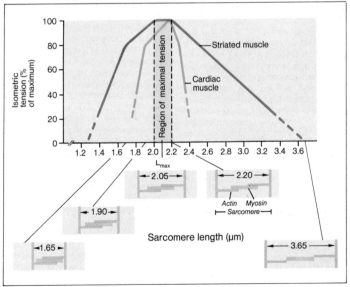

C. Isometric tension relative to sarcomere length

formed, and tension again decreases. The L-T relationship is also influenced by the electrolyte composition of the ICF; changes in Ca^{2+} concentration alter the shape and amplitude of the L-T curve. This influence permits fine modulation of muscle response and perhaps has its greatest significance in cardiac muscle (homeometric regulation).

Resting tension is developed when the fibril is stretched while in the resting state. Tension increases nonlinearly: when a skeletal muscle fiber is pulled to 125% of its resting length (L_{max}), resting tension is only 3% of the active tension; at lengths greater than 130%, the PEC contributes increasingly to resting tension and resists stretch; the limit to stretch occurs at 165% of resting length, at which point the filaments are pulled almost completely apart. Resting tension shows *hysteresis*: for a given fiber length, tension is greater during the stretch phase than during the relaxation phase.

The L-T curve is the basis for the Frank-Starling law of the **heart** ($\rightarrow$ p. 168). Clinically, it is not possible to measure sarcomere length in the heart, and it becomes necessary to estimate length and tension. For length, approximate substitutes are: end-diastolic volume or pressure, venous or atrial pressure, ventricular filling pressure; for tension, they are: stroke work or volume, cardiac output. End-diastolic tension is resting tension; stroke volume is active tension. Preload is end diastolic pressure, afterload is aortic pressure.

Differences between skeletal and cardiac muscle ($\rightarrow$ p. 43): (a) striated muscle can be stretched to greater fiber lengths ($\rightarrow$ p. 39, **C**); (b) the parallel elastic component of the heart resists lengthening and generates a much greater resting tension for a given fiber length ($\rightarrow$ **E**); (c) skeletal muscle normally operates on the plateau of the L-T vurve ($\rightarrow$ p. 39, **C**).

Heart muscle has no plateau; it operates on the ascending limb below its L_{max}; this furnishes the heart with a reserve; it allows the tension to increase when the heart is stretched by a greater diastolic filling ($\rightarrow$ p. 170); (d) when sarcomere length varies from L_{max}, active tension falls off more rapidly for the heart ($\rightarrow$ p. 39, **C**); a 10% increase from L_{max} causes a 50% fall in active tension; further stretch is resisted by an exponential increase in resting tension ($\rightarrow$ **E**); (e) series compliance is greater in the heart; (f) the heart has a longer action potential ($\rightarrow$ pp. 29, 43) and a slower onset of active state. Since tetanus can develop only if the refractory period is short, the heart cannot be tetanized.

The **Force-velocity** (F-V) curve ($\rightarrow$ **F**) relates the speed of contraction to the force generated. If shortening is slow, work is done and little heat is produced; if rapid, less work and more heat are produced. The maximum force or tension (P_0) is produced when there is no shortening, as when pulling or pushing; force depends on the rate of make-and-break of cross-bridges in the myofilaments. Maximum velocity of shortening (V_{max}) occurs when the work performed is minimal as when throwing or in playing an instrument. The F-V curve demonstrates that light weights can be lifted faster than heavy weights. The energy liberated (work and heat) is greater in isotonic than in isometric contraction. The difference between isotonic and isometric contraction is the heat of shortening and becomes smaller as the speed of shortening increases. **Contractility** can be described by the V_{max}.

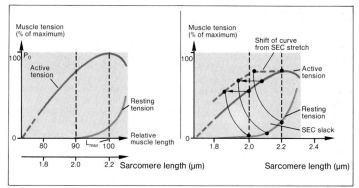

D. Active and passive tension of the muscle

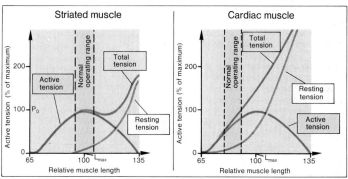

E. Length-tension curves for striated and cardiac muscle

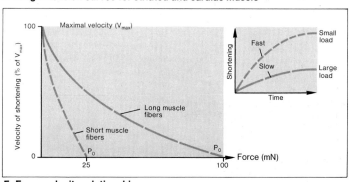

F. Force-velocity relationship

Smooth Muscle

Smooth muscle (SM) includes all the different types of muscle that do not have cross-striations. Because of technical experimental difficulties, SM has not been as thoroughly studied as have cardiac and skeletal muscles. Nevertheless, SM are of great clinical significance since they operate the viscera and control blood flow by influencing vascular resistance.

There are several different kinds of smooth muscle and no single description suffices for all of them. All SM contain filaments of F actin ($\rightarrow$ p. 32). Although a form of myosin is also present, thick filaments are not common and there is no organization into sarcomeres. The tubular system and triads of striated muscle do not occur in SM. There is no true resting membrane potential; the membrane potential is often unstable. It can vary rhythmically at a slow frequency and low amplitude with variations of only a few mV; the muscle is therefore in a constant state of partial contraction. Superimposed on the low amplitude waves are irregularly occurring **spikes** with a duration of c. 50 ms and a variable amplitude that may exceed 0 mV and become positive. A spike is followed after 150 ms by a slowly rising and falling contraction that reaches its peak 500 ms after the start of the spike. In some SM, the spike has a plateau that is similar to phase 2 and 3 of repolarization in the cardiac action potential ($\rightarrow$ p. 24). As in other muscles, the membrane potential is determined to a large extent by the K^+ gradient. The role of Ca^{2+} in modulating excitation-contraction coupling is not clearly defined.

Two major types of SM may be described: (1) SM of the *viscera*, such as intestine, ureter, bladder, uterus, shows a number of cellular bridges that connect the muscle fibers and form a **syncytium**. Cells show a spontaneous depolarization or pacemaker activity that is transmitted to adjoining cells. In this way, a contraction wave may spread slowly along a length of tissue or may die out before it reaches the end. This activity is intrinsic and is largely independent of neural stimulation. The muscle is under constant tension or **tone**. Stretch of the organ may also stimulate contraction. The muscle is unusally sensitive to circulating humoral agents. (2) SM of **blood vessels** (multi-unit SM) is not a syncytium, and contractions do not spread as in visceral muscle. This SM is more closely controlled by neural stimuli. The transmitter tends to persist at its site of release and elicits a sustained response. In general, **neural supply** *to the SM* is dual and antagonistic ($\rightarrow$ pp. 50–51). In some SM, acetylcholine relaxes and norepinephrine constricts, whereas in others the reverse is true.

Length-tension curve. SM shows both active and resting tension as do other muscles. However, the tension that can be generated at a given initial length varies. If SM is kept stretched, the tension at that length progressively decreases. In effect, the L-T curve is shifted so that peak tension develops at longer lengths. This property is **plasticity** and may be exemplified by the urinary bladder. As the bladder fills and is stretched, tension develops, falls after a short time and relieves the urgency, and then rises again as the bladder continues to fill.

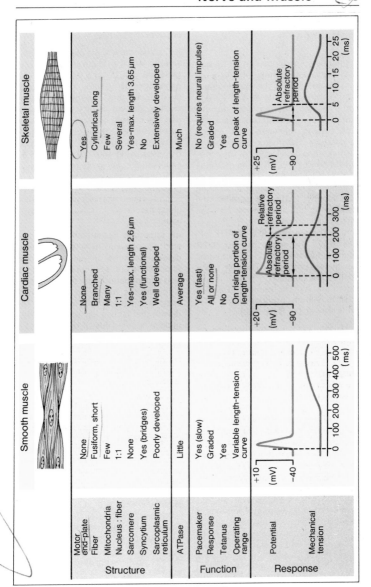

		Smooth muscle	Cardiac muscle	Skeletal muscle
Structure	Motor end-plate	None	None	Yes
	Fiber	Fusiform, short	Branched	Cylindrical, long
	Mitochondria	Few	Many	Few
	Nucleus : fiber	1:1	1:1	Several
	Sarcomere	None	Yes-max. length 2.6 µm	Yes-max. length 3.65 µm
	Syncytium	Yes (bridges)	Yes (functional)	No
	Sarcoplasmic reticulum	Poorly developed	Well developed	Extensively developed
	ATPase	Little	Average	Much
Function	Pacemaker	Yes (slow)	Yes (fast)	No (requires neural impulse)
	Response	Graded	All or none	Graded
	Tetanus	Yes	No	Yes
	Operating range	Variable length-tension curve	On rising portion of length-tension curve	On peak of length-tension curve
Response	Potential			
	Mechanical tension			

A. Structure and function of 3 classes of muscle

Energy for Muscle Contraction

The ultimate energy source for muscle contraction is glycogen, which makes up 0.5% to 1.0% of the muscle mass. The direct energy source is **ATP** ($\rightarrow$ **A**). In the muscle fibril, an Mg^{2+}-dependent ATPase is associated with the thick myosin filaments. The energy liberated by ATP hydrolysis is used in the make-and-break cycle of the actin-myosin bond ($\rightarrow$ p. 36) (one molecule ATP for each cycle). ATPase activity is greatest when muscle is shortening most rapidly. Hydrolysis of ATP, and therefore contraction of muscle, can proceed anaerobically; O_2 is not essential. The reaction: $ATP \rightarrow ADP + P_i$ + energy. **Regeneration of ATP** occurs simultaneously with its hydrolysis by two processes: The first, using creatine phosphate (CrP), is rapid but limited ($\rightarrow$ **A**). The second, oxidative phosphorylation, produces more ATP but is slower ($\rightarrow$ **B**; p. 182).

1. Muscle contains the high-energy compound **CrP**. It is possible to regenerate ATP by transferring the high-energy P bond from CrP to the ADP, which is formed during contraction. The reaction is anaerobic and requires no utilization of O_2. However, since muscle contains only c. 5 µmol/g of ATP and c. 20 to 30 µmol/g of CrP, the process is limited. The combined ATP and CrP stores allow c. 50 to 100 contractions before they are depleted. They nevertheless have an important practical value since they permit short bursts of muscular activity without the limitation of slow regenerative processes. For example, a sprinter can race at maximum effort without the necessity to breathe; his muscles operate just as efficiently for short distances as when he breathes. In addition, holding his breath furnishes the advantage of a more rigid skeletal structure for the leverage required in running. The CrP that is used in this process is restored from the energy generated during anaerobic glycolysis.

2. The second restorative process is the aerobic breakdown of **glycogen** and **glucose** to CO_2 and water with the generation of 38 mol ATP/mol glucose. When the O_2 supply is insufficient to support aerobic phosphorylation processes, anaerobic glycolysis takes place. *Anaerobic glycolysis* generates 4 mol of ATP/1 mol glucose. However, when glucose is the starting material, the *net* gain is only 2 mol ATP because of two obligatory hexose phosphorylation steps. When glycogen is the starting material, only one hexose phosphorylation is needed and the net gain is 3 mol ATP.

An important consequence of anaerobic glycolysis in muscles is the production of large amounts of *lactate*. This lactate diffuses into the blood but also accumulates in the muscle, where it lowers the pH to the point where enzyme reactions are inhibited. The reason that lactate accumulates is that muscle is incapable of metabolizing it anaerobically; it must be transported to the liver ($\rightarrow$ p. 227) or to the heart ($\rightarrow$ p. 174) for disposition. When the period of exertion is ended, extra O_2 is utilized to consume the excess lactate. The amount of extra O_2 needed is proportional to the extent to which the capacity of aerobic processes was exceeded. This extra O_2 represents the **oxygen debt** and is measured by determining O_2 utilization after exercise until a basal rate of O_2 consumption is reached. This value may be as much as six times greater than the basic O_2 consumption; the individual is capable of exerting himself six times more with the O_2 debt than without it. Anaerobic gain of energy is only possible for a limited period of time.

For **longer lasting exercise** energy must be obtained **aerobically**. The power of the long term performance (in top athletes ca. 370 W) is thus determined by the rate of oxygen supply and of the aerobic breakdown of glucose (and fat).

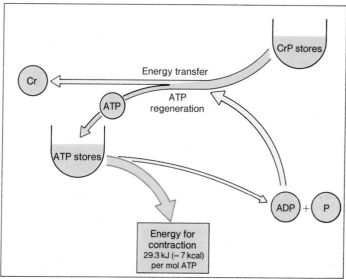

A. Energy sources of muscle

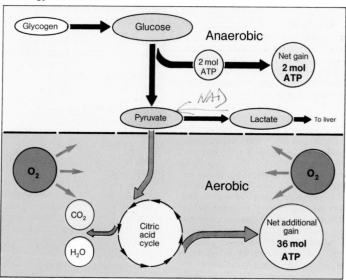

B. Aerobic and anaerobic metabolism of glucose

Exercise

The total energy output of the body determines its metabolic rate; it is the sum of **external work, heat, and stored energy**. External work is done when a mass is moved (work = force × × distance). Muscular activity is limited by both force and velocity of contraction (force-velocity curve) ($\rightarrow$ p. 40). The balance between these two depends upon the nature of the exertion. In static effort (isometric) ($\rightarrow$ p. 38), force is large and speed is low; no external work is done and energy appears almost exclusively as heat. In dynamic work (isotonic) ($\rightarrow$ p. 38), speed is high and force is low; output is limited by the availability of energy since contraction ceases when energy stores are depleted ($\rightarrow$ p. 44). At rest, muscles need 30 ml O_2/min/kg tissue but at maximum dynamic activity the need may rise to 3000 ml O_2/min/ kg. To meet these needs, there are adjustments in the cardiovascular and respiratory systems.

Distribution of blood flow ($\rightarrow$ **A**; p. 162–163). In exercise, vascular resistance increases in the kidney and GI tract; blood is shunted to the muscles where the resistance falls. Cerebral blood flow does not change. At the onset of exercise, muscle vessels dilate. The dilation is maintained by *local* changes during exercise: temperature and pCO_2 increase, pO_2 falls, lactate accumulates ($\rightarrow$ p. 44) and acidosis develops. In the *heart*, both contractility and automaticity are enhanced; the volume of blood ejected with each heart beat (= stroke volume) and the number of heart beats per minute increase. The minute volume of the heart (output/min = stroke volume × beats/min) may rise from a resting value of c. 6 l/min to as much as 35 l/min. During heavy work, the maximum attainable heart rate varies with sex and training but declines with age. At 20 years, the maximal rate is c. 200 beats/min ($\rightarrow$ **B**) but at 65 years it is only 160 beats/min.

During exercise, systolic blood pressure rises more than diastolic. During recovery, cardiac output subsides but because vasodilation in muscles persists, systolic pressure may fall below the initial level. In **capillaries**, the steady state of fluid exchange ($\rightarrow$ p. 144) is altered. There is a 10 to 100-fold increase in the number of open capillaries. Pressures change so that filtration occurs throughout the capillary; resorption is minimal. Hemoconcentration occurs and viscosity increases; there are more red cells in each milliliter of blood reaching the tissue. Lymph flow increases to carry away the larger volume of filtered water. **Pulmonary ventilation** ($\rightarrow$ p. 74) increases from a resting level of 5 to 7 l/min to well above 100 l/min. Ventilation increases in parallel with elevated O_2 consumption as a result of an increase both in respiratory rate and in tidal volume. Tidal volume, normally 10% of the vital capacity, rises to 50% at the expense of the inspiratory reserve ($\rightarrow$ **C**). Increased ventilation in exercise is always less than the maximum voluntary ventilation so that respiration is not the main limiting feature in exercise. At the **tissues**, extraction of O_2 from blood (AV difference) may triple. The decreased pH and elevated temperature shift the dissociation curve for hemoglobin ($\rightarrow$ p. 89) to the right and more O_2 is given up.

Training. The chief distinction of the trained athlete is the ability to exercise without raising his blood level of lactate. Without training, lactate rises at only 40% of the maximum O_2 consumption; with training, lactate is unchanged even at 90% of maximum O_2 consumption.

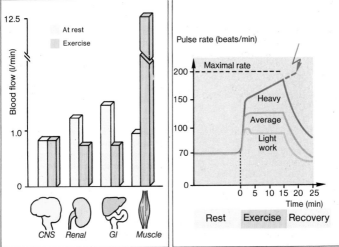

A. Blood flow response to exercise

B. Work load and pulse rate

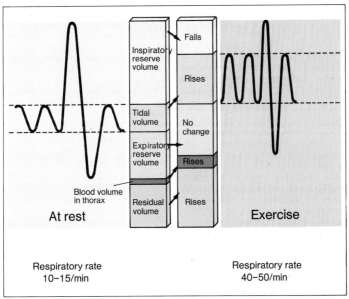

C. Respiration and exercise

Organization of Autonomic Nervous System

In animal evolution, before development of a centralized integrated nervous system, segmental organization was achieved by localization of nerve cells in ganglia. Ganglia persist in vertebrates, receiving fibers from both brain and spinal cord. Paired ganglia occur in chains along the spinal column; other ganglia, paired or unpaired, occur at various locations. These ganglia and their associated nerve fibers comprise the autonomic nervous system (ANS). The ANS is arranged on the same principles as the somatic nervous system (SNS) and is based functionally on the reflex arc with afferent and efferent limbs. **Afferent** fibers mediate visceral sensation, pain, and reflexes from blood vessels ($\rightarrow$ p. 167), lungs ($\rightarrow$ p. 92), and GI tract ($\rightarrow$ p. 197). **Integration** can occur in the spinal cord. Respiration and cardiovascular functions are integrated in the medulla. The hypothalamus ($\rightarrow$ p. 270) coordinates the ANS with the SNS via its connections to thalamus and cortex. **Efferent** fibers activate involuntary functions served by smooth muscle, heart, and glands.

The ANS is comprised of two functionally and anatomically distinct divisions ($\rightarrow$ **A**): sympathetic (symp) (adrenergic) and parasympathetic (p-symp) (cholinergic). In general, sympathetic activity accomplishes quick mobilization for response to emergencies (fight-or-flight). Metabolic stores are released and utilized and circulation is increased. In contrast, parasympathetic activity accomplishes restoration, digestion, and recovery.

Many tissues are dually innervated. The tissue response to the two divisions may be antagonistic (symp stimulates the heart, p-symp suppresses it) or parallel (salivary secretion is stimulated by symp and p-symp). Some organs are not dually innervated

(blood vessels have predominantly symp fibers).

Sympathetic efferent fibers originate in the **thoracic** and **lumbar** regions of the spinal cord. Preganglionic fibers either end in the paravertebral ganglionic chain (short fibers) or pass through the chain to unpaired ganglia (celiac and mesenteric ganglia) ($\rightarrow$ pp. 50–51). The preganglionic neurotransmitter is acetylcholine (AcCh) ($\rightarrow$ p. 52). From the ganglia, relatively long unmyelinated fibers innervate the end organs. The postganglionic neurotransmitter is norepinephrine (NE) ($\rightarrow$ p. 54).

Parasympathetic efferent fibers originate in nuclei of the cranial nerves (III, VII, IX, and X) and the sacral division of the spinal cord ($\rightarrow$ **A**). P-symp ganglia are located near the effector organ so that preganglionic fibers tend to be long and postganglionic fibers short. In both fibers, the neurotransmitter is AcCh.

The adrenal medulla ($\rightarrow$ p. 56) is essentially an autonomic ganglion. It receives cholinergic preganglionic fibers. Its cells liberate epinephrine (Ep) and norepinephrine (NE) ($\rightarrow$ p. 56), which are distributed by the blood to the effector organs. The symp postganglionic fibers to the sweat glands liberate AcCh instead of NE.

When the autonomic fiber to an effector organ is cut, the effector becomes increasingly sensitive to circulating neurotransmitter. Similarly, after chronic pharmacologic blockade is removed, a rebound sensitivity may be observed.

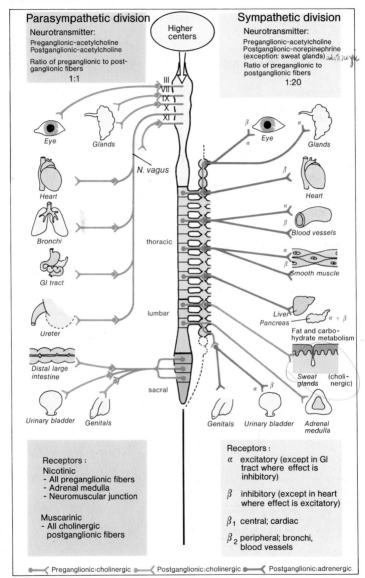

Parasympathetic division

Neurotransmitter:
Preganglionic-acetylcholine
Postganglionic-acetylcholine

Ratio of preganglionic to postganglionic fibers
1:1

Higher
centers

Sympathetic division

Neurotransmitter:
Preganglionic-acetylcholine
Postganglionic-norepinephrine
(exception: sweat glands)
Ratio of preganglionic to
postganglionic fibers
1:20

III
VII
IX
X
XI

Eye
Glands

β Eye α
α
Glands

N. vagus

Heart

β Heart

Bronchi

α
β Blood vessels

GI tract

thoracic

α
β Smooth muscle

Ureter

lumbar

Liver
Pancreas $\alpha + \beta$
Fat and carbo-
hydrate metabolism

Distal large
intestine

sacral

Sweat (choli-
glands nergic)

Urinary bladder Genitals

Genitals Urinary bladder Adrenal
medulla
α β

Receptors :
Nicotinic
- All preganglionic fibers
- Adrenal medulla
- Neuromuscular junction

Muscarinic
- All cholinergic
  postganglionic fibers

Receptors :
α excitatory (except in GI
tract where effect is
inhibitory)

β inhibitory (except in heart
where effect is excitatory)

β_1 central; cardiac

β_2 peripheral; bronchi,
blood vessels

Preganglionic:cholinergic   Postganglionic:cholinergic   Postganglionic:adrenergic

A. Survey of autonomic nervous system

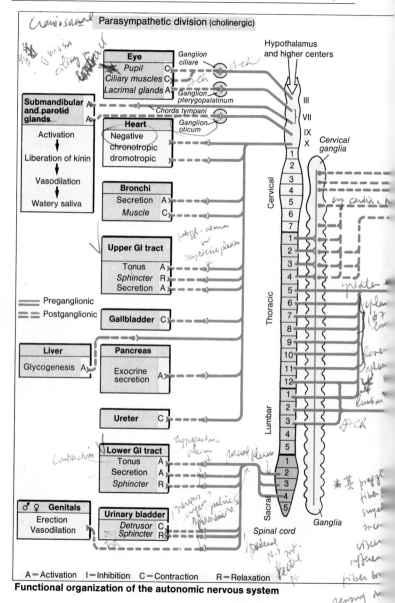

Parasympathetic division (cholinergic)

Eye
Pupil — C
Ciliary muscles — C
Lacrimal glands — A

Ganglion ciliare

Hypothalamus and higher centers

Ganglion pterygopalatinum

Submandibular and parotid glands — A

Activation
↓
Liberation of kinin
↓
Vasodilation
↓
Watery saliva

Chorda tympani

Heart
Negative chronotropic dromotropic

Ganglion oticum

III
VII
IX
X

Cervical ganglia

1
2
Cervical
3
4
5
6
7

Bronchi
Secretion — A
Muscle — C

Upper GI tract
Tonus — A
Sphincter — R
Secretion — A

1
2
3
4
5
Thoracic
6
7
8
9
10
11
12

═══ Preganglionic
═ ═ Postganglionic

Gallbladder — C

Liver
Glycogenesis — A

Pancreas
Exocrine secretion — A

1
2
Lumbar
3
4
5

Ureter — C

Lower GI tract
Tonus — A
Secretion — A
Sphincter — R

1
2
Sacral
3
4
5

Ganglia

Spinal cord

♂ ♀ **Genitals**
Erection
Vasodilation

Urinary bladder
Detrusor — C
Sphincter — R

A = Activation I = Inhibition C = Contraction R = Relaxation

Functional organization of the autonomic nervous system

Sympathetic division (preganglionic cholinergic, postganglionic mostly adrenergic)

α - adrenergic	β - adrenergic	
Eye	**Eye**	
D *Pupil*	Ciliary muscle for distance accommodation	
Submandibular gland	**Heart**	**Sweat glands**
A Viscous salivary secretion	Positive chronotropic dromotropic inotropic	V—A Sympathetic cholinergic fibers
V—C **Piloerection**	R β₂ **Bronchi**	**Blood vessels (muscle)**
		V—D Sympathetic cholinergic vasodilation is of questionable physiologic importance
Upper GI tract	**Upper GI tract**	
C *Sphincter*	R *Tonus*	
	R **Gallbladder**	
Pancreas	**Pancreas**	**Cholinergic**
I Insulin secretion		**Adrenal medulla**
I Exocrine secretion	A Insulin secretion	A Secretion

Preganglionic fiber

Blood vessels		
C **Spleen**	V—D *Skin, muscles and others*	
Blood vessels	**Fat cells**	
C *Skin, skeletal muscles, GI-tract and others*	V—C Lipolysis	
	Liver	♂ **Genitals**
	Gluconeogenesis	Ejaculation
Urinary bladder	**Urinary bladder**	
C *Sphincter*	R *Detrusor*	
C *Pregnant uterus*	R *Nonpregnant uterus*	

Ganglion coeliacum

Ganglion mesentericum sup. et inf.

D = Dilatation   V = Segmental innervation

Neurotransmitters: Acetylcholine (AcCh)

Acetylcholine (AcCh) is the neurotransmitter at all autonomic preganglionic nerve endings, at parasympathetic postganglionic endings, at the neuromuscular junction ($\rightarrow$ p. 30), at several synapses in the central nervous system and in some sympathetic postganglionic endings.

Synthesis of AcCh occurs in the cytoplasm of the nerve endings. Acetylcoenzyme A (AcCoA) is synthesized in the mitochondria, and its acetyl group is transferred to choline in the presence of the enzyme choline acetyltransferase (ChAT). ChAT is synthesized in the soma of the neuron and is transported by axoplasmic flow to the nerve endings. Choline itself cannot be synthesized in nerves and must be accumulated from the extracellular fluid (ECF) by a sodium-dependent active membrane transport process. This is the rate-limiting step in AcCh synthesis. Plasma choline concentration is constant at 5 to 10 μmol $\times l^{-1}$. Transport of choline and therefore synthesis of AcCh can be blocked by *hemicholinium*. When AcCh is broken down in the synaptic cleft, 50% of the choline that is produced can be recovered by reuptake into the nerve endings; recovery is increased by nerve stimulation.

Storage and release. AcCh is stored in vesicles in the nerve endings. The total amount stored remains constant because synthesis of AcCh varies to match loss when AcCh is released into the synaptic cleft. In the resting nerve, there is a continuous slow release of vesicle content by exocytosis. Each vesicle releases the same quantum of AcCh, c. 4000 molecules, but a single quantum is insufficient to fire a postsynaptic action potential (AP) ($\rightarrow$ p. 28). However, when an AP in the presynaptic neuron reaches the nerve ending, Ca^{2+} is mobilized and several hundred quanta are released. These combine with receptors in the postsynaptic neuron to cause a change in membrane conformation, which increases permeability to Na^+, K^+, and Ca^{2+}. This action generates an EPSP ($\rightarrow$ p. 28). When sufficient quanta are released to raise the EPSP above the threshold potential ($\rightarrow$ p. 24), the postsynaptic neuron generates and propagates an AP. Termination of AcCh action occurs by hydrolysis of AcCh by cholinesterase (AcChE) but also by diffusion of AcCh into the ECF and by reuptake into the presynaptic nerve ending. Since nerve impulses can be transmitted at frequencies of several hundred per second, hydrolysis of AcChE must be very fast (ms). True AcChE occurs in neurons and in the neuromuscular junction at sites that are close to the AcCh receptor. Pseudo-AcChE occurs in plasma, liver, and kidney. Inhibitors of AcChE are effective only against true AcChE.

There are two classes of receptors for AcCh: (1) **Nicotinic** receptors occur at autonomic ganglia, motor endplates ($\rightarrow$ p. 30), and probably also in the central nervous system at Renshaw cells in the spinal cord. These receptors respond to nicotine in the same way as to AcCh. At low concentrations, both nicotine and AcCh stimulate the postsynaptic receptor; at high concentrations, they inhibit it ($\rightarrow$ p. 30). (2) **Muscarinic** receptors occur at all cholinergic effector cells. These receptors respond to muscarine and similar drugs, in addition to AcCh. Stimulation of muscarinic receptors in the heart is inhibitory (slowing of heart rate) and in the GI tract is excitatory (increased motility).

Nicotinic receptors in skeletal muscle can be blocked by *curare* ($\rightarrow$ p. 30); muscarinic receptors in heart, smooth muscle, and in other sites can be blocked by *atropine*.

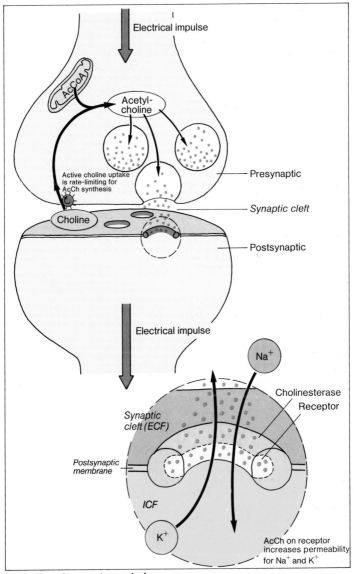

A. Cholinergic neurotransmission

Neurotransmitters: Catecholamines (NE and Ep)

Norepinephrine (NE) is the neurotransmitter at sympathetic postganglionic nerve endings and at some synapses in the central nervous system, especially in the hypothalamus. In the adrenal medulla ($\rightarrow$ p. 56), both epinephrine (Ep) and NE are released into the blood stream for humoral transmission to reinforce neurotransmission.

Synthesis and storage. Unmyelinated sympathetic postganglionic fibers are characterized by varicosities or swellings along their terminal fibers, which are sites for synaptic contact. The membranes of these swellings can actively transport tyrosine (Tyr) for synthesis of NE via L-dopa and dopamine. The conversion of Tyr to L-dopa by Tyr-hydroxylase is the rate-limiting reaction; it is activated by Na^+ and Ca^{2+} and can be retarded by NE (negative feedback, $\rightarrow$ p. 218). When NE is discharged from the neuron, the decrease in intracellular concentration of NE stimulates the enzyme reaction to synthesize more NE. By this means, neuronal stores of NE are kept constant. The NE is stored in large granules in combination with a protein, enzymes, and ATP in a micellar ($\rightarrow$ p. 204) complex.

Release of NE into the synaptic cleft occurs by exocytosis when an action potential (AP) reaches the nerve ending. Several drugs can influence storage and release. Release is prevented by an ATPase; PGE_2, a prostaglandin, inhibits release by stimulating the ATPase. In contrast, when an AP reaches the nerve ending, it increases influx of Na^+ and Ca^{2+}, which inhibit the ATPase and permit release of the NE. Simultaneously, Na^+ stimulates synthesis of new NE.

Adrenergic receptors. Two classes of receptors, α and β, are characterized by their sensitivity to adrenergic agonists. β are sensitive to isoproterenol (IPR) $\geq$ Ep > NE; α are sensitive to NE > $\geq$ Ep $\geq$ IPR. Stimulation of α generally causes an excitatory response such as

vasoconstriction or muscular contraction (ureter, uterus, sphincter, splenic capsule). In pancreas α inhibit insulin and exocrine secretion. α stimulation tends to reduce the ratio cAMP/cGMP ($\rightarrow$ p. 222) and influences Ca^{2+} influx at the effector cell. There are two types of β; both types appear to operate via cAMP (increased cAMP/cGMP ratio). β_2 tend to increase Ca^{2+} efflux and are inhibitory; they relax smooth muscle in bronchi, uterus, GI tract, and in blood vessels of skeletal muscle. They are most sensitive to IPR and Ep. In contrast, β_1 tend to increase Ca^{2+} influx and are excitatory; they stimulate lipolysis and activate insulin secretion. In the heart, β_1 cause positive inotropic and chronotropic ($\rightarrow$ p. 150) responses. (β_1 in the heart do not appear to be uniform; although Ep and NE are nearly equipotent in causing inotropic responses, Ep has greater chronotropic action than NE.)

Termination of transmitter action occurs by four mechanisms ($\rightarrow$ **A**): (**1**) Excess NE in the synaptic cleft diffuses into capillaries (overflow). (**2**) NE in the cleft is inactivated by catechol-O-methyltransferase (COMT). (**3**) The most important process is **reuptake** into the presynaptic nerve ending by a Na-dependent active transport process. As much as 70% of NE in the cleft can be recovered by this mechanism. Reuptake can be inhibited by a variety of agents. (**4**) Free NE in the cell is inactivated by mitochondrial monoamine oxidase (MAO). End products of metabolism are excreted in the urine; they vary according to the enzyme involved and the site of inactivation.

Feedback. The presynaptic membrane of adrenergic fibers contains α-receptors. These respond to NE in the synaptic cleft and inhibit release of more NE from the granules. Their affinities for adrenergic transmitters differ from those of α-receptors in the postsynaptic membrane ($\rightarrow$ p. 57, A).

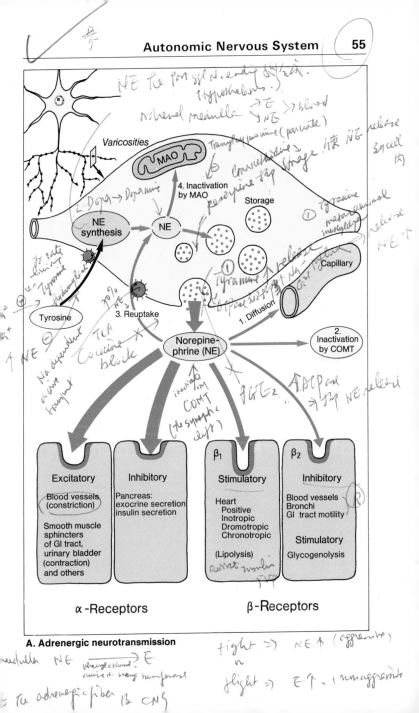

A. Adrenergic neurotransmission

Humoral Transmitters: Adrenal Medulla

The adrenal medulla is a neuroendocrine transducer. It releases both epinephrine (Ep) and norepinephrine (NE) (catecholamines) in response to autonomic stimulation by sympathetic preganglionic fibers ($\to$ pp. 50 to 51). The neurotransmitter at these fiber endings is AcCh. The adrenal medulla normally functions at a low basal level of activity. Increased secretion is recruited in stressful or emergency situations to support the NE-dependent functions of the autonomic sympathetic division.

Synthesis. As in sympathetic nerves ($\to$ p. 54), the adrenal medulla accumulates tyrosine and converts it to NE via dopa and dopamine. It contains an additional enzyme, phenylethanolamine - N - methyltransferase, which methylates NE to generate Ep. Activity of this enzyme is enhanced by cortisol ($\to$ p. 240). The ratio Ep:NE depends upon activity of the transmethylase and differs according to species. The ratio correlates roughly with the fight-or-flight response of animals; nonagressive, hunted animals (rabbit) have relatively greater Ep, whereas aggressive, hunting animals (tiger, lion) have greater NE.

Storage. Like all water-soluble hormones, Ep and NE are stored intracellularly in membrane-delineated vesicles or granules. These contain Mg^{2+}, protein (11%), lipid (7%), and 1 mol ATP for every 4 mol of catecholamine. There is a slow continuous leak out of these vesicles, but the concentration of catecholamine in the cytoplasm is held to a minimum by the action of mitochondrial MAO ($\to$ p. 54). (In contrast, AcCh is tightly held in its vesicles and does not leak into the cytoplasm. Instead, there is a continuous release of vesicles into the extracellular fluid ($\to$ p. 52).

Release. Stimuli for release are sympathetic discharge, exercise, cold, heat, hypoglycemia, pain, hypoxia, and fright or anxiety. The hypothalamus ($\to$ p. 270), partly through its pain and hunger centers, is intimately involved with neural control of the adrenal medulla. Discharge of AcCh in the medulla changes membrane permeability of medullary cells, which allows fusion of the vesicles with the membrane (exocytosis) and discharge of both Ep and NE. The vesicle is the all-or-none unit of neurotransmitter action.

Actions. In response to an emergency, Ep and NE **mobilize energy stores** for increased muscular activity. **Lipolysis** and **glycogenolysis** ($\to$ p. 226) are stimulated. Simultaneously, **insulin** secretion is turned off ($\to$ p. 229). This action partially reduces uptake of glucose by muscle and shunts it to the central nervous system, where glucose uptake is not influenced by insulin. In the **muscle**, stimulation of cAMP activates phosphorylase from its β form to its α form to increase glycogenolysis. In the **heart**, a positive inotropic effect ($\to$ p. 150) increases cardiac rate, blood pressure, blood flow and cardiac output. Blood is shunted from the splanchnic to the muscle circulation ($\to$ p. 47). At the same time that these catabolic functions are stimulated, catecholamines, by a stimulatory action in the *hypothalamus* (CRH $\to$ ACTH $\to$ Cortisol) ($\to$ p. 240), release hormones that initiate reactions to restore the original preemergency state. These hormones reach their peak c. 4 h after the emergency.

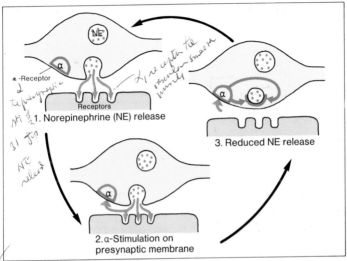

A. Feedback control of NE release

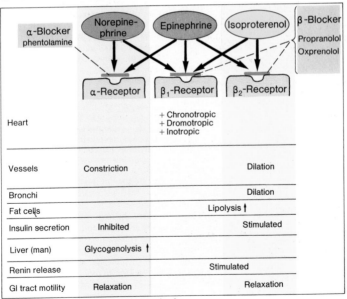

B. Adrenergic stimulation and blockade

Composition and Function of the Blood

The **blood volume** of the adult human comprises 6 to 8% of body weight; 1 l of blood contains 0.46 l red blood cells in males, 0.41 l in females. This value which may be also expressed as a percent (46% in males) is the **hematocrit**. In $1 mm^3$ ($= 1 \mu l$) of blood there are 5×10^6 red blood cells (**erythrocytes**) in males, 4.5×10^6 in females, 4 to 6×10^3 **leukocytes**, and 0.15 to 0.3×10^6 **thrombocytes**. A normal *differential count* of leukocytes includes about 67% granulocytes, 27% lymphocytes, and 6% monocytes.

The fluid phase of the blood is the **plasma**; it has an osmolar concentration of c. 290 mOsm/l and contains 65–80 g protein/l. The various **proteins** in the plasma include albumin (55%), α_1-globulin (4%), α_2-globulin (8%), β-globulin (12%), γ-globulin (18%), and fibrinogen (3%) ($\rightarrow$ p. 64–67). When the blood clots, fibrinogen is used up; the remaining fluid is **serum**. Serum and plasma are similar except for the fibrinogen content.

Function of blood. Transport of dissolved and particulate matter (O_2, CO_2, nutrients, metabolites, etc.); *transport of heat* for heating and cooling; **transmission** of signals (hormones); **buffering** the body fluids (acid-base balance); **defense** ($\rightarrow$ p. 60, 62).

Functions of blood proteins. Immune defense reactions; maintenance of colloid osmotic (= oncotic) pressure; transport of water-insoluble substances (lipids $\rightarrow$ lipoproteins). Substances that are bound to proteins cannot be excreted by the kidneys and are ineffective osmotically.

Erythrocytes (red blood cells, RBC) are synthesized in bone marrow. Iron, and the vitamins B_{12} and folic acid, are some of the prerequisites. In the fetus, erythrocytes are produced also in the spleen and liver. Immature RBC in the marrow are nucleated but lose their nuclei when they reach the bloodstream. Their dimensions (7.5×2 μm) are such that they are deformed in the small capillaries.

The chief function of the RBC is to carry O_2 and CO_2 between lungs and tissues; it is accomplished by hemoglobin (Hb) (male 160 g/l blood, female 145 g/l). The high **intra**cellular Hb concentration (c. 300 g/l) is a large component of the cellular osmolarity and requires that the intracellular electrolyte concentration be maintained at a lower level than that in the plasma. To accomplish this, the RBC membrane contains an active transport system for Na^+ and K^+. The chemical energy (ATP) needed for this purpose comes from anaerobic glycolysis ($\rightarrow$ p. 45, B).

Production of RBCs is influenced primarily by hormones. O_2 deficiency in the tissues (cellular hypoxia) is the initiating event in the production and release of **erythropoietin**, which stimulates red cell synthesis in the bone marrow. More than 90% of erythropoietin is produced in the kidney, the rest mainly in the liver. When the red cell mass increases and corrects the hypoxia, synthesis of erythropoietin subsides after some hours.

Lactate sensitive cells have been found near the juxtaglomerular apparatus ($\rightarrow$ p. 139, A). In hypoxia lactate production rises because of increased anaerobic glycolysis ($\rightarrow$ p. 44). Therefore, it has been speculated that the lactate sensitive cells produce erythropoietin. Another influence on erythropoiesis is the CNS, which can stimulate the marrow to discharge stored RBC.

The life span of the RBC is c. 120 days. Aged RBC are removed from the blood in sinuses of the spleen and are degraded. The fragments are broken down in the reticuloendothelial system (RES) of the spleen, liver, bone marrow, etc. When the membrane of the RBC ruptures (hemolysis), Hb is set free and is metabolized to globin and bilirubin ($\rightarrow$ p. 202). The iron of Hb is recycled. In spherocytic anemia, the *osmotic fragility* of the RBC is greater than normal, and its life span is reduced.

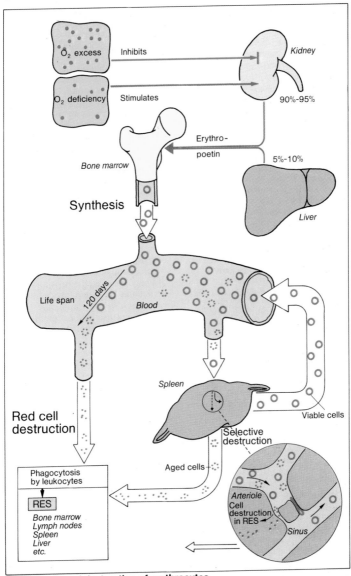

A. Synthesis and destruction of erythrocytes

Immune Defense Mechanisms
Blood Groups

The body has three immune mechanisms for protection against invasion by foreign high molecular weight substances (foreign proteins, bacteria, viruses): (1) a nonspecific cellular (→ p. 62), (2) a specific cellular (→ p. 62), and (3) a specific humoral defense. **Cellular immunity** is mediated by lymphocytes. It accounts for delayed reactions (rejection of transplants; poison ivy) and depends for its development on the presence of the thymus during maturation: T-lymphocytes. **Humoral immunity** is mediated by antibodies (immunoglobins = Ig) in the blood.

Ig are principally α-, β- and γ-globulins; they are manufactured in plasma cells (→ p. 62) of the RES (lymph nodes, bone marrow, spleen, and liver). The immunoglobins are divided into five general categories. In order of decreasing concentration in the blood, they are: IgG, major antibacterial and antiviral activity; IgA, in saliva, tears etc.; IgM, RBC agglutinins etc.; IgD, function not known; IgE, parasitic infections; hay fever etc.

When microorganisms invade a tissue, they initiate a disease process and they act as **antigens**. In response, the body produces **antibodies** to the invaders to inactivate them. After recovery from the disease, a later exposure to the same microorganisms stimulates a much more rapid antibody response that inactivates the invader before it can initiate a disease process (→ **A**). **Immunization** is a means of stimulating the antibody response without the necessity of first having the disease. In **active immunization**, antibodies are stimulated by injecting nonpathogenic bacteria or attenuated pathogens. If the disease has already become established, serum containing antibodies from animals that have had the same disease is injected (**passive immunization**). Antibodies can also be produced to foreign substances of high molecular weight that enter the body by inspiration (grass pollen) or by ingestion (lobster, strawberries). Even substances of low molecular weight can react with natural proteins as haptens and stimulate antibodies to the combination. When the exposure is repeated, an **allergic response** (e.g., asthma, hives) can result. The antigen-antibody reaction can liberate varying amounts of histamine; a severe reaction can lead to edema and shock (anaphylaxis).

The membranes of the **erythrocytes** have antigenic properties, called **agglutinogens** (→ **B**). The most important of these, the A and B agglutinogens, determine the four major **blood types**: A, B, AB, 0. Individuals with type A in the RBCs always have an antibody to B (anti-B agglutinin) in the circulation and vice versa. When plasma from a type A individual is mixed with type B red cells, the RBC will clump (agglutinate) and then hemolyze. RBCs of type 0 have both anti-A and anti-B agglutinins in serum, while those of type AB have no agglutinins (→ **B**). Before transfusing blood, cross matching should be performed to test for compatibility. Donor's cells and recipient's plasma are mixed on a slide. If they are incompatible, agglutination and hemolysis will be observed.

The Rh (rhesus) system is highly complex and has a large number of antigenic determinants. In the Western world, 85% of the population are Rh$^+$, i.e., they are homozygous or heterozygous for the Rh gene. The 15% who are rh$^-$ have a 50% chance of forming antibodies after a single transfusion of Rh$^+$ blood and an 80% chance after repeated transfusion. Similarly, in an rh$^-$ mother, an Rh$^+$ fetus can stimulate maternal antibodies to the fetal red cells (→ **C**). Depending on the magnitude of the reaction in the mother, various degrees of hemolysis and jaundice in the fetus may result. If hemolysis is extreme, fetal damage may be irreversible.

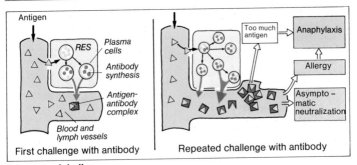

A. Immunoglobulins

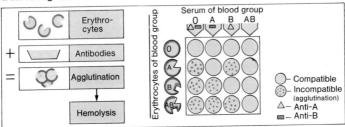

B. AB0 blood group compatibilities

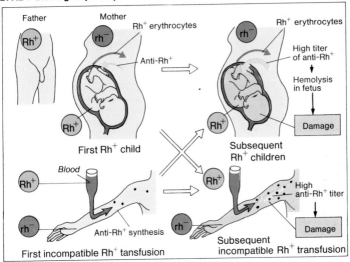

C. Rh sensitization (pregnancy or transfusion)

Cellular Defense Mechanisms

In man there are normally 4000 to 6000 leukocytes (WBC) per mm³ (µl) of blood. These include (→ p. 58) granulocytes (polymorphonuclear leukocytes) (PMN), lymphocytes (LMC), and monocytes (MNC). The PMNs include neutrophils, basophils, and eosinophils.

The PMNs are formed in the bone marrow. They have a life span in the circulation of c. 30 h. They (and the RES) (→ p. 58–60) are involved in the **unspecific cellular defense mechanisms** against bacteria and viruses, foreign particles, and under certain circumstances, endogenous substances (e.g., RBC fragments). In infections, their production increases dramatically. They contain lysosomal hydrolytic enzymes, histamine and peroxidase. They are attracted to an infected area by bacterial products (*chemotaxis*) or by changes in electric potential of damaged cells (*galvanotaxis*). They can migrate through capillary walls (*diapedesis*) to extravascular spaces where they ingest the bacteria (**phagocytosis**) (→ **A**). Peroxides are generated to kill the bacteria. As a second line of defense, the MNC, which are also phagocytic, follow PMNs into the infected area. Eosinophils phagocytize antigen-antibody complexes (→ p. 60) and are mobilized in allergies.

LMCs enter the blood stream via the lymphatics (→ **B, C**) and are destroyed in the RES. Two populations of LMC develop from the stem cells: those dependent on the thymus (**T-lymphocytes**) and those independent of the thymus (**B-lymphocytes**). The B cells mature into plasma cells (→ p. 60). Development of cellular immunity in T cells requires the thymus in early life; humoral immunity (→ p. 60) does not. An initial exposure of small LMCs to antigens is followed by production of **cellular (sessile) antibodies**. Immunoblasts, a larger cell form, are formed from small LMCs and play a role in this process.

Humoral immune response can result in inactivation of the invader by antigen antibody complex formation within minutes. In contrast, cellular reactions may take days (e.g., skin test for TB). The immune systems of an organism are adapted in the perinatal period to distinguish between autologous proteins of the individual, which should not elicit an immune response, and foreign proteins, which should. The organism's own proteins produce an "**immunologic tolerance**" early in life; proteins to which the individual is exposed in later life are not protected by this tolerance and stimulate production of antibodies. The newborn is immunologically deficient in the first few months and receives immunologic protection from the mother's blood before birth (IgG) and from her milk. This humoral immune protection must suffice until the intrinsic immune systems develop and mature.

The cellular immune system is the reason for frequent *rejection of tissue transplants* such as kidney or heart (*host-vs.-graft reaction*) since the donor's tissues contain histocompatibility antigens that are recognized as foreign proteins by the recipient; exceptions occur in identical twins. The recipient may be given immunosuppressive compounds to produce a transient tolerance of the host tissue. A reaction may also occur when an immune competent tissue produces antibodies to the recipient (*graft-vs.-host reaction*). The cellular immune system plays a role in destruction of autogenous cells that deviate genetically (e.g., cancer cells). This process is advantageous but may be perverted to produce immune response to the individual's tissues and to produce a number of **autoimmune** reactions and disease states.

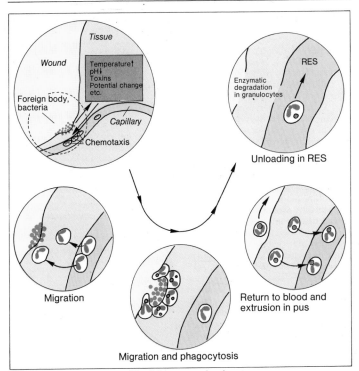

A. Granulocytes - nonspecific defense mechanisms

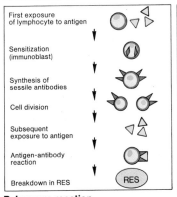

**B. Immune reaction -
   specific defense mechanism**

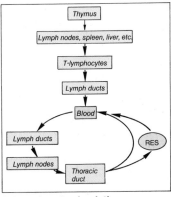

C. Lymphocyte circulation

Hemostasis

Injury of a blood vessel initiates a series of reactions involving *platelets* (*thrombocytes* or TMC), *plasma*, and *tissue factors*; it leads to plugging of the vessel. This process is **hemostasis** and is adapted to prevent or minimize blood loss. Damage to the vascular endothelium; e.g., when the vessel is punctured or cut, ($\rightarrow$ **A**), brings blood into contact with subendothelial collagen fibers. **Platelets** at the site of injury are influenced to stick together (aggregation) and to stick to the injury (adhesion). This plug of TMCs is a **white thrombus** and stops the leak of blood provisionally. During this phase, the TMCs change form (viscous metamorphosis) and liberate serotonin (5-HT), ADP, and other substances. The 5-HT stimulates constriction of the vessel and serves to reduce blood loss. The ADP attracts other platelets. In addition, the inner layer of the vessel rolls inward and reduces the diameter of the cut end. At the same time, the actual blood clotting processes that lead to a proper clot are set into motion by two mechanisms: (a) the **extrinsic pathway** ($\rightarrow$ p. 67), which is comprised of tissue factors set free when tissue is damaged, and (b) the **intrinsic pathway** ($\rightarrow$ p. 67), which begins when blood factor XII contacts the collagen fibers. Each pathway can activate factor X, either singly or in combination. **Factor X** in turn activates the sequences that convert **prothrombin** (factor II) to **thrombin** and allow **fibrin** to be formed from **fibrinogen** (factor I) ($\rightarrow$ **B**). **Phospholipids** ($\rightarrow$ **B**) are liberated from the TMC (thrombocyte factor 3, TF3) or from damaged tissues; these are essential both for activation of the intrinsic pathway and for effectiveness of the activated factor X (= Xa). The fibrin that is formed polymerizes to form a loose mesh of fibers. These fibers enmesh TMC and RBC and condense to form the definitive "mixed (red)clot" ($\rightarrow$ **B**). This clot undergoes three subsequent stages: (1) **retraction** (mediated by contractile proteins of TMC) (2) **organization**, during which fibroblasts proliferate, and (3) **scar formation**. These are followed by complete recovery and regeneration of the endothelium.

Vitamin K is essential for the formation of the plasma clotting factors prothrombin (II), VII, IX, and X. After synthesis of their protein chains these factors are carboxylated at some N-terminal glutamyl residues ($\rightarrow$ γ-carboxyglutamyl). Vit K is a cofactor of this enzymatic step. The γ-carboxyglutamyl residues bind to Ca^{2+} which in turn combine with the phospholipids involved in clotting. Vitamin K is normally supplied by the bacteria of the gut. Therefore *vitamin K deficiency* occurs mainly if the gut flora is damaged, as for example by orally administered antibiotics.

Factor		half-life in vivo (h)
I	fibrinogen	96
II	prothrombin	72
III	thromboplastin, -kinase, tissue factor	
IV	ionized Ca^{2+}	
V	accelerator globulin	20
(VI	no longer used)	
VII	proconvertin	5
VIII	antihemophilic globulin (A)	12
IX	plasma thromboplastic component (PTC), Christmas f.	24
X	Stuart-Prower factor	60
XI	plasma thromboplastin antecedent (PTA)	48
XII	Hageman factor	60
XIII	fibrin-stabilizing factor	120

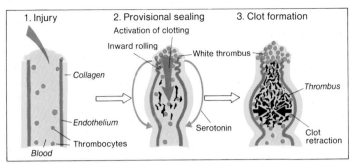

A. Hemostasis

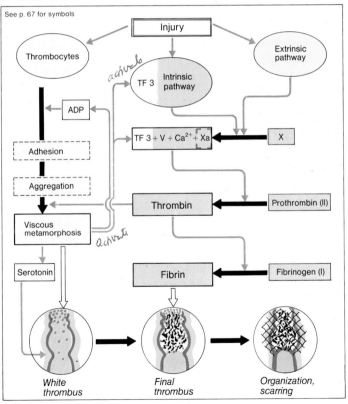

B. Hemostasis

Blood Clotting and Fibrinolysis

The clotting mechanism ($\rightarrow$ **A**) involves a series of reactions in a **cascade** that ends with the formation of a fibrin clot. The cascade starts with the aggregation of platelets and the activation of the **intrinsic pathway**. Its initial reaction is conversion of inactive factor **XII** to active factor XIIa. kallikrein (activated from prekallikrein by XIIa) and other proteins are cofactors. The activation can occur when blood is brought into contact with collagen underlying an endothelial injury, the walls of a test tube, or other surfaces. The requirements for the intrinsic pathway are the plasma factors including Ca^{2+} and the thrombocytic factor, TF3 ($\rightarrow$ **A**). In the case of widespread damage involving tissues as well as blood vessels, the **extrinsic pathway**, involving Ca^{2+}, tissue factors, and plasma factor VII, comes into play.

Both pathways, either individually or together, activate factor **X** ($\rightarrow$ p. 64) to factor Xa. Factor Xa catalyzes conversion of prothrombin to thrombin in the presence of phospholipids (from either platelets or tissues), factor V, and Ca^{2+} ($\rightarrow$ **A**). **Thrombin** has three major effects: (1) it catalyzes conversion of fibrinogen to fibrin for clot formation, (2) it activates the fibrin-stabilizing factor XIII to XIIIa ($\rightarrow$ **A**), and (3) it influences the thrombocytes during hemostasis ($\rightarrow$ p. 65, **B**). Fibrin is formed as a loose weave of single, monomeric strands that polymerize. Stabilization of fibrin (fibrin$_i$) is under the influence of factor XIIIa ($\rightarrow$ **A**).

Once the clotting cascade is initiated, mechanisms must operate to prevent clotting from spreading throughout the intravascular system ($\rightarrow$ **A**, bottom). **Plasmin** or **fibrinolysin** prevent such spread and dissolve any clots that do form. Plasmin is derived from inactive plasminogen in a reaction that requires various blood and tissue factors, including factor XIIa and kallikrein ($\rightarrow$ **A**). A number of **antiplasmin** substances can hinder fibrinolysis. These include ϵ-aminocaproic acid and aprotinin. A feedback system of sorts also operates: the fiber fragments that result from the action of fibrinolysis inhibit the action of thrombin on fibrin formation and slow the clotting sequence ($\rightarrow$ **A**).

Pathologic defects in the clotting can lead to *bleeding tendencies*. Such may come about by (a) congenital deficiencies of essential plasma factors, e.g., deficiency of factor VIII accounts for hemophilia A, (b) acquired deficiencies, as in liver disease or vitamin K deficiency ($\rightarrow$ p. 64), (c) increased utilization of clotting factors, (d) thrombocytopenia, (e) specific peripheral vascular diseases, and (f) excessive fibrinolysis and others.

Hemostatic mechanisms can be inhibited by **drugs** at various levels:

1. **Removal of Ca^{2+}** (applicable *only in vitro*): Ca^{2+} can be bound to citrate, oxalate or EDTA. Since Ca^{2+} is an essential component of many clotting reactions, its removal prevents clotting.

2. **Heparin** (applicable in vivo and in vitro), acting with **antithrombin III**, inhibits thrombin formation and factors IXa, Xa, XIa, and XIIa: its action is monitored by measuring the so-called *partial thromboplastin time* (PTT). Heparin is mainly used for initial parenteral anticoagulant medication because of its almost immediate effect.

3. The **oral anticoagulants** (*Dicumarol, Warfarin sodium* etc.) (applicable only in vivo) inhibit vitamin K ($\rightarrow$ p. 64). These drugs exert their effect only after a *latent period* of at least 12 to 24 h which is a consequence of the half-life of the factors involved ($\rightarrow$ table on p. 64). The action of oral anticoagulants is monitored mostly by measuring the *one-stage prothrombin time* (PT or Quick time).

4. Platelet aggregation can be inhibited by **Aspirin, Dextran 40** etc.

5. If a clot has already been formed, but not longer than a few hours ago (for example in acute pulmonary embolism), activators of the fibrinolytic system ($\rightarrow$ **A**) like **streptokinase** may be used in *thrombolytic therapy*.

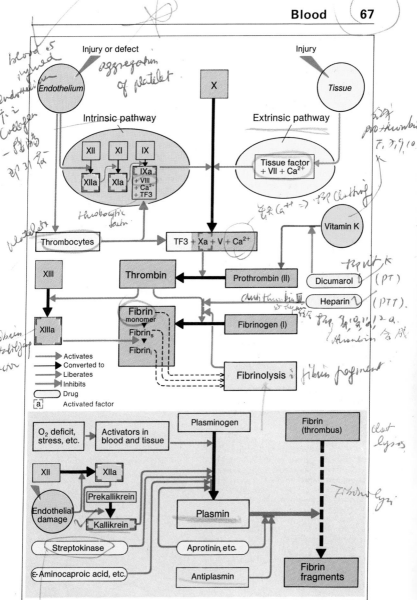

A. Blood clotting and fibrinolysis

Functions of Respiration

Respiration, in the broadest sense, is the exchange of gases between an organism and its invironment, and utilization of O_2 and production of CO_2 (cellular respiration or oxidative metabolism, → p. 182). The **respiratory rate** at rest is 12 to 16 breaths/min. Since intake is c. 500 ml of air/breath, intake is 6 to 8 l air/min. Approximately 250 ml O_2 are inspired and 200 ml CO_2 are expired/min (→ p. 82). On exertion, the respiratory rate can increase to c. 100 l/min (the **respiratory limit**, → p. 81, B).

In a mixture of gases, the **partial pressure** of each gas (the pressure that each gas exerts) equals the total pressure of the gas mixture times the relative fraction of the individual gas. The partial pressures of the individual gases add up to the total gas pressure. At sea level, air has a barometric pressure of 101,3 kPa (= 760 Torr or mmHg).

Composition of air is (fractional concentrations, → p. 6) $O_2 = 0.209$, $CO_2 = 0.0003$, $N_2 = 0.781$, inert gases = 0.009. Partial pressures are: $pO_2 = 21$ kPa (= 158 Torr), $pCO_2 = 0.03$ kPa (= 0.23 Torr). pH_2O varies but averages 0.76 kPa (= 5.7 Torr). At the alveoli, pH_2O has a constant value of 6.27 kPa (= 47 Torr) at 37 °C.

Values for partial pressures at different levels in the respiratory environment are shown in the figure. As the inspired gas flows through the respiratory channels, it becomes saturated with water. pH_2O rises to a maximum constant value at the alveoli. In the circulating blood, the sum of all partial pressures (Σp) is less than the barometric pressure for two reasons: (1) at the cells, pO_2 falls faster than pCO_2 rises and (2) the CO_2 binding curve is steeper than the O_2 binding curve (→ p. 89). Σp is therefore lower than atmospheric pressure by c. 7.17 kPa (= 54 Torr). This explains why gas pockets in the body do not persist. Whenever a gas gains access to a body cavity (after abdominal or pulmonary surgery), the pressure difference allows the gas to be resorbed.

In unicellular organisms, diffusion paths for respiratory gases are short and gaseous exchange takes place along the gaseous concentration gradient. In multicellular organisms, c-however, the diffusion paths are too long for rapid delivery of O_2 and removal of CO_2. These organisms have developed a transport system to carry respiratory gases between the lungs and the distant cells. O_2 in the inspired air reaches the pulmonary alveoli (ventilation) where it diffuses into the blood. O_2 is carried in blood partly in dissolved form but this quantity is inadequate to supply tissue needs. Instead, chemical combination between O_2 and hemoglobin (Hb) in the red blood cells allows much larger quantities of O_2 to be carried. The blood vessels decrease in size until they reach the tissues where they have the magnitude of capillaries (→ p. 143). At this level, the diffusion paths for the gases are sufficiently small for gas exchange. O_2 diffuses to the cells and CO_2 diffuses to the capillaries. CO_2 is carried to the lungs by the veins. In the alveolar capillaries, the reverse process takes place; CO_2 diffuses out of the blood and O_2 diffuses in.

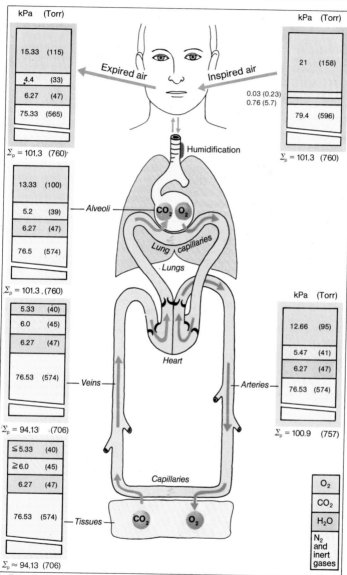

kPa	(Torr)
15.33	(115)
4.4	(33)
6.27	(47)
75.33	(565)

$\Sigma_p = 101.3$ (760)

Expired air   Inspired air

kPa	(Torr)
21	(158)
0.03	(0.23)
0.76	(5.7)
79.4	(596)

$\Sigma_p = 101.3$ (760)

Humidification

kPa	(Torr)
13.33	(100)
5.2	(39)
6.27	(47)
76.5	(574)

$\Sigma_p = 101.3$ (760)

Alveoli CO_2 O_2

Lung capillaries

Lungs

kPa	(Torr)
5.33	(40)
6.0	(45)
6.27	(47)
76.53	(574)

$\Sigma_p = 94.13$ (706)

Heart

Veins

kPa	(Torr)
12.66	(95)
5.47	(41)
6.27	(47)
76.53	(574)

Arteries

$\Sigma_p = 100.9$ (757)

kPa	(Torr)
≤5.33	(40)
≥6.0	(45)
6.27	(47)
76.53	(574)

$\Sigma_p \approx 94.13$ (706)

Capillaries

Tissues CO_2 O_2

O_2
CO_2
H_2O
N_2 and inert gases

A. Respiration

Respiratory Mechanics

The lungs are elastic structures and as such have a tendency to recoil to their smallest possible volume. However, the lungs normally are expanded and adapt to the contours of the chest wall. The lining of the lung cavity, the pleura, has two layers; one is applied to the lungs (pleura pulmonalis) and the other to the chest cavity (pleura parietalis) ($\rightarrow$ **B**). The space between these two layers, the pleural space, is filled with a thin film of fluid that is *in*elastic and not compressible. The lung remains in contact with the chest wall as long as the pleural space is not disturbed. If the chest wall is opened, air enters the pleural space and the lungs collapse ($\rightarrow$ p. 73, B). The **intrapleural (= intrathoracic) pressure** (P_{pl}) ($\rightarrow$ **B**) is normally negative at $-0.3\,kPa$ ($= -3\,cmH_2O$ relative to atmospheric). Normal inspiration lowers the pressure to -0.6 kPa ($= -6\,cmH_2O$) and expands the lungs. Strong inspiration can lower P_{pl} to $-4\,kPa$ ($-40\,cmH_2O$). On strong expiration the P_{pl} may become slightly positive (above atmospheric).

The driving force for ventilation is the pressure difference between the atmosphere and the **intrapulmonic pressure** (P_{pul}) ($\rightarrow$ **B**). For inspiration, P_{pul} in the alveoli must be below the external atmospheric pressure, for expiration it must be above. Relative to atmospheric pressure, P_{pul} is negative on inspiration and positive on expiration ($\rightarrow$ **B**). These pressure gradients are established when the lung volume is increased on ispiration and decreased on expiration by action of the diaphragm and thorax.

Inspiration is an active process. Muscular contraction (v.i.) increases the volume of the chest, the lungs inflate, and P_{pul} falls so that air flows into the lungs. At the end of quiet inspiration, the lungs and chest recoil to their original positions at the beginning of inspiration. Thus, quiet expiration is largely a passive process but it can be assisted by the respiratory muscles.

Respiratory muscles. The **diaphragm** exerts a **direct influence** on lung volume by contracting (inspiration) and relaxing (expiration). It accounts for 75% of the volume change in quiet inspiration. It moves as much as 7 cm on deep inspiration. **Indirect influences** on inspiration are exerted when the thorax enlarges by contraction of the **scalene**, and **external intercostal muscles** and other accessory muscles. Expiration occurs chiefly by passive collapse of the thoracic cage from its elasticity and mass. It is assisted indirectly by contraction of the abdominal muscles, which increases intra-abdominal pressure and by contraction of the internal intercostal muscles.

The external intercostal muscles run downward and forward from rib to rib ($\rightarrow$ **A**). The attachment of the muscle to the upper rib (B) is closer to the pivot point (A) than the attachment to the lower rib (C') is to its joint (A'); the lower rib has a longer lever arm (A – B < A' – C'). Therefore, when the muscles contract, they *elevate* the lower rib, which pivots on its joint at the vertebra. This action pushes the sternum outward and increases the anterior-posterior diameter of the rib cage.

The internal intercostal muscle are oppositly oriented ($\rightarrow$ **A**), and their contraction causes the ribs to pivot in the opposite direction. As the rib cage is pulled inward, intra abdominal pressure is elevated and assists in pushing the diaphragm upward.

The diaphragm or the external intercostal muscles alone can maintain adequate ventilation at rest.

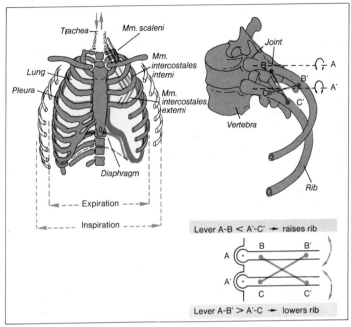

A. Respiratory musculature

Trachea
Mm. scaleni
Lung
Pleura
Mm. intercostales interni
Mm. intercostales externi
Diaphragm
Expiration
Inspiration

Joint
B
B'
A
A'
C'
Vertebra
Rib

Lever A-B < A'-C' → raises rib

A
B
B'
A'
C
C'

Lever A-B' > A'-C → lowers rib

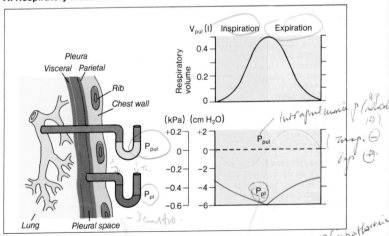

B. Intrapleural and intrapulmonic pressures

Pleura
Visceral   Parietal
Rib
Chest wall
P_{pul}
P_{pl}
Lung
Pleural space

V_{pul} (l)   Inspiration   Expiration
Respiratory volume
0.4
0.2
0

(kPa) (cm H$_2$O)
+0.2   +2
0   0
-0.2   -2
-0.4   -4
-0.6   -6
P_{pul}
P_{pl}

Artificial Respiration

Artificial respiration is necessary when spontaneous ventilation is insufficient or fails completely. Artificial respiration should always be attempted because the heart continues to function even after respiratory stimuli have ceased. Transection of the spinal cord above C_3 produces total paralysis of respiration since the phrenic nerves arise between C_3 and C_5 and the intercostals arise at their corresponding thoracic segments. Lack of oxygenation for seconds causes loss of consciousness and for only a few minutes causes irreversible damage to the CNS and death ($\rightarrow$ p. 90).

Mouth-to-mouth respiration is an emergency measure until spontaneous respiration can be restored ($\rightarrow$ **A**). In all attempts at artificial respiration, the airway should be cleared. The subject is turned on his back with his nose held closed. The first aid assistant blows into his mouth. This process raises the P_{pul} ($\rightarrow$ p. 71) above atmospheric pressure and inflates both the lungs and the chest. Breaking the mouth-to-mouth contact allows expiration as the chest collapses passively by elastic recoil. Pressure applied to the chest can increases the rate and completeness of expiration. This process is repeated at c. 15 cycles/min. Success of the procedure can be recognized by a change in the skin color from blue (cyanosis) to normal pink. This technique has the advantage that it expands the lungs. In the prone pressure method, expiration is forced by pressing on the rib cage. Passive recoil draws air into the lungs for the inspiratory phase. However, since the lungs are not expanded, this method is not as effective as mouth-to-mouth respiration.

Mechanical positive pressure respiratory assistance is used in anesthesia during surgery when spontaneous respiration is deliberately paralyzed by curarization. Inspiration is achieved by a pump ($\rightarrow$ **A**). The conventional respirator used for treatment of chronic respiratory insufficiency, as may occur after bulbar poliomyelitis, is the "iron lung" ($\rightarrow$ **A**). The patient is enclosed to his neck in an airtight chamber. Air is pumped out of the chamber to lower the pressure and to permit inspiration. Pumping air into the chamber allows expiration. One deficiency of these systems is that they hinder venous return to the heart ($\rightarrow$ p. 170), but this can be avoided if expiration as well as inspiration is assisted.

Pneumothorax

Pneumothorax occurs when air enters the pleural space ($\rightarrow$ p. 71). In *open pneumothorax*, when the chest wall is penetrated, the lung on the open side collapses by elastic recoil and does not contribute to ventilation ($\rightarrow$ **B**). Air moves in and out of the pleural space when the subject breathes. Gas exchange in the healthy lung is also compromised because (1) air is exchanged in part with the collapsed lung instead of with the external atmosphere, (2) the weight of the collapsed lung prevents full ventilation, and (3) atmospheric pressure in the pleural space presses the mediastinum into the healthy side. Respiration is stimulated but distress can be severe. In *tension pneumothorax* ($\rightarrow$ **B**), a flap of tissue over the puncture acts as a valve allowing entry but not exit of air from the pleural space. The P_{pl} rises above atmospheric pressure. Hypoxia stimulates vigorous ventilation, which builds up higher pressures, up to 2.7 to 4 kPa ($= 20$ to 30 Torr.). The mediastinum shifts to the unaffected side. Cardiac filling ($\rightarrow$ p. 148) is reduced and peripheral veins distend. Cyanosis develops, and the condition may become fatal if pressure is not relieved by withdrawing air. However, if the hole becomes sealed, P_{pl} stabilizes, the healthy lung retains its function, and anoxia does not develop. After 1 to 2 weeks, the air pocket is completely absorbed ($\rightarrow$ p. 68). *Closed pneumothorax* may develop spontaneously, especially, in emphysema when the lung ruptures through the visceral pleura. In this circumstance, air escapes into the pleural space.

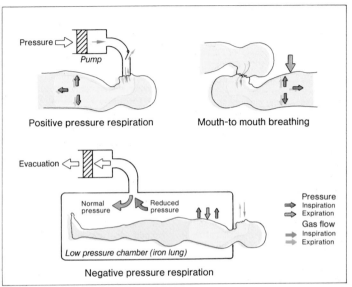

Positive pressure respiration

Mouth-to mouth breathing

Negative pressure respiration

A. Artificial respiration

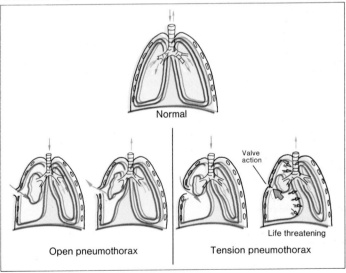

Normal

Open pneumothorax

Tension pneumothorax

B. Pneumothorax

Respiratory Volumes and the Spirometer

The **spirometer** (→ **A**) is an apparatus for measuring pulmonary air volumes. It consists of an inverted chamber in a water seal. A subject breathes into and out of the chamber, which moves an indicator to depict magnitude of the corresponding volume change. The indicator moves upward during inspiration, downward during expiration. These movements are recorded as the **spirogram** on a drum that moves at a fixed rate and allows calculations of ventilatory rates as well as volumes.

At expiration, the thorax comes to rest at the **resting expiratory level**. Quiet breathing at rest allows a ventilatory exchange of c. 0.5 l, the **tidal volume** (V_T). An active maximal inspiratory effort allows intake of air in excess of the tidal volume, the **inspiratory reserve volume** (IRV). After passive expiration, an additional active maximal expiratory effort yields the **expiratory reserve volume** (ERV). These two volumes, IRV and ERV, are in reserve and come into use during exertion when the tidal volume becomes inadequate and additional gas exchange becomes necessary. Even after maximal expiration, some air remains in the lungs: the **residual volume** (RV). The **vital capacity** (VC) is the air expired after maximal inspiration. VC = TV + IRV + ERV = c. 4.5 l but it varies with age. **Total lung capacity** (TLC) (c. 6 l) is the sum of the VC and the RV. The **functional residual capacity** is the sum of the ERV and the RV (→ **A**; p. 76). These volumes and capacities vary according to age, sex, and physical training. The value for vital capacity might indeed have a normal range between 2.5 and 7 l in the absence of pulmonary disease. Absolute values, therefore, have much less significance in describing disease states than the changes that these values undergo in an individual patient. In clinical practice the most important volume is the RV, which must be determined by other means than the spirometer since it is not included in the spirogram. The RV is necessary to measure pulmonary diffusion and TLC. With the spirograph, **dynamic function tests** can be performed. The product of tidal volume (V_T) and respiratory frequency (f) yields the respiratory rate (l × min^{-1}): $\dot{V}_T = V_T \times f$ (→ p. 81, B). This measurement is useful in evaluating the course of diseases involving the respiratory muscles (e.g., myasthenia gravis). A number of measurements can be made from a fast or forced expiratory trace as a function of time. The **forced expiratory volume** (FEV_1) (→ p. 81, C) is the fraction of the vital capacity that can be expired in 1 s. Other FEV times are sometimes used as well. The **forced vital capacity** (FVC) is the volume that can be expired rapidly after a full inspiration. The VC performed slowly may be normal, but the FVC and FEV_1 can be greatly reduced, especially in diseases such as asthma, in which bronchial constriction increases frictional resistance of the airways. The ratio FEV_1/VC (relative FEV_1, → p. 81, C) can be used to distinguish between *obstructive* (e.g., asthma, emphysema) and *restrictive* (e.g., pulmonary edema or inflammation, scoliosis) *disease*. It is > 0.80 at 20 years, > 0.75 at 40 years, and > 0.70 at 60 years. In restrictive disease, FEV_1 and VC are both reduced, but the ratio FEV_1/VC remains normal. The ratio is reduced in obstructive disease (→ p. 81, C). The spirograph serves also to measure compliance (→ p. 78) and O_2 utilization (→ p. 82).

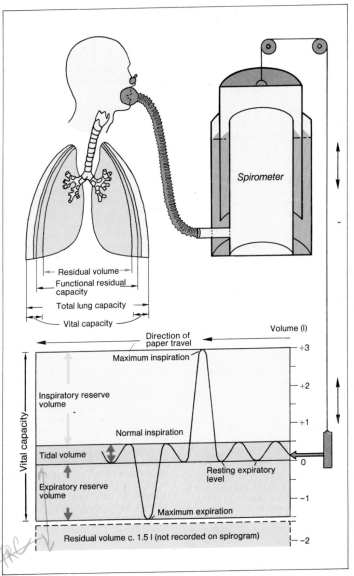

Spirometer

Residual volume
Functional residual capacity
Total lung capacity
Vital capacity

Volume (l)

Direction of paper travel

Maximum inspiration

+3

Inspiratory reserve volume

+2

+1

Normal inspiration

Vital capacity

Tidal volume

0

Resting expiratory level

Expiratory reserve volume

−1

Maximum expiration

Residual volume c. 1.5 l (not recorded on spirogram)

−2

A. Measurement of lung volumes

Dead Space and Residual Volume

Gas exchange in the respiratory system occurs only in the alveoli. That part of the ventilation that is effective in ventilating the alveoli is the **alveolar ventilation** ($\dot{V}_A$); the part that does not ventilate the alveoli is the **dead space ventilation** ($\dot{V}_D$). The **dead space** is the total volume of the airways that conduct inspired air to the alveoli; it does not take part in gas exchange. The mouth, nose, pharynx, trachea, and bronchi comprise the **anatomic dead space** (ADS), which has a volume equivalent to the body weight in pounds expressed in ml (c. 150 ml). The ADS functions as an air conduit in which the air is simultaneously freed of dust particles, humidified, and warmed before it reaches the alveoli. It also contributes to speech as a sound box that determines characteristics of the voice ($\rightarrow$ p. 304). The ADS is normally equivalent to the **functional dead space** (V_D); however, in pathology, when gas exchange in some alveoli is restricted, the V_D exceeds the ADS ($\rightarrow$ p. 82).

The V_D can be calculated from the CO_2 of expired air, the CO_2 of alveolar gas, and the tidal volume (V_T) by means of the **Bohr equation** ($\rightarrow$ **A**). The tidal volume, V_T, is comprised of the gas that equilibrated in the alveoli, V_A, and the gas that did not (dead space), V_D ($\rightarrow$ **A**, I, III). In each of these three volumes there is a corresponding equivalent fractional CO_2 concentration ($\rightarrow$ **A**, II): $F_{E_{CO2}}$ in V_T, $F_{A_{CO2}}$ in V_A and the original concentration of the inspired air, $F_{I_{CO2}}$ in V_D. The product of each volume component and its corresponding fractional CO_2 concentration gives the volume of CO_2 for each. The CO_2 volume in the expiratory volume ($F_{E_{CO2}} \times V_T$) equals the sum of the CO_2 volume in V_A and V_D (II). However, the term $F_{I_{CO2}}$ is small enough to be disregarded (III).

To determine V_D, three magnitudes must be measured: (1) V_T is determined with the spirometer, (2) $F_{A_{CO2}}$ is determined in the terminal portion of the expired air i.e., in the alveolar air, and (3) $F_{E_{CO2}}$. When $V_T = 0.5$ l, $F_{A_{CO2}} = 0.06$ l/l (6 vol%) and $F_{E_{CO2}} = 0.045$ l/l (4.5 vol%), $V_D = 0.15$ l. The ratio V_D/V_T is an index of wasted ventilation (at rest c. 0.2 to 0.3).

Residual volume and **functional residual capacity** ($\rightarrow$ p. 74) must be measured indirectly because they are not represented on the spirograph. A test gas, such as N_2, must be used ($\rightarrow$ **B**). The fractional concentration of N_2 in the lungs, $F_{L_{N2}}$, is constant at c. 0.80 l/l. The subject rebreathes several times from a bag of known volume (V_B) containing a N_2-free gas. N_2 that is in the lungs is expired into the bag until its distribution is equilibrated between the lungs and the bag volume. In this process the volume (amount) of N_2 remains nearly unchanged, but its concentration in the expired air will decrease by virtue of the larger volume of distribution. Thus, the amount of N_2 at the beginning of the determination (N_2 that was only in the lungs) is the same amount that ultimately is distributed between lung and bag volume at the end of the determination. The pulmonary residual volume, V_L, can then be calculated ($\rightarrow$ **B**). V_B and $F_{L_{N2}}$ are known values and the fractional volume of N_2 in the bag at the end of the determination, $F_{X_{N2}}$, need be determined. If the procedure is begun at the extreme expiratory level, V_L measures residual volume (c. 1.5 l); if begun at the resting expiratory level, V_L measures functional residual capacity (c. 3 l) ($\rightarrow$ p. 74). The value that is clinically significant is the fraction of the total lung capacity ($\rightarrow$ p. 74) that the residual volume represents (normally 25%). In obstructive pulmonary disease (e.g., emphysema), this fraction may be more than 55% and indicates the severity of the disease process.

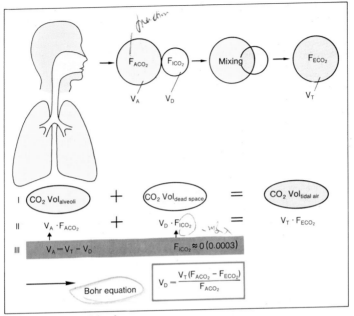

I $\boxed{CO_2 \ Vol_{alveoli}}$ $+$ $\boxed{CO_2 \ Vol_{dead\ space}}$ $=$ $\boxed{CO_2 \ Vol_{tidal\ air}}$

II $V_A \cdot F_{ACO_2}$ $+$ $V_D \cdot F_{ICO_2}$ $=$ $V_T \cdot F_{ECO_2}$

III $V_A = V_T - V_D$ $F_{ICO_2} \approx 0 \ (0.0003)$

Bohr equation

$$V_D = \frac{V_T (F_{ACO_2} - F_{ECO_2})}{F_{ACO_2}}$$

A. Measurement of dead space

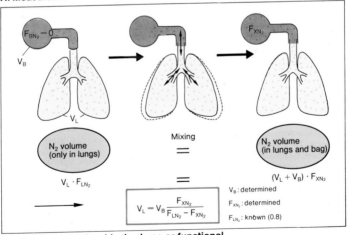

N$_2$ volume
(only in lungs)

$=$

Mixing

$=$

N$_2$ volume
(in lungs and bag)

$V_L \cdot F_{LN_2}$ $(V_L + V_B) \cdot F_{XN_2}$

$$V_L = V_B \frac{F_{XN_2}}{F_{LN_2} - F_{XN_2}}$$

V_B : determined
F_{XN_2} : determined
F_{LN_2} : known (0.8)

**B. Measurement of residual volume or functional
residual capacity**

Pressure-Volume curves
Work of Breathing

The pressure-volume (PV) curve is determined by measuring the pressure in the airway (intrapulmonic pressure, P_{pul}) ($\rightarrow$ p. 71) at different stages of chest inflation (V_{pul}) during a respiratory cycle. The relationship is shown graphically by plotting V_{pul} against P_{pul} ($\rightarrow$ **A**).

To determine the pressure-volume characteristics, V_{pul} at the resting expiratory level is set at zero relative to the atmospheric pressure (P_B) ($\rightarrow$ **A**, **a**). From this starting point, small measured volumes of air are inspired ($+V_{pul}$) or expired ($-V_{pul}$). At the end of each stepwise stage, the valve to the spirometer is closed and P_{pul} is measured. (Respiratory muscles must be relaxed.) During these measurements, V_{pul} is respectively reduced or enlarged if compared to the volume originally measured by means of the spirometer ($\rightarrow$ **A**, diagonal arrows). Under these static conditions, the pressure – volume curve for **lung and thorax** can be determined ($\rightarrow$ **A**, **c**, **a**, **b**). Inspiration of a volume generates a rise in pressure ($V_{pul} > 0$) ($\rightarrow$ **A**, **b**), expiration generates a fall in pressure ($V_{pul} < 0$) ($\rightarrow$ **A**, **c**). These pressures arise from the passive elastic recoil ($\rightarrow$ **A**, blue arrows) of the lungs and the chest wall, which attempt to restore the original resting position ($\rightarrow$ **A**, **a**). The recoil forces become greater the more V_{pul} deviates from 0.

The slope of the static pressure curve describes the static **compliance** ($\rightarrow$ **B**), or stretchability, of the lungs and thorax. Compliance will be different at different volumes; stretchability will naturally be less when the lungs are stretched maximally. The largest compliance is observed in the range between $V_{pul} = 0$ to 1 l, the normal respiratory range.

Compliance is determined by the change in lung volume per unit change in airway pressure ($\Delta V_{pul}/\Delta P_{pul}$) (normal = 1 ml/Pa or 0.1 l/cm H_2O). The compliance can also be determined separately for the thorax ($\Delta V_{pul}/\Delta P_{pl}$), and the lungs ($\Delta V_{pul}/\Delta (P_{pul} - P_{pl})$), where P_{pl} = the

intrapleural pressure ($\rightarrow$ p. 71). Compliance is low in obstructive pulmonary disease, pulmonary fibrosis, and pulmonary congestion. In a similar manner, the PV diagram can also be determined during maximal active participation of the respiratory muscles ($\rightarrow$ **A**, red and green curves), the maximal inspiratory and expiratory curves (MIC, MEC). At an extreme expiratory position ($V_{pul} \ll 0$), the expiratory muscles can produce only very small changes in pressure ($\rightarrow$ **A**, **g**); in contrast, when V_{pul} is positive, the pressure may exceed 15 kPa (= 150 cm H_2O) ($\rightarrow$ **A**, **e**). The reverse relationships are found in regard to the inspiratory muscles ($\rightarrow$ **A**, **d**, **f**).

If the PV relationships are measured during breathing, different values are obtained during the two phases of inspiration and expiration (**hysteresis**, $\rightarrow$ **C**). The "driving pressure gradient" on the abscissa of this plot represents, for example, the pressure difference between the mouth and the atmosphere during artificial positive pressure respiration ($\rightarrow$ p. 73, A). The areas enclosed by the two curves ($A_{R_{insp.}}$, $A_{R_{exp.}}$) represent the *work of breathing* done *against frictional resistance* to air flow and to chest movement during inspiration or expiration respectively. They have the dimension pressure times volume ($Pa \times m^3 = J$, $\rightarrow$ p. 6). The hatched area ($\rightarrow$ **C**) is the *elastic work* (A_{elast}) required to stretch the lungs and chest wall. Total **work of inspiration** is $A_{R_{insp.}} + A_{elast}$ (pink + hatched area). Total **work of expiration** is $A_{R_{exp.}} - A_{elast.}$ (greenish *minus* hatched area). Because at rest the elastic energy stored during inspiration is larger than the expiratory work against frictional resistance ($A_{elast.} > A_{R_{ex}}$), expiration does not need extra energy. In forced respiration the greenish area becomes larger than the hatched area ($A_{R_{exp.}} > A_{elast.}$) because muscle activity is needed (1) for faster air flow and (2) for decreasing V_{pul} below the resting respiratory level. Work may increase up to 20 times above its resting value.

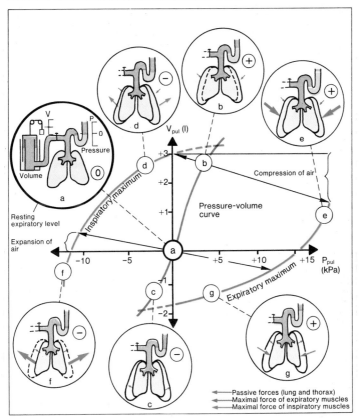

A. Pressure-volume curves (lung and thorax)

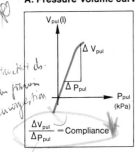

B. Static compliance

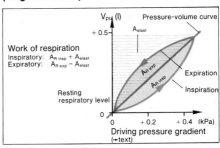

C. Dynamic pressure-volume curves

Within figure A:

V / Pressure / Volume / 0

Resting expiratory level

Expansion of air

Inspiratory maximum

Compression of air

Pressure-volume curve

V_{pul} (l)

$+3$ $+2$ $+1$ -1 -2

-10 -5 $+5$ $+10$ $+15$ P_{pul} (kPa)

Expiratory maximum

a b c d e f g

Passive forces (lung and thorax)
Maximal force of expiratory muscles
Maximal force of inspiratory muscles

Within figure B:

V_{pul} (l)

ΔV_{pul}

ΔP_{pul}

P_{pul} (kPa)

$$\frac{\Delta V_{pul}}{\Delta P_{pul}} = \text{Compliance}$$

Within figure C:

V_{pul} (l)

Pressure-volume curve

$+0.5$

A_{elast}

Work of respiration
Inspiratory: $A_{R\,insp} + A_{elast}$
Expiratory: $A_{R\,exp} - A_{elast}$

$A_{R\,exp}$

$A_{R\,insp}$

Expiration

Inspiration

Resting respiratory level

0 $+0.2$ $+0.4$ (kPa)

Driving pressure gradient
(→text)

Surface Tension of the Alveoli

Compliance of the lungs depends to a large extent on the low **surface tension** of the film of fluid that lines the alveolar surface. This surface, across which gas exchange between air and blood takes place, has a total area of c. 80 m². Gas exchange is effective only as long as the alveoli remain open.

The influence of surface tension can be demonstrated by measuring the intrapulmonary pressure while attempting to fill a fully collapsed lung. When air is used to fill the lung there is, especially at the outset, a high resistance to opening, just as in attempting to inflate a new ballon. In a newborn baby during the first breath of life this resistance is overcome only by an intrapleural pressure ($\rightarrow$ p. 71) of as much as -6 kPa ($= -60$ cmH$_2$O) whereas in later life only c. -0.6 kPa ($= -6$ cm H$_2$O) are necessary for expansion of the lungs. In contrast, when the lung is filled with a saline solution, the resistance is only one fourth as large because saline reduces the surface tension to nearly zero. The phenomenon is defined by the Laplace relationship ($\rightarrow$ p. 142): the tension in a chamber wall is proportional to the pressure times the radius. When a gas bubble is surrounded by a fluid, the surface tension (γ) (dimensions: N/m) of the fluid increases the pressure (Δp) in the bubble. $\Delta p = 2\gamma/\text{radius}$. The γ is normally a constant for a given fluid (for plasma $= 10^{-3}$ N/m), so that Δp is influenced by the radius, becoming larger as the air space becomes smaller.

These relationships may be exemplified by a soap film across the opening of a cylinder ($\rightarrow$ **A**, **1**), where r is large and pressure is small. To blow a bubble, the surface must become spherical with a smaller radius; this requires a large "opening pressure" (Δp) ($\rightarrow$ **A**, **2**). As the bubble enlarges (r in-

creases), ($\rightarrow$ **A**, **3**), less pressure is required to enlarge it still further until, finally, the walls of the bubble are no longer able to contain the pressure and the bubble breaks. The alveolus responds similarly, but rupture at high pressures is resisted by the tissue elasticity.

Two other features are apparent: (1) below a critical pressure, the bubble collapses ($\rightarrow$ **A**, **2**) and (2) when two bubbles of different radii are connected, there is a pressure gradient from the smaller to the larger bubble ($\rightarrow$ **A**, **4**); gas will flow to the larger bubble at the expense of the smaller bubble.

In the alveoli, these effects are resisted by a *surface-active* subst*ance* or **surfactant**, a lipoprotein film in the alveoli that lowers surface tension; it is more effective in smaller alveoli. In some newborn infants, surfactant is insufficiently effective (*hyaline membrane disease*); many alveoli are collapsed, and the opening pressure cannot be achieved. Surfactant is a complex of protein and lipids, of which the chief component is dipalmitoyl lecithin. Two types of cells line the alveoli; one of these, the granular pneumocytes, secrete surfactant by exocytosis. Surfactant deficiency contributes to the development of pulmonary edema in selected conditions and to the development of pulmonary abnormalities after bronchial occlusion or after breathing air with a high pO$_2$ ($\rightarrow$ p. 96).

[handwritten annotations:]

Laplace law -

Tension = Pressure × radius

$\delta P = \dfrac{2T}{r}$

constant 10^{-3}

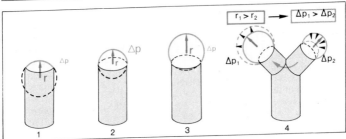

A. Surface tension (soap bubble model)

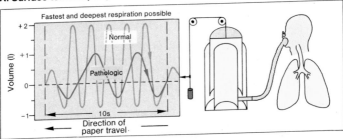

Fastest and deepest respiration possible

Normal

Pathologic

10s

Direction of paper travel

B. Respiratory limit (see p. 68)

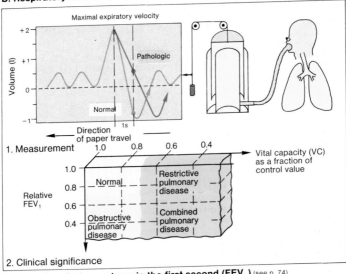

Maximal expiratory velocity

Pathologic

Normal

1s

Direction of paper travel

1. Measurement

Vital capacity (VC) as a fraction of control value

Relative FEV₁

Normal | Restrictive pulmonary disease

Obstructive pulmonary disease | Combined pulmonary disease

2. Clinical significance

C. Forced expiratory volume in the first second (FEV₁) (see p. 74)

Gas Exchange in the Lung

For gas exchange between alveoli and blood, the lungs must be ventilated. However, not all of the tidal volume (V_T) reaches the alveoli. The volume reaching the alveoli (V_A) is less than V_T because of the dead space (V_D) in the upper airway. Thus, $V_T = V_A + V_D$ ($\rightarrow$ p. 77). The total ventilatory rate, $\dot{V}_T$ (l/min), $= V_T \times f$ (f = frequency). Similarly, the alveolar ventilatory rate, $\dot{V}_A$, and the dead space ventilatory rate, $\dot{V}_D$, can be derived from V_A and V_D. At rest normally, $\dot{V}_A$ amounts to c. 70% of $\dot{V}_T$ (c. 350 ml/breath). In rapid shallow breathing, f is increased, V_T is decreased but $\dot{V}_T$ remains unchanged; $\dot{V}_D$ will increase ($= V_D \times f$) because the magnitude of V_D is fixed by the anatomy of the upper airway. As a consequence, the functionally significant value, $\dot{V}_A$, will decrease.

Example: Normal $V_T = 0.5$ l, $f = 15$ breaths /min, $V_D = 0.15$ l, $V_A = 0.35$ l. Then $\dot{V}_T = = 7.5$ l/min, $\dot{V}_D = 2.25$ l/min, $\dot{V}_A = 5.25$ l/min. In rapid shallow ventilation: $V_T = = 0.375$ l, $f = 20$ breaths/min, $V_D = 0.15$ l (constant), $V_A = 0.225$ l. Then $\dot{V}_T = 7.5$ l/min, $\dot{V}_D = 3$ l/min (elevated), $\dot{V}_A = 4.5$ l/min (reduced). Because of the lower alveolar ventilatory rate ($\dot{V}_A$), gas exchange decreases. The same consequences follow when the dead space increases ($\rightarrow$ p. 94).

The inspired air contains an O_2 fraction of 0.21 (21 vol%) ($F_{I_{O2}}$) and a CO_2 fraction of 0.0003 ($F_{I_{CO2}}$). In the expired air, the fractions are for O_2, 0.17 ($F_{E_{O2}}$) and for CO_2, 0.035 ($F_{E_{CO2}}$). The rate of inspiration of $O_2 = \dot{V}_T \times F_{I_{O2}}$ and of expiration $= \dot{V}_T \times F_{E_{O2}}$. The difference of the rates, $\dot{V}_T (F_{I_{O2}} - F_{E_{O2}})$, is the rate of **oxygen utilization** ($\dot{V}_{O2}$). **CO_2 formation** ($\dot{V}_{CO2}$) is calculated from $\dot{V}_T (F_{E_{CO2}} - F_{I_{CO2}})$. At rest, $\dot{V}_{O2}$ is c. 0.3 l/min and $\dot{V}_{CO2}$ is c. 0.26 l/min. Both $\dot{V}_{O2}$ and $\dot{V}_{CO2}$ increase c. tenfold on heavy exertion. Normally, when exercise is mild, $\dot{V}_{O2}$ is 1 l/min, when moderate, 1.5 l/min, and when severe, 2 l/min. The

relationship $\dot{V}_T / \dot{V}_{O2}$ is constant in mild and moderate exercise (c. 25); in severe exercise it increases to c. 35. In the lungs, $\dot{V}_{CO2} / \dot{V}_{O2}$ is the **respiratory exchange rate** (R); in the cells, it is the **respiratory quotient** (RQ) (range 0.75 to 1.0 depending on diet) ($\rightarrow$ p. 184). R = RQ only at steady state; in activity during breath holding, R (lungs) and RQ (cells) have quite different values.

Partial pressures *in the alveoli* are ($\rightarrow$ **A**): O_2, 13.33 kPa ($= 100$ Torr) and CO_2, 5.33 kPa ($= 40$ Torr); *in venous blood*, they are: O_2, 5.33 kPa ($= 40$ Torr) and CO_2, 6.13 kPa ($= 46$ Torr). Thus, there is an O_2 gradient from alveolus to blood of 8 kPa ($= 60$ Torr) and a CO_2 gradient from blood to alveolus of 0.8 kPa ($= 6$ Torr). *These gradients constitute the driving force for diffusion of gas across the alveolar membrane.* The diffusion path from the alveolus is c. 1 µm to the plasma and an additional 1 µm to the red cell. The path is short enough to allow equilibration of gas pressure during the time it takes for blood to flow through the alveolar capillaries (< 1 s) ($\rightarrow$ **A**).

Gaseous exchange ($\rightarrow$ **B**, **1**) can be *hindered* in several ways: by reducing flow in the alveolar capillaries ($\rightarrow$ **B**, **2**); by thickening of the alveolar membrane (diffusion barrier) ($\rightarrow$ **B**, **3**); and by reducing ventilation of alveoli ($\rightarrow$ **B**, **4**). In the first two cases ($\rightarrow$ **B**, **2**, **3**), the functional dead space is enlarged ($\rightarrow$ p. 76); in the second and third cases, blood is insufficiently oxygenated/arterialized.

Insufficient oxygenation can also occur by an arteriovenous shunt (extra-alveolar). A physiologic AV shunt also occurs: the bronchial veins of the lung and thebesian veins of the heart carry unsaturated (venous) blood to the oxygenated blood in the left ventricle of the heart. For this reason, pO_2 falls from 13.33 kPa ($= 100$ Torr) in the pulmonary capillaries to 12.66 kPa ($= 95$ Torr) in the aorta (changes in pCO_2 are analogous but opposite).

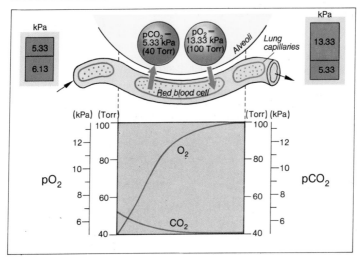

A. Alveolar gas exchange

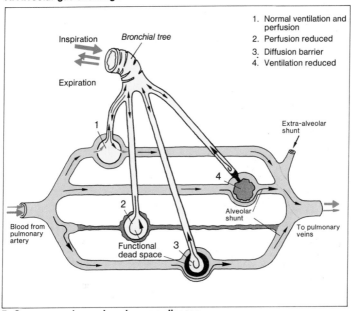

1. Normal ventilation and perfusion
2. Perfusion reduced
3. Diffusion barrier
4. Ventilation reduced

B. Gaseous exchange in pulmonary disease

CO$_2$ Transport in Blood

The main end products of cellular metabolism are CO$_2$ and H$_2$O. The CO$_2$ that is formed in the cells dissolves in cell water and diffuses through the ECF to the venous blood. Transport and delivery of CO$_2$ to the lungs is accomplished by the red blood cells (RBC). At rest, transport accounts for c. 260 ml or 11.7 mmol CO$_2$/min (1 mol = 22.26 l). At a cardiac output of 5 l/min, 2.3 mmol or 52 ml CO$_2$/l blood are taken up in the tissues and delivered to the alveoli ($\rightarrow$ "A-V difference" in the table on p. 86).

Dissolved CO$_2$. CO$_2$ is 20 times more soluble in blood than is O$_2$; there is always more CO$_2$ in solution than O$_2$. The dissolved CO$_2$ in the plasma equilibrates by diffusion with the cell water in the RBC. The pCO$_2$ of arterial blood supplying the cells is 5.33 kPa (= 40 Torr). At the cells, CO$_2$ diffuses into the blood and raises the pCO$_2$ to 6.27 kPa (= 47 Torr) in venous blood.

HCO$_3^-$. The CO$_2$ diffusing into the RBC ($\rightarrow$ **A**, blue arrow) is rapidly converted to HCO$_3^-$. Carbonic anhydrase (CA) plays a decisive role in catalyzing the reaction: CO$_2$ + H$_2$O $\rightleftarrows$ HCO$_3^-$ + H$^+$ ($\rightarrow$ p. 131, A). In the absence of CA, the reaction rate is slow; the enzyme accelerates the reaction so that the short contact time with RBCs in the capillaries (c. 1 s) is sufficient for the conversion of all of the CO$_2$ to H$^+$ and HCO$_3^-$. The H$^+$ is buffered within the RBC, chiefly by hemoglobin (Hb); the HCO$_3^-$ diffuses out of the RBC into the plasma along its concentration gradient.

Chloride shift. About 70% of the HCO$_3^-$ formed in the RBC diffuses into the plasma. It is not accompanied by cations since neither protein nor intracellular K$^+$ diffuses readily out of the RBC. To maintain electric neutrality, an anion (Cl$^-$) must diffuse into the cell to replace the HCO$_3^-$ that diffuses out. Thus, RBC in venous blood contain more Cl$^-$ than RBC in arterial blood. The intracellular Cl$^-$ increases osmolality and motivates water movement into the cells, which increases their volume.

Carbamino compounds ($\rightarrow$ **A, B**). Some of the CO$_2$ combines with free NH$_2$ groups of Hb in the RBC to form carbamino compounds. A much smaller amount (c. 1/10) forms carbamino bonds with plasma proteins. The reaction is Hb$-$NH$_2$ + CO$_2$ $\rightleftarrows$ Hb$-$NH$-$COO$^-$ + H$^+$.

Buffering of H$^+$ ions

Both the carbamino and the CA reactions generate H$^+$, a portion of which lowers the pH of blood from 7.40 (arterial) to 7.36 (venous). A much larger portion is buffered, principally by Hb. The binding capacity of reduced Hb for H$^+$ is greater than of oxy-Hb ($\rightarrow$ **B**). When O$_2$ is given up at the tissues, oxy-Hb is converted to reduced Hb; thus, Hb has its greatest buffering capacity just at the site where H$^+$ production is greatest. By buffering H$^+$, Hb maintains a low concentration of H$^+$ in the RBC, drives the reaction CO$_2$ + H$_2$O $\rightleftarrows$ HCO$_3^-$ + H$^+$ to the right, and allows more CO$_2$ to be taken up (*Haldane effect*).

In the pulmonary capillaries, all of these reactions are reversed by the lowered pCO$_2$ ($\rightarrow$ **A**, red arrows). HCO$_3^-$ diffuses from plasma back into the RBC, accepts H$^+$, and forms CO$_2$ and H$_2$O. This reaction is supported by Hb, which becomes oxygenated in the lung and liberates H$^+$ in the process (*Haldane effect*) ($\rightarrow$ **B**). The lower pH in the RBC reverses the carbamino bonding. CO$_2$ is thus generated by two processes. It diffuses along its concentration gradient to the alveoli where the pCO$_2$ is lower and exits with the expired air. At the same time, the RBC give up the extra Cl$^-$ and H$_2$O and shrink in volume.

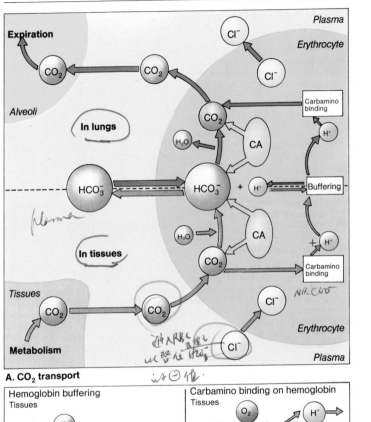

A. CO₂ transport

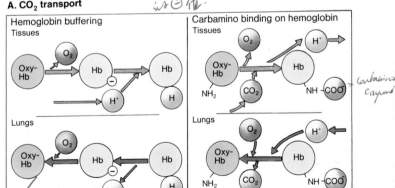

B. Buffering and carbamino binding in erythrocytes

CO_2 Binding and Distribution in Blood

The **blood content of CO_2** in mixed venous blood, both as dissolved and as chemically bound CO_2, is 23 to 24 mmol/l, in arterial blood 22–23 mmol/l. The actual content at any time is more or less dependent on the partial pressure of CO_2 (pCO_2). The distribution of CO_2 in arterial blood and in venous blood is shown in the table (note that content of plasma and RBC is given per l *blood*!).

Distribution of CO_2 in blood (approx. values in mmol/l *blood* §):

	Dissolved	HCO_3^-	Carbamino	Total
Arterial plasma*)	0.7	13.2	0.1	14.0
Arterial RBC**)	0.5	6.5	1.1	8.1
Arterial blood	1.2	19.7	1.2	22.1
Venous plasma*)	0.8	14.3	c. 0.1	15.2
Venous RBC**)	0.6	7.2	1.4	9.2
Venous blood	1.4	21.5	1.5	24.4
A-V difference blood	0.2	1.8	0.3	2.3
(% of total A-V diff.)	(9%)	(78%)	(13%)	

*) 0.55 l/l blood;   **) 0.45 l/l blood;   §) 1 mmol = 22.26 ml CO_2

The amount of **dissolved CO_2** in blood is linearly dependent on the pCO_2. The amount is given by $\alpha \times pCO_2$ where α represents the solubility coefficient; for CO_2 in plasma at 37°C, $\alpha = 0.22$ mmol $\times l^{-1} \times kPa^{-1}$. The curve for dissolved CO_2 is therefore a straight line ($\rightarrow$ **A**, green curve). In contrast, the curve for **chemically combined CO_2** relative to pCO_2 is curvilinear. The shape of this curve is determined by the limited buffer sites of Hb and by the limited number of carbamino bonds that can be formed with the available Hb. The curves for **total CO_2** represent the sum of dissolved and of bound CO_2 ($\rightarrow$ **A**, red and blue curves). Two curves are shown, one for CO_2 binding when Hb is fully saturated with O_2 (100% HbO_2) and one when Hb is fully desaturated (0% HbO_2). The capacity to bind CO_2, i.e., to bind H^+ during formation of HCO_3^- and carbamino compounds, depends on the state of Hb ($\rightarrow$ p. 84). For a given pCO_2, blood fully saturated with O_2 binds *less* CO_2 than O_2-free blood ($\rightarrow$ **A**, red and violet curves). The physiologic meaning of this difference is twofold. (1) venous blood (low in O_2, high

in CO_2) can carry more CO_2 away from the tissues. (2), in the pulmonary capillaries, when Hb is saturated with O_2, release of CO_2 is facilitated by formation of HbO_2 (Haldane effect). Venous blood is 70% saturated with O_2 (70% HbO_2). The dissociation curve corresponding to this extent of saturation lies between the two curves for 0% and 100% saturation. The pCO_2 for venous blood is 6.27 kPa (= 47 Torr) ($\rightarrow$ **A**, point v). In arterial blood, Hb saturation is close to 97% and the pCO_2 is 5.33 kPa (= 40 Torr) ($\rightarrow$ **A**, point a). The line joining points a and v represent the dissociation curve for CO_2 under physiologic conditions.

The relationship between HCO_3^- and dissolved CO_2 varies according to the pH of the blood. At pH 7.4, in plasma, the ratio HCO_3^-/CO_2 is 20:1. These values compare with the normal pH in the RBC of 7.2, which establishes a ratio of 12:1 ($\rightarrow$ pp. 98–107).

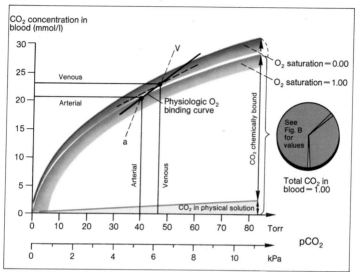

A. CO₂ binding curve

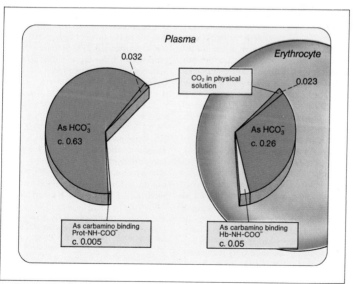

B. CO₂ distribution in arterial blood

O_2 Binding and Transport in Blood

Hemoglobin (Hb) is a protein (MW 64 800) with four subunits, each containing a **heme** moiety. Heme is a complex of porphyrin and Fe (II). Each of the four Fe (II) combines reversibly with one O_2 molecule. The process is **oxygenation**, not oxidation. The O_2 *dissociation curve* ($\rightarrow$ **A**, red curve) has a sigmoid shape because of varying affinities of the heme groups for O_2. When O_2 binds to the first heme group, affinity of the second is enhanced; binding to the second enhances the third, etc. When fully saturated with O_2, 1 g Hb can carry 1.38 ml O_2. Full saturation occurs at a pO_2 of 20 kPa; O_2 carrying capacity cannot be increased significantly by elevating pO_2 further. The average normal concentration of Hb in blood is 150 g/l; therefore, the **O_2 capacity** of blood, the maximal potential O_2 concentration = 0.207 l or 9.24 mmol O_2/l blood (1 mol O_2 = 22.4 l).

O_2 content is the concentration of O_2 actually in the blood. The fraction O_2 *content*/O_2 *capacity* represents **O_2 saturation** of Hb. Because of the return of bronchial venous blood to the left ventricle ($\rightarrow$ p. 82), saturation of arterial blood is 0.97 at a pO_2 of 13.3 kPa (= 100 Torr). Saturation of mixed venous blood is 0.75 at a pO_2 of 5.33 kPa (= 40 Torr), but pO_2 varies locally in the various organs according to their metabolic activity.

The **O_2 dissociation curve** is sigmoid, rises steeply at low values of pO_2 but becomes almost flat at high values of pO_2, when it approximates to the O_2 capacity. The curve can be displaced by various influences. Since the amount of available Hb affects O_2 content, the curve may be shifted upward in Hb excess and downward in Hb deficiency. This shift can be seen only if *concentration* of O_2 or of Hb-O_2 ($\rightarrow$ **A, B**, yellow and purple curve), but not if O_2 *saturation* ($\rightarrow$ **C**) is plotted against pO_2. In contrast, shifts of the curve to the left or right occur without changing the O_2 capacity. A **shift to the left** increases the slope of the rising portion of the curve because the affinity of Hb for O_2 has been increased ($\rightarrow$ **A**, blue curve). In effect, (a) for a given pO_2, more oxy-Hb is formed or (b) for a given O_2 content, the pO_2 is less. A **shift to the right** represents decreased affinity of Hb for O_2. Three principal factors influence the affinity: pH (Bohr effect), pCO_2 as such, temperature, 2,3-diphosphoglycerate (DPG).

1. At the tissues, pCO_2 is increased and pH is decreased. Both facts reduce affinity of Hb, allowing more O_2 to dissociate. In the lung capillaries CO_2 is given off, the pH rises and Hb can carry more O_2 (affinity increases). The effect of pH is known as the **Bohr effect**. A physiologic dissociation curve ($\rightarrow$ **B**) can be made for O_2 (as for CO_2, $\rightarrow$ p. 86) in order to take into account changes of pH and pCO_2 in venous blood compared to arterial blood.

2. Reduced temperature shifts the O_2 dissociation curve to the left.

3. 2,3-DPG is generated by glycolysis; it binds to Hb but not to oxy-Hb. High concentrations of DPG therefore reduce affinity of Hb for O_2 and shift the dissociation curve to the right.

Other O_2-carrying pigments, **myoglobin** in muscles and **fetal Hb**, differ in detail from Hb ($\rightarrow$ **C**). They have a steeper rising portion to the curve (shift to the left). Other gases can bind with Hb. Carbon monoxide (CO) has a much steeper dissociation curve and, even in low concentrations, displaces large amounts of O_2 from Hb. Methemoglobin is formed when ferrous iron is oxidized to ferric. Met-Hb has no affinity for O_2.

The amount of **dissolved O_2** is very small and is linearly dependent on the pO_2 ($\rightarrow$ **A**, orange line). The amount is given by $\alpha \times pO_2$ where α represents the solubility coefficient. For O_2 in plasma at 37 °C $\alpha = 0.01$ mmol $\cdot l^{-1} \cdot kPa^{-1}$. In arterial blood ($pO_2 = 12.66$ kPa) dissolved O_2 amounts to ca. 0.13 mmol/l. This value is only 1/70 of the O_2 combined with Hb.

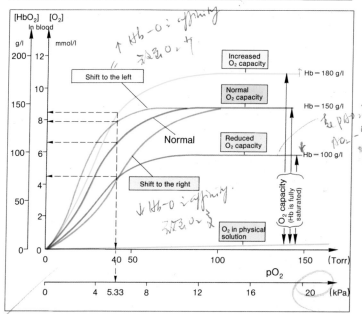

A. O₂ binding curve

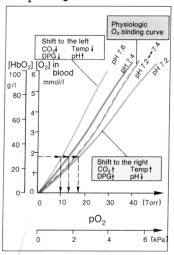

B. O₂ binding curve (detail)

C. Other binding curves

O_2 Deficit – Hypoxia

Anoxia, the complete absence of O_2, does not actually occur clinically, but the term is often used. **Hypoxia** occurs when there is a deficiency of O_2 at the tissues. It occurs at high altitudes and in a number of common diseases. Hypoxia can develop because only a limited cylindric space surrounding a capillary can be adequately supplied with O_2 by diffusion processes. The radius of such a cylinder (or the volume subtended) depends upon the pO_2, the O_2 diffusion coefficient, the O_2 utilization of the tissue, and the rate of capillary flow. For a maximally functional muscle, 20 μm is a realistic radius for the cylinder (diameter of capillary lumen is 6 μm); tissue cells at a distance greater than 20 μm from the capillary will be undersupplied with O_2 (hypoxia). ($\rightarrow$ **A, 4**). Four types of hypoxia are recognized clinically:

Hypoxic hypoxia ($\rightarrow$ **A, 1**) occurs when the O_2 availability to the RBC from the atmosphere is reduced (low pO_2). **Causes**: (a) low atmospheric pO_2, as at high altitudes ($\rightarrow$ p. 96); (b) hypoventilation, as in paralysis of respiratory muscles, in depression of respiratory control in the medulla, in high external pressure on the thorax ($\rightarrow$ p. 94), in airway obstruction and in atelectasis (collapsed lung); (c) alveolar-capillary diffusion block, as in pneumonia or fibrosis; (d) ventilation/perfusion imbalance, as in emphysema.

Anemic hypoxia ($\rightarrow$ **A, 2**) occurs because the O_2-carrying capacity of the blood is reduced. The arterial pO_2 is normal, but the amount of Hb available to carry O_2 is reduced. At rest, hypoxia due to anemia is rarely severe, but in exercise it can become restrictive. **Causes**: (a) reduced erythrocyte count from blood loss, decreased production or increased destruction of RBC; (b) reduced Hb concentration in spite of normal erythrocyte count (hypochromic anemia), as can result from Fe deficiency; (c) synthesis of abnormal Hb,

as in sickle cell anemia; (d) reduced binding of O_2, as in carbon monoxide poisoning ($\rightarrow$ p. 88) or chemical alteration of Hb (methemoglobinemia).

Ischemic or stagnant hypoxia ($\rightarrow$ **A, 3**) occurs during shock, heart failure, or intravascular obstruction. The pO_2 at the lungs and the Hb concentration are normal, but delivery of O_2 to the tissue is inadequate, O_2 extraction is increased, and the local pO_2 falls to low levels. The arteriovenous difference for O_2 is increased as a consequence of the increased extraction.

Histotoxic hypoxia ($\rightarrow$ **A, 5**) occurs when the tissues cannot utilize O_2 for oxidative processes. Delivery of O_2 is adequate, but the cells cannot make use of the O_2 supplied to them. This circumstance occurs in cyanide poisoning, for example, in which cytochrome oxidase (among others) is inactivated.

Effects of hypoxia depend upon the tissue affected because tissue sensitivities to hypoxia differ. In general, the brain is most sensitive ($\rightarrow$ **B**). Anoxia can cause unconsciousness in 15s, irreversible cell damage in c. 2 min, and cell death in 4 to 5 min. Lesser degrees of hypoxia, as in heart failure or chronic pulmonary disease, may be manifested clinically by confusion, disorientation, and bizarre behavior.

Respiratory stimulation occurs reflexly as **hyperpnea** (increased rate and depth of respiration) or **tachypnea**, a rapid shallow respiration (increased rate). **Dyspnea** is the awareness of shortness of breath and depends in part on the respiratory reserve. **Cyanosis** occurs when reduced Hb in the capillaries exceeds 50 g/l. Since reduced Hb has a dark color, nail beds, lips, ear lobes, and areas where the skin is thin take on a dusky purplish coloration. Development of cyanosis depends on concentration of total and reduced Hb and on state of circulation. It does not appear in all hypoxic states.

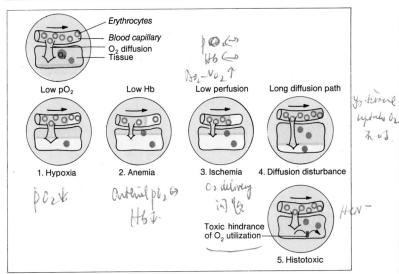

A. Types of hypoxia

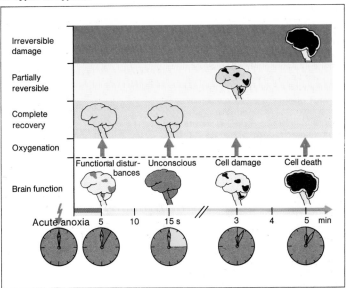

B. Effects of anoxia on the CNS (brain)

Control of Respiration

Respiration is controlled in the CNS; voluntary respiration is governed by the cortex and automatic respiration by structures in the medullopontine region ($\to$ **A**). The spinal cord integrates the efferent output of the two systems. Regulation of respiration adjusts ventilation to maintain appropriate body levels of O_2 and CO_2. The level of CO_2 influences the pH of the ECF and thereby the affinity of hemoglobin for O_2. The level of O_2 can be critical for cellular function especially in the CNS ($\to$ p. 90).

Receptors. There are several sensors for afferent input to the CNS. **Peripheral chemoreceptors** are found in the carotid and aortic bodies. In man, the primary O_2 sensory organ is the **carotid body**. Impulses from these sensors begin to increases as pO_2 falls below 66.5 kPa ($= 500$ Torr) and increase more rapidly below c. 13.3 kPa ($= 100$ Torr). Output of impulses cannot be sustained below 4 kPa ($= 30$ Torr). The increased ventilatory response to a fall in pO_2 is potentiated by a rise in pCO_2 or in concentration of H^+. Responses to pCO_2 are linear above 5.3 kPa ($= 40$ Torr) and to H^+ from pH 7.7 to 7.2. **Central chemoreceptors** are found in the medulla. They are the primary sensors for changes in pCO_2 and respond to corresponding changes in pH of the ECF or cerebrospinal fluid (CSF). **Peripheral mechanoreceptors** occur in the upper airway and in the lungs. They are of several types and have various functions. In the lungs, the principal receptor are the **pulmonary stretch receptors** (PSR) of the Hering-Breuer reflex. Inflation of the lung stretches the PSR and initiates impulses that are carried to the CNS by large myelinated fibers in the vagus (X). They increase respiratory time and reduce frequency. They are also involved in reflexes that bring about bronchoconstriction, tachycardia, and vaso-

constriction. Control of automatic breathing by the central nervous system is governed by so-called centers in the pons and the medulla. These centers modulate the depth of inspiration and the cutoff point that terminates inspiration. The medullary center is important for establishing respiratory rhythm and for the Hering-Breuer reflex, which inhibits inspiration when the lung is stretched.

Other inputs to the medullary center include: **proprioceptors** ($\to$ p. 258), which coordinate muscular activity with respiration; **body temperature**, which increases respiratory rate during fever; **baroreceptors** ($\to$ p. 166), which send afferents to the medullary center as well as to the cardioinhibitory area in the medulla; in the reverse direction, respiratory activity affects blood pressure and pulse rate; the effects are small; **higher CNS centers** (cortex, hypothalamus, limbic system), which influence respiration during anxiety, pain, sneezing, yawning, etc. **Voluntary breath holding** inhibits automatic respiration until the **breakpoint** is reached when the rise in pCO_2 and fall in pO_2 override the voluntary inhibition. The breakpoint can be delayed by previous hyperventilation.

In **chronic CO_2 retention**, the medullary center becomes insensitive to changes in pCO_2 so that pO_2 becomes the chief respiratory drive. In this state, if the pO_2 is raised by breathing 100% O_2, the respiratory drive may be abolished resulting in coma and death. To avoid this event, patients with chronically elevated pCO_2 should receive only O_2-enriched air rather than 100% O_2 ($\to$ p. 96).

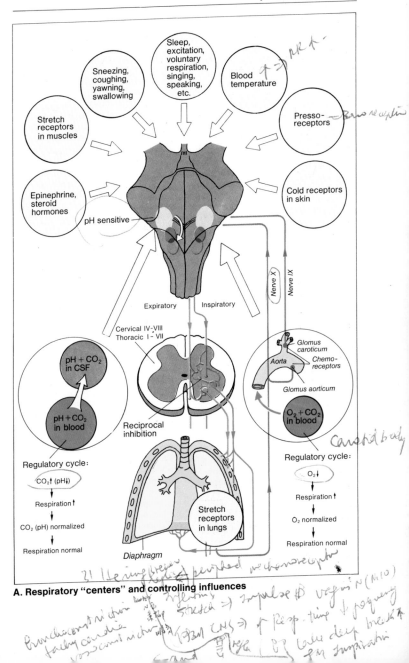

A. Respiratory "centers" and controlling influences

Respiration at Elevated Barometric Pressure (Diving)

The barometric pressure increases 98 kPa ($= 1$ at $= 735$ Torr) for each 10 m depth in the ocean. Swimming below the water surface is possible when the upper airway is lengthened by a snorkel. However, there are two limits to the depth at which **snorkel diving** is practical:

1. The snorkel increases total dead space ($\to$ **B**). When the total dead space ($\to$ pp. 76, 82) approximates the ventilatory volume, fresh air no longer reaches the alveoli. Reducing the diameter of the snorkel to reduce its dead space volume is not practical because the airway resistance increases.

2. Water pressure on the chest increases inspiratory effort. The maximum inspiratory pressure is c. 11 kPa (112 cm H_2O) ($\to$ p. 79) so at a depth of 112 cm, inspiration is impossible and the chest is fixed in expiration ($\to$ **B**).

To permit ventilation at greater depths, respiration must be assisted by an **apparatus** that adjusts the pressure at which gas is delivered from a gas bottle to the pressure that exists at the diving depth. The elevated barometric pressure that exists at depth increases the amount of dissolved N_2 and other gases in the plasma by elevating their partial pressure ($\to$ **A**). At 60 m, seven times more N_2 is dissolved in plasma than at sea level. If return to sea level is gradual, the pN_2 is reduced by diffusion and expiration of N_2; no ill effects occur. If return is not slow enough, the relatively rapid decompression allows N_2 bubbles to form in plasma and in tissues. In the tissues, bubbles cause severe pain; in the plasma, they may occlude small vessels and may cause damage to the heart or CNS. The effect of decompression on the blood is similar to opening a bottle of carbonated drink. Treatment for decompression sickness (the bends) is immediate recompression and slower decompression.

Although N_2 is physiologically nonreactive at atmospheric pressure, higher levels of pN_2 are toxic and depress the nervous system in the same manner as common anesthetic gases. At a pressure of 4 to 5 at (30 to 40 m depth), N_2 produces euphoria; at greater depths and at longer bottom times it produces narcosis ("rapture of the deep"), which resembles alcoholic intoxication. These effects can be prevented by substituting helium for N_2, since helium has a much weaker anesthetic effect. The normal fractional O_2 concentration in air (0.21 l/l or 21 Vol%) becomes fatal at a depth of 100 m (325 ft) because pO_{2insp} rises to the absolutely toxic value of 220 kPa ($\to$ p. 96).

Holding the breath during a dive without apparatus leads to elevation of pCO_2 because the CO_2 produced in the tissues is not eliminated by expiration. When the pCO_2 rises sufficiently, chemoreceptors are stimulated and the breakpoint is reached ($\to$ p. 92). It is possible to delay the rise in pCO_2 by hyperventilation prior to diving. This process lowers the level of pCO_2 at the start of a dive ($\to$ **C**) and allows a longer period of time before pCO_2 rises to the breakpoint. Figure **C** shows partial pressures for CO_2, O_2, and during hyperventilation and during a dive at 10 m depth for 40 s. The barometric pressure during the dive increases alveolar pCO_2, pO_2, and pN_2. The gases diffuse from alveoli to blood. When pCO_2 rises to the breakpoint, the diver returns to the surface. During the ascension, pO_2 in blood falls because barometric pressure is reduced and because tissues continue to utilize O_2. Eventually, alveolar exchange ceases. It is possible by hyperventilation to prolong the rise in pCO_2 to such an extent that pO_2 falls to intolerable levels during decompression and the diver loses consciousness before he can surface ($\to$ **C**, dotted line).

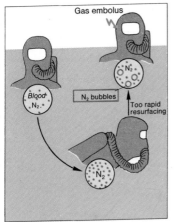

A. Diving with equipment

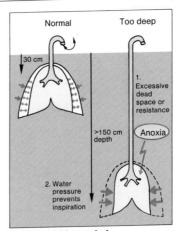

B. Diving with snorkel

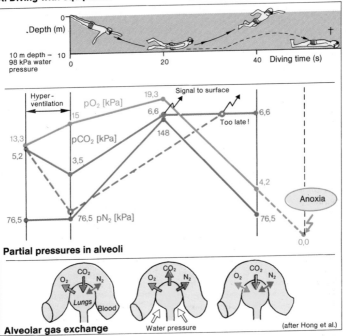

Partial pressures in alveoli

Alveolar gas exchange

(after Hong et al.)

C. Diving unassisted

Respiration at High Altitude

At sea level, the mean barometric pressure (p_B) is 101.3 kPa ($= 760$ Torr). O_2 comprises 21% of the air and has a partial pressure ($pO_{2 insp}$) at sea level of 21.33 kPa ($= 160$ Torr) ($\rightarrow$ **A**, Column 1, blue curve). pO_2 in alveoli ($pO_{2 alv}$) is c. 13 kPa ($= 100$ Torr) ($\rightarrow$ **A**, column 2). At increasing altitude, p_B falls and therefore $pO_{2 insp}$, and $pO_{2 alv}$ also fall ($\rightarrow$ **A**, column 1 and 2). Even at 3000 m (10,000 ft) pO_2 is low enough (8 kPa = 60 Torr) to stimulate an increase in ventilation via the chemoreceptors ($\rightarrow$ p. 92). At 4000 m (13,000 ft) ($\rightarrow$ **A**, column 2, dashed line), the $pO_{2 alv}$ falls below a critical level of 4.7 kPa ($= 35$ Torr) and hypoxia ($\rightarrow$ p. 91) becomes severe. Progressively increasing stimulation of the chemoreceptors allows a gain in ventilation ($\rightarrow$ **A**, column 4) and makes it possible to reach still higher altitudes before the critical value of $pO_{2 alv}$ is reached at 7000 m (23,000 ft) ("altitude gain") ($\rightarrow$ **A**, column 3). Above this altitude, consciousness is invariably lost, However, breathing O_2 from a gas bottle allows higher elevations to be achieved ($\rightarrow$ **A**, column 1, green curve). The critical $pO_{2 alv}$ is displaced to 12,000 m (39,500 ft) ($\rightarrow$ **A**, column 3), but if V_T is also increased, the limit can be pushed to 14,000 m (46,000 ft). The p_B becomes the limiting factor when breathing 100% O_2. At an altitude of 10,400 m (34,000 ft), p_B falls below 25 kPa ($= 187$ Torr). At this pressure, it is not possible to achieve a $pO_{2 alv}$ of 13.3 kPa ($= 100$ Torr) because of the contribution of the partial pressures of CO_2 and water in the alveoli. This altitude cannot be exceeded without development of hypoxia. Modern airliners fly below this limiting p_B so that, in case of depressurization of the cabin, ventilation can be sustained with an O_2 mask and survival is still feasible. At the critical $pO_{2 alv}$ (4.7 kPa = 35 Torr) or the critical p_B of 16.3 kPa ($= 122$ Torr), a pressurized suit is essential

for survival. For space travel, pressurized cabins are essential.

The maximal increase in ventilation rate that can be stimulated by the chemoreceptors during O_2 deficiency is only approximately three times the resting rate. This value is relatively small when compared to the increase that can occur during exercise at sea level ($\rightarrow$ p. 46). The explanation for this difference is that hyperventilation at elevated altitudes appreciably reduces pCO_2 in blood and produces a respiratory alkalosis ($\rightarrow$ p. 104), which becomes partially compensated, although plasma levels of HCO_3^- and pCO_2 remain lower than normal. In this state, the CO_2 and pH receptors have a low firing rate and do not permit the full hyperventilatory response to develop.

During **acclimatization** to high altitude there is an increase in ventilatory rate over the first few days; after this time the rate declines somewhat but requires years at altitude to return to its original level. Erythropoietin secretion ($\rightarrow$ p. 58) increases to stimulate a surge in red cell production; when arterial pO_2 rises, erythropoietin synthesis subsides. In the high Andes (5,500 m = 18,000 ft), the inhabitants have polycythemia and low $pO_{2 alv}$ but survive quite normally.

O_2 toxicity

If $pO_{2 insp}$ rises above normal (> 22 kPa) *hyperoxia* develops. The pO_2 may be high due to an increased O_2 concentration (O_2 therapy) or due to an elevated total pressure (p_B) at a normal O_2 content (diving, $\rightarrow$ p. 94). O_2 toxicity is a function of $pO_{2 insp}$ (critical level c. 40 kPa or 300 Torr) and time of exposure. *Lung damage* with a decrease of surfactant ($\rightarrow$ p. 80) develops after days if $pO_2 \approx 70$ kPa (0.7 at) and within 3 to 6 h if $pO_2 \approx 200$ kPa (2 at). *Coughing* and *pain during breathing* are the main symptoms. At $pO_2 > 220$ kPa (2.2 at), corresponding to 100 m depth during diving with compressed air, *convulsions* and *unconsciousness* develop. Premature babies become *blind* if exposed to a $pO_{2 insp} \geqslant 40$ kPa for a longer period.

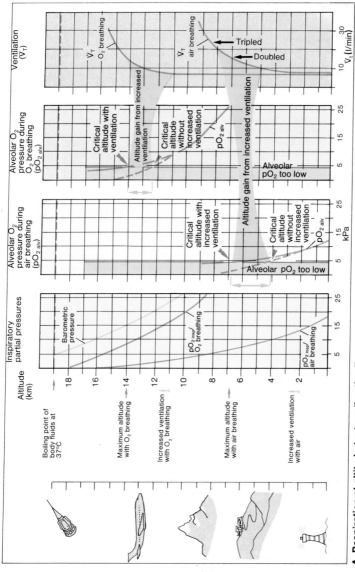

A. Respiration at altitude (not acclimatized)

pH — Buffers — Acid-Base Balance

The hydrogen ion concentration or pH of plasma is normally maintained within a narrow range. Values for pH outside the range of 7.0 to 7.8 are incompatible with life. Acid-base balance is achieved by homeostatic mechanisms that resist a change from the normal plasma pH of 7.4. These mechanisms depend upon a variety of constituents of the body fluids, **buffers**, which partially neutralize the acids and bases that arise from diet and metabolism. All buffers in a fluid compartment are in equilibrium at a given pH. Except for local deviations in the microenvironment, the pH is similar in all fluid compartments (ECF, ICF, etc.). In the blood, the **total buffering capacity** is normally c. 48 mEq/l; it represents the sum of the **buffer bases** HCO_3^-, Hb^-, HbO_2^-, HPO_4^{2-}, etc.

Among the buffers, the pair **CO_2/ HCO_3^-** is unusual because its two components can be *varied independently* by different organs. CO_2 is produced continually as an end product of metabolism; its concentration or partial pressure (pCO_2) in the blood is controlled principally by the **lungs**. By varying the ventilation, pCO_2 can be kept constant even when it is produced in excess or when it is consumed in buffering base. HCO_3^- is controlled by the kidney. The supply of HCO_3^- at any one time is limited, but when it is consumed in buffering acid, the **kidney** can regenerate more. When it is formed in excess by buffering alkali, the kidney can increase its excretion.

For any buffer pair, the pH is a function of the ratio of the alkaline form and the acid form. At pH = 7.4, the ratio HCO_3^-/CO_2 = 20/1 and reflects the normal plasma levels of HCO_3^- (= 24 mmol/l) and of CO_2 (= 1.2 mmol/l). An increase in the relative concentration of the alkaline form increases the ratio and raises the pH; a relative increase of the acid form lowers the ratio and the pH.

Plasma and cell pH can be changed by a variety of influences that influence the ratio HCO_3^-/CO_2 either directly or indirectly:

1. **Production of acid** by normal metabolic processes. Most of this acid takes the form of CO_2, which is eliminated in the lungs. A smaller amount of acid, **fixed acid**, is derived from sulfur- and phosphorus-containing proteins and from fat and carbohydrate (lactate, acetoacetate, β-hydroxybutyrate, ketoacids, etc.). Acid production is normally c. 50 to 80 mmol/day and is matched by an equivalent renal excretion of acid ($\rightarrow$ p. 130).

2. **Production of base** by metabolism of plant substances (as in herbivores) leads to a rise in HCO_3^- and raises the pH.

3. **Change in CO_2 concentration** (pCO_2). An increase in pCO_2 lowers blood pH. Changes in pCO_2 are brought about by factors that influence primarily the effectiveness of respiration (pulmonary exchange of CO_2) or, to a lesser extent, the rate of metabolic processes (cellular generation of CO_2).

4. **Change in HCO_3^- concentration.** An increase in HCO_3^- raises the blood pH. Changes in HCO_3^- are brought about primarily by factors that influence generation of base or its loss ($\rightarrow$ pp. 132, 102).

Anion gap. The sum of plasma cations always matches exactly the sum of anions. However, in routine clinical laboratory analyses the sum of cations (Na^+ and K^+) is more than the sum of anions (Cl^- and HCO_3^-). This cation excess represents the anions that are not usually measured, the anion gap; it is normally less than 17 mEq/l. Increases in Cl^- or HCO_3^- can reduce the gap, but increases in other anions, such as lactate, produce metabolic acidosis with a larger than normal anion gap.

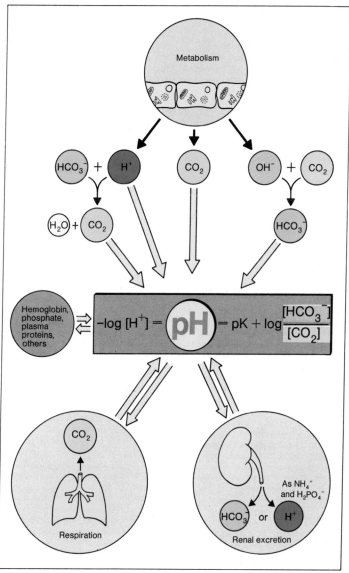

A. Regulation of pH in ECF

Bicarbonate / CO_2 Buffer

In any buffered solution, the pH is determined by the **relative** concentrations of the buffer pair. In bicarbonate buffers, the concentration of $CO_2 = \alpha \times pCO_2$ where α = the solubility coefficient of CO_2. According to the Henderson-Hasselbalch equation ($\rightarrow$ **A**), the pH of a bicarbonate buffer is 7.4 when $HCO_3^-/CO_2 = 20/1$ ($\rightarrow$ **A**). Changing the absolute concentrations of HCO_3^- and CO_2 will not change the pH as long as the ratio remains 20/1. Example: pH = 7.4 if $HCO_3^-/CO_2 = 24/1.2$ mmol/l (normal) or 10/0.5 (low) or 34/1.7 (high), since the ratio in each case is 20/1.

When H^+ is added to a HCO_3^- buffer in a **closed** system ($\rightarrow$ **A**, left side), the reaction $H^+ + HCO_3^- \rightleftarrows CO_2 + H_2O$ is driven to the right. Addition of H^+, 2 mmol/l, reduces HCO_3^- from 24 to 22 mmol/l and raises pCO_2 from 1.2 to 3.2 mmol/l. At this ratio, the pH falls to 6.93. In contrast, the body is an **open** system ($\rightarrow$ **A**, right side, **B**) in which the concentration of CO_2 is continuously regulated by respiration. Thus, when pCO_2 is elevated to 3.2 mmol/l, the excess is blown off in the lungs. As a result, addition of acid or base to the open system changes only the HCO_3^- concentration. Addition of H^+, 2 mmol/l, lowers HCO_3^- to 22 but leaves CO_2 normal at 1.2 mmol/l. The ratio 22/1.2 = 18.3 at a pH of 7.36! Respiratory adjustments therefore restrict pH changes to a much narrower range than would occur in a closed system. Normal metabolic processes continuously generate CO_2 at a rate of 15,000 to 20,000 mmol/day but the pCO_2 in arterial plasma is maintained constant by respiratory activity. To these amounts, H^+ is added continuously at a rate of c. 60 mmol/day. **Neutralization of H^+** adds CO_2 to the plasma ($HCO_3^- + H^+ \rightarrow CO_2 + H_2O$), but the excess CO_2 is eliminated each time the blood passes through the lungs (high CO_2 gradient in the alveoli), and both pCO_2 and pH are kept relatively constant ($\rightarrow$ **B**, left side). Quantitatively, this compensation is trivial; for example, if H^+ production should double, the excess CO_2 (assuming all other buffers were inoperative) would be only 60 mmol/day or 0.3 % of the normal daily load. On the other hand, the HCO_3^- consumed for buffering these 60 mmol/d of fixed acid is quantitatively much more important. Per day it constitutes c. 1/6 of the HCO_3^- amount of the ECF. Renal excretion of H^+ is therefore necessary. For each H^+ ion excreted one HCO_3^- ion is generated ($\rightarrow$ p. 133, C 1).

In principle, **addition of base** operates similarly. The reaction is $OH^- + CO_2 \rightleftarrows HCO_3^-$. The concentration of HCO_3^- increases and changes the buffer pair ratio, and the pH becomes more alkaline. In this case, initially the concentration of CO_2 is relatively lower than normal. Accordingly, the rate of elimination of CO_2 declines (lowered alveolar CO_2 gradient) and the pCO_2 builds up to restore the normal ratio ($\rightarrow$ **B**, right side).

The HCO_3^-/CO_2 buffer accounts for about half the buffering capacity of blood. **Nonbicarbonate buffers** are predominantly intracellular and make up approximately half the buffering capacity. The nonbicarbonate buffers operate as in a closed system; the total concentration of the buffer pair remains constant when pH changes occur. In the blood, hemoglobin is the chief nonbicarbonate buffer. Buffering capacity of the blood may be reduced in case of large losses of blood or in anemia. Nonbicarbonate buffers not only cooperate with the HCO_3^-/CO_2 buffer in buffering fixed acids or bases. They become the *only effective buffers in respiratory acid-base disturbances* where prolonged changes of plasma CO_2 are the primary events ($\rightarrow$ p. 104).

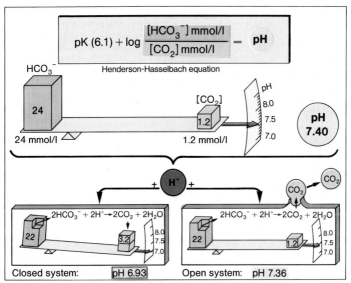

A. Bicarbonate buffer: open and closed systems

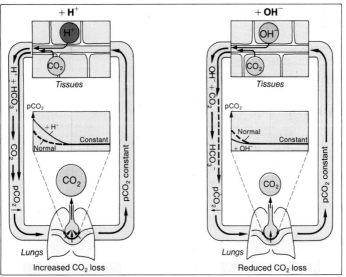

B. Bicarbonate as a blood buffer: open system

Metabolic Disturbances in Acid-Base Balance

Acid-base balance refers to the homeo-static mechanism that resist changes in total body acid and maintain pH constant. **Normal acid-base parameters of arterial plasma** (although measured in blood) are as follows (for RBC values see table on p. 86):

	Females	Males
pH	7.40 ± 0.015	7.39 ± 0.015
pCO$_2$	5.07 ± 0.3 kPa ($= 38$ Torr)	5.47 ± 0.3 kPa ($= 41$ Torr)
HCO$_3^-$	24 ± 2.5 mmol/l	24 ± 2.5 mmol/l

At steady state:

1. H$^+$ production − HCO$_3^-$ production = H$^+$ excretion − HCO$_3^-$ excretion = = 60 mmol/day
2. CO$_2$ production = CO$_2$ excretion = = 15,000 to 20,000 mmol/day

For equation 1, the primary contributions are H$^+$ production as phosphoric and sulfuric acids (end products of protein metabolism) and renal excretion of H$^+$ ($\rightarrow$ p. 130). On the other hand, a diet high in vegetable matter generates large amounts of HCO$_3^-$ and accounts for the alkaline urine that herbivores excrete.

When the pH falls below 7.35, **acidosis** occurs; when it rises above 7.45, **alkalosis** occurs. A primary change in pCO$_2$ is a **respiratory** disturbance; a primary change in HCO$_3^-$ is a **metabolic** disturbance. Metabolic disturbances most often represent a discrepancy between production and excretion of H$^+$. The acid-base disturbance may be **compensated** or **uncompensated**.

Metabolic acidosis ($\rightarrow$ **A**) (low pH, low HCO$_3^-$) may occur from (1) renal failure to excrete the normal acid load; (2) ingestion of acid (e.g., NH$_4$Cl); (3) excess endogenous production of acid as occurs in diabetes mellitus and starvation (incomplete metabolism of fats, ketoacids); (4) anaerobic production of lactate (incomplete oxidation of carbohydrate); (5) increased production of phosphates and sulfates; (6) loss of HCO$_3^-$ as in diarrhea, renal tubular acidosis, or use of carbonic anhydrase inhibitors ($\rightarrow$ p. 128). In all of these events, a gain of H$^+$ produces an effect equivalent to a loss of HCO$_3^-$. The H$^+$ is neutralized (H$^+$ + HCO$_3^-$ $\rightleftarrows$ $\rightleftarrows$ H$_2$O + CO$_2$) and the CO$_2$ is expired. The buffer pair HCO$_3^-$/CO$_2$ is low in HCO$_3^-$ and the pH is acid. Chemoreceptors in the medulla ($\rightarrow$ p. 92) stimulate *hyperventilation* to blow off the relative excess of CO$_2$ and restore the buffer *ratio* toward normal (**respiratory compensation**). Simultaneously, the *kidney secretes more* H$^+$, forms more HCO$_3^-$, and *synthesizes more ammonia* to neutralize H$^+$ ($\rightarrow$ pp. 130). **Metabolic alkalosis** (high pH, high HCO$_3^-$) may occur from (1) intake of alkaline substances (HCO$_3^-$); (2) metabolism of organic anions such as lactate and citrate to CO$_2$ and H$_2$O; (3) loss of H$^+$ from vomiting, or from increased renal excretion in K$^+$ deficiency ($\rightarrow$ p. 134). Chemical responses resemble those in metabolic acidosis except that they are opposite in direction. Hypoventilation may occur as a respiratory compensation, but the extent is limited because when CO$_2$ begins to accumulate respiration is stimulated again ($\rightarrow$ p. 102). Furthermore, hypoventilation would lead to hypoxia. The *kidneys* excrete HCO$_3^-$ and reduce the excretion of H$^+$.

Metabolic alkalosis is common in the clinic. The elevated pH reduces serum calcium and may lead to tetany. Irritability of the nervous system increases and may produce convulsions. ECG changes resemble hypokalemia. Alkalosis is especially hazardous to patients receiving digitalis.

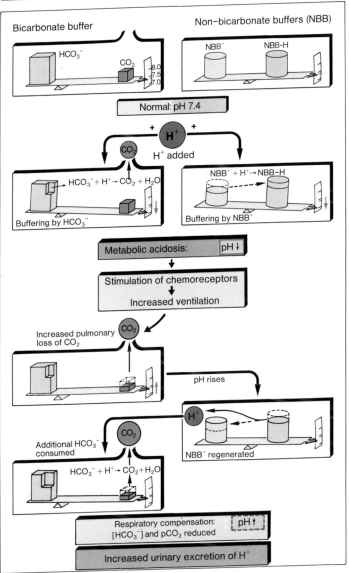

Bicarbonate buffer   Non−bicarbonate buffers (NBB)

HCO_3^- CO_2 NBB⁻   NBB-H

Normal: pH 7.4

H^+ added

$HCO_3^- + H^+ \rightarrow CO_2 + H_2O$ NBB⁻ + H^+ → NBB−H

Buffering by HCO_3^- Buffering by NBB⁻

Metabolic acidosis: pH ↓

Stimulation of chemoreceptors

Increased ventilation

Increased pulmonary loss of CO_2 CO_2

pH rises

H^+

Additional HCO_3^- consumed CO_2

$HCO_3^- + H^+ \rightarrow CO_2 + H_2O$ NBB⁻ regenerated

Respiratory compensation: pH ↑
$[HCO_3^-]$ and pCO_2 reduced

Increased urinary excretion of H^+

A. Metabolic acidosis

Respiratory Disturbances in Acid-Base Balance

In metabolic disturbances of acid-base balance, the concentrations of bicarbonate and of the bases of the non-bicarbonate buffers change in parallel; in respiratory disturbances, changes in the concentrations are dissociated. The HCO_3^-/CO_2 buffer is no longer effective because in respiratory disturbances the change in pCO_2 is the *cause* and not the *result* as it is in metabolic disturbances ($\rightarrow$ pp. 100 to 103).

Respiratory acidosis ($\rightarrow$ A) (low pH, high pCO_2) occurs whenever the ability of the lungs to eliminate CO_2 is impaired, as in (1) reduction of pulmonary tissue (tuberculosis, pneumonia), (2) reduced ventilation (poliomyelitis, barbiturate overdosage), and (3) abnormalities of the rib cage (scoliosis).

In these states, decreased respiratory exchange elevates the pCO_2 and lowers the pH. These changes are followed by a very small increase in HCO_3^-. The H^+ excess is buffered by the intracellular non-bicarbonate buffers (NBB) at the expense of the salt (base) form of the buffer pairs $(NBB^- + H^+ \rightleftarrows NBB - H)$. As a result, the combined total of bicarbonate concentration and NBB bases remains unchanged, unlike the situation in metabolic disturbances. Nevertheless, the rise in HCO_3^- is by far not enough to account fully for the high pCO_2 so that the ratio HCO_3^-/CO_2 remains below normal (lower pH, acidosis). **Renal excretion of H^+** becomes increasingly important to compensate for the basic pathology that accounts for the reduced pulmonary exchange.

The kidneys have a remarkably large potential capacity to excrete H^+ in acidosis ($\rightarrow$ pp. 130–134). In the process of H^+ secretion, HCO_3^- formation in the proximal renal tubules is also increased; the same mechanism

that eliminates acid restores HCO_3^- (renal compensation). As HCO_3^- retention progresses, the blood pH approaches its normal value and the NBB shift their equilibria (NBB$-$H $\rightarrow$ NBB$^- + H^+$). A portion of the retained HCO_3^- is used to neutralize the H^+, which is generated by this shifting equilibrium.

With the development of acidosis, excretion of **ammonium (NH_4^+)** and **titratable acidity** (TA) increase. TA is determined by measuring the equivalents of NaOH required to titrate urine back to the pH of blood. Normal daily excretion of H^+ is 40 to 60 mmol as NH_4^+ and 20 to 40 mmol as TA. In contrast, when acidosis develops, NH_4 production can account for excretion of as much as 300 to 400 mmol of H^+ daily. Because of the 1 to 2 day delay in renal metabolic response to acidosis, *acute* respiratory acidosis is poorly compensated in comparison to chronic respiratory acidosis. In the latter, HCO_3^- may increase 1 mmol for each 10 mmHg rise in pCO_2 above normal. Uncorrected respiratory acidosis can cause severe damage to the central nervous system.

Respiratory alkalosis (high pH, low pCO_2) occurs in hyperventilation as may develop in O_2 deficiency or from psychic causes. The HCO_3^- is variably decreased depending upon the magnitude of renal compensation. The NBB shift their equilibrium and liberate H^+, which reacts with HCO_3^- to lower its concentration (NBB$-$H $\rightarrow$ NBB$+H^+$; $H^+ + HCO_3^- \rightarrow CO_2 + H_2O$). The kidney reduces its excretion of H^+ and allows more HCO_3^- to escape into the urine. As a broad generalization, pCO_2 and renal tubular resorption of HCO_3^- change in parallel.

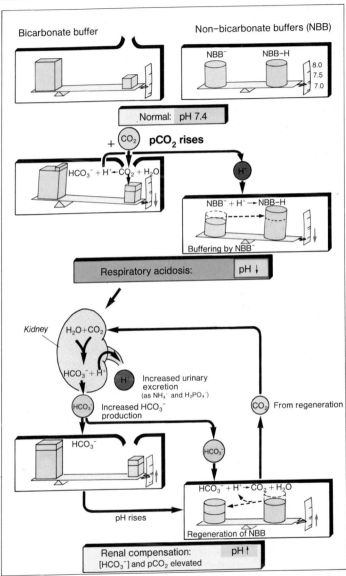

Bicarbonate buffer

Non−bicarbonate buffers (NBB)

NBB⁻ NBB−H

8.0
7.5
7.0

Normal: pH 7.4

+ CO_2 **pCO₂ rises**

$HCO_3^- + H^+ \rightarrow CO_2 + H_2O$

H⁺

$NBB^- + H^+ \rightarrow NBB-H$

Buffering by NBB⁻

Respiratory acidosis: pH ↓

Kidney

$H_2O + CO_2$

$HCO_3^- + H^+$

H Increased urinary
excretion
(as NH_4^+ and $H_2PO_4^-$)

HCO_3^- Increased HCO_3^-
production

CO_2 From regeneration

HCO_3^-

HCO_3^-

$HCO_3^- + H^+ \rightarrow CO_2 + H_2O$

pH rises

Regeneration of NBB

Renal compensation:
[HCO_3^-] and pCO₂ elevated pH ↑

A. Respiratory acidosis

Determination of Acid-Base Status

In the Henderson-Hasselbalch equation pH $= pK_a + \log [HCO_3^-]/[CO_2]$. The pK_a of CO_2/HCO_3^- = 6.1 at physiologic conditions. The concentration of $CO_2 = \alpha \times pCO_2$. The solubility coefficient, α, is also a constant (= 0.22 mmol $\cdot$ l$^{-1} \cdot$ kPa^{-1}). The remaining values, pH, HCO_3^-, and pCO_2 vary interdependently; when one is held constant the other two vary proportionally. Plotting log pCO_2 against pH yields a straight line ($\rightarrow$ **A**). The blue line shows data from a solution of HCO_3^- at 24 mmol/l, with no other buffers. Changing pCO_2 changes the pH without influencing HCO_3^-. A family of parallel lines can be drawn to describe the relationships for other $[HCO_3^-]$. The scale of the coordinates is adjusted so that the slope of the line is 45° (HCO_3^- titration curve).

In blood, the presence of additional buffers (NBB) influences the titration curve; when pCO_2 is changed, the pH changes much less than when HCO_3^- is the only buffer. The slope of the titration curve is therefore steeper ($\rightarrow$ **B**, green and red lines). A change in blood pCO_2, for example from 40 ($\rightarrow$ **B**, d) to 20 mmHg ($\rightarrow$ **B**, d) is followed by a change in the **actual** $[HCO_3^-]$ in the same direction ($\rightarrow$ **B**, point c on red line; c lies now on the interrupted line which represents a lower $[HCO_3^-]$ than the yellow line). The "**standard**" $[HCO_3^-]$, however, which is always measured after equilibration with a pCO_2 of 5.33 kPa, does not change.

The **Siggaard-Andersen nomogram** ($\rightarrow$ **C**) is used for clinical evaluation of acid-base balance. Log pCO_2 is plotted on the y axis and pH on the x axis. A horizontal reference line is drawn from the y axis at the normal pCO_2 (= 5.33 kPa = 40 Torr). Along this line are values for HCO_3^- which represent the family of HCO_3^- titration curves in Fig. **A**. To any point on this line, the 45° line

may be redrawn along which the actual HCO_3^- value can be found. On the upper part of the nomogram, the curve has two scales, one for hemoglobin (Hb) and one for **total buffer bases** (TBB). The TBB are composed chiefly of HCO_3^-, Hb, and other proteins. A line drawn from a given value for Hb to the point pCO_2 = 40; pH = 7.4 describes the HCO_3^- titration curve in the presence of NBB. When Hb is low, as in anemia, the slope of this line falls to some extent.

To use the nomogram, arterial blood is drawn and measured for pH. It is then equilibrated with CO_2 at two different pressures, and the two corresponding pH values are plotted. *Example*: Fig. **C**, point A: pH at pCO_2 of 10 kPa (= 75 Torr) and point B: pH at pCO_2 of 2.67 kPa (= 20 Torr). The pH of the blood before equilibration is plotted along this line and the pCO_2 corresponding to that pH is read from the y axis ($\rightarrow$ **C**, green line for normal acid-base values).

In Fig. **C**, the green capitals show normal values whereas the red letters represent the respective values in an abnormal state: a and b are the equilibrated values. The original pH of 7.2 indicates acidosis. The standard HCO_3^- for this curve (d) of 13 mmol/l is low. The pCO_2 of the original blood (4 kPa = 30 Torr) is also low and represents a partial respiratory compensation by hyperventilation. A 45° line from c to e shows the actual HCO_3^- in the original blood to be lower (11 mmol/l) than the standard HCO_3^-. This difference reflects the decreased pCO_2. The TBB can be read from the upper scale at g (normal TBB = 48 mEq/l when Hb = 15 g%). Base excess (BE) is shown where the titration line crosses the lower curve on the nomogram. BE is the amount of acid or base needed to titrate 1 liter of blood to normal at a pCO_2 of 5.33 kPa (= 40 Torr). The BE is (+) in alkalosis and (−) in acidosis.

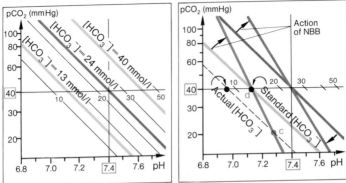

A. pCO₂/pH nomogram (without NBB)   **B. pCO₂/pH nomogram (with NBB)**

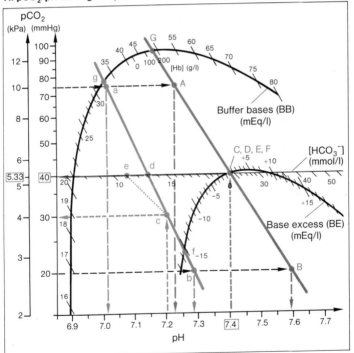

C. Siggaard-Andersen nomogram

Kidney: Anatomy and Function

The primary function of the kidney is to maintain constancy of the extracellular fluid (ECF) (→ p. 125). It achieves constancy both of ECF **volume** and of **osmolality** by balancing intake and excretion of salt and water; it achieves constancy of blood and cellular pH by adjusting excretion of K^+, H^+, and HCO_3^- to their intake and to respiration and metabolism (→ p. 98–104). In addition to these basic responsibilities, the kidney conserves nutrients and eliminates end products of metabolism. It also has numerous metabolic functions (e.g., gluconeogenesis, peptide hydrolysis) and is a source of hormones (e.g., renin, erythropoietin, D-hormone).

The functional unit of the kidney is the **nephron**; 1.2 million nephrons make up each human kidney. The nephron may be thought of as a funnel: blood that reaches the kidney is filtered in the **glomerulus** (→ **A, B**); in the long stem (tubule) of the nephron, most of the filtered fluid and solutes are returned to the blood by a process of **resorption** (→ p. 112). The fraction that is not resorbed remains in the tubule and is excreted as the terminal urine.

Subunits of the Nephron

1. The **renal corpuscle** (→ **A, 1, 2; B**) is made up of Bowman's capsule and the **glomerulus**, which contains the filtering surface of the nephron. The **afferent** arteriole brings blood to the nephron and breaks up into several capillaries, which comprise the **glomerular tuft**. The capillaries rejoin to form the **efferent** arteriole, which carries blood to a second capillary network around the tubular cells (→ p. 110). The glomerular capillaries invaginate Bowman's capsule like fingers pushed into an inflated balloon. In this way, two layers are formed within the capsule: a visceral layer (→ **B, 2**), which is applied closely to the capillaries, and a parietal (→ **B, 1**) layer, which forms the outer surface of the corpuscle. The capsular space between the two layers collects the glomerular filtrate.

The filtering surface has several layers (→ **B**): the parietal layer of the capsule contains the **podocytes**. Their appendages, the **pedicels**, interdigitate closely. The slit-like spaces between the pedicels are covered by the **slit membrane**. The other face is made up of the capillary endothelium; spaces between its cells are 50 to 100 nm. A three-part **basal membrane** lies between the podocyte layer and the capillary layer. The gaps in the filtration surface allow all components of blood to pass except blood cells and proteins having a molecular weight greater than c. 70,000.

2. The **proximal tubule** has both a convoluted (→ **A, 3**) and a straight (**pars recta**) (→ **A, 4**) segment. Cellular characteristics include: a **brush border** on the luminal surface, elaborate infolding of the basal surface membrane to form a complex **basal labyrinth**, and numerous mitochondria near the basal surface (→ **A** and p. 14.

3. The **loop of Henle** has a thin descending limb (→ **A, 5**) and a thin and thick ascending limb (→ **A, 6**). Long loops of Henle dip into the medulla and represent c. 20% of all nephrons; the remaining nephrons have shorter loops.

4. The **macula densa** is a group of specialized cells at the point of contact between tubule and glomerulus. (→ p. 138).

5. The **distal tubule** (→ **A, 7**) begins at the macula densa and connects with the collecting ducts (→ **A, 8**). Its cells have no brush border and fewer mitochondria than those of the proximal tubule.

6. The **collecting ducts** make final modifications to the urine and conduct it to the renal papillae and pelvis for excretion.

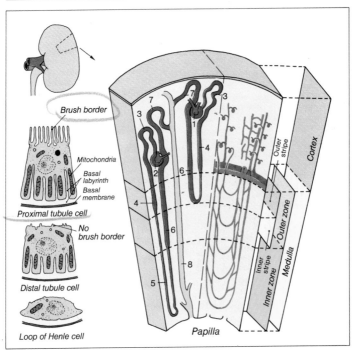

Brush border

Mitochondria
Basal labyrinth
Basal membrane

Proximal tubule cell

No brush border

Distal tubule cell

Loop of Henle cell

Cortex

Outer stripe

Outer zone

Inner stripe

Inner zone

Medulla

Papilla

A. Functional anatomy of the kidney

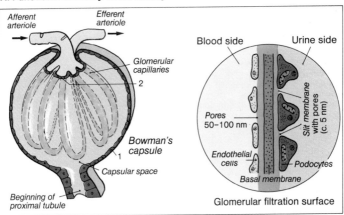

Afferent arteriole

Efferent arteriole

Glomerular capillaries
2

Bowman's capsule

Capsular space

Beginning of proximal tubule

Blood side   Urine side

Pores 50–100 nm

Slit membrane with pores (c. 5 nm)

Endothelial cells

Podocytes

Basal membrane

Glomerular filtration surface

B. Glomerulus

Renal Circulation

Blood reaches the kidney through the short, high-pressure, large-diameter **renal-artery**. The **arcuate arteries** ($\to$ **A, 1a, 2a**), a major division of the renal artery, form a boundary between the **cortex** and **medulla**. Distribution of blood to these two regions differs. Total renal **plasma flow** (**RPF**) in man is 500 ml/min in the cortex, 120 ml/min in the outer medulla, and 25 ml/min in the inner medulla ($\to$ p. 109, **A**). These relationships may be altered in pathologic states. Total renal **blood** flow (**RBF**) in man is c. 25% of the cardiac output or c. 1,2 l/min.

The **microcirculation of the kidney** is unusual; it consists of two capillary networks ($\to$ **A, 2; B**). The first is a high-pressure net ($\to$ **B**) between the afferent and efferent arterioles of the glomerulus ($\to$ p. 108). Changes in resistance of these vessels influence flow to the glomerulus and affect the glomerulus filtration pressure. Blood in the efferent arteriole flows to the second capillary network, a low-pressure net surrounding the tubules ($\to$ **B**). The kidney contains two classes of nephrons; the peritubular capillary net is adapted differently to each of these classes ($\to$ **A, 2**): **cortical nephrons** have short loops of Henle and a limited capacity to conserve Na^+; around these tubules, the capillaries form a network as in any other tissue; (2) **juxtamedullary nephrons** have long loops of Henle and a greater capacity to conserve Na^+; around these nephrons, the capillaries break up into long (40 mm) loops or slings that parallel the loops of Henle. The capillary loops, the **vasa recta**, are the only blood supply to the medulla. Changes in distribution of blood to these two network influences the extent of salt excretion. Medullary blood flow also influences osmolar concentration of the medullary ECF; increased flow reduces the concentration and limits capacity of the kidney to conserve water ($\to$ p. 122).

Autoregulation of RBF.

When the systemic blood pressure is varied from 10.6 to 26.6 kPa (= 80 to 200 Torr), only trivial changes are observed in GFR ($\to$ p. 114) and RBF. ($\to$ **C**). This phenomenon is **autoregulation** ($\to$ p. 162) and occurs even in a denervated kidney. It is influenced largely by changes in afferent arteriolar resistance ($\to$ **B**). When the systemic systolic pressure falls below 10.6 kPa (= 80 Torr) autoregulation is no longer effective and both GFR and RBF fall off sharply ($\to$ **C**).

Measurement of RPF can be accomplished by means of the **Fick principle**: the rate of removal of a test substance from the blood by an organ is a measure of the blood flow to that organ. The full expression of the Fick principle is: blood flow = rate of removal/(arterial − venous concentration). It is applicable to any organ, provided an appropriate test substance is used. Concentration of the test substance in the blood entering and leaving the organ must be determined. For the kidney, the concentration of the test substance, p-aminohippurate (PAH), in venous blood leaving the kidney is *assumed* to be negligible (at low arterial concentrations renal venous concentration of PAH is c. 1/10 of the arterial value). Only the arterial concentration is measured. Thus $RPF = U_{PAH} \times \dot{V}_u / P_{PAH}$. U_{PAH} is the urinary PAH concentration and $\dot{V}_u$ is urinary flow rate. The term $U_{PAH} \times \dot{V}$ is the rate of removal from blood or the rate of excretion into the urine; P_{PAH} is the arterial concentration. If P_{PAH} is too high PAH excretion ($\to$ p. 112) is saturated, and $U_{PAH} \times \dot{V}_u/P_{PAH}$ becomes much lower than RPF.

Renal blood flow, RBF, is calculated from $RPF/(1\text{-Hematocrit})$ ($\to$ p. 58).

Renal O_2 consumption in man is c. 18 ml/min. Because of the large RBF, the arteriovenous difference for O_2 is small (14 ml/l blood). O_2 consumption of the cortex is more than 20 fold higher than that of the inner medulla. In the cortex, energy metabolism is oxidative and uses fatty acids and other substances as substrates. In the medulla, it is anaerobic, O_2 utilization is low, and glucose is the chief substrate.

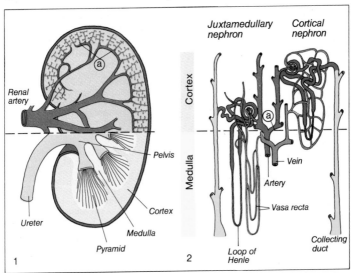

A. Vascular arrangement in kidney

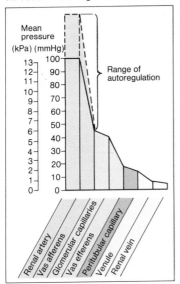

B. Blood pressure gradients in renal vessels

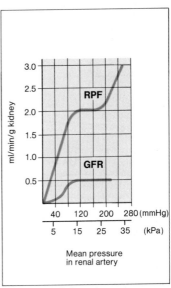

C. Autoregulation of renal blood flow and filtration

Transport Processes in the Nephron

The final urine is formed by one or more of the following mechanisms: (1) **filtration** at the glomerulus ($\rightarrow$ **B**), (2) **tubular resorption** (either active or passive) ($\rightarrow$ **A, w, x, y**), (3) **active tubular excretion** from plasma to urine ($\rightarrow$ **A, z**), and (4) **tubular secretion** from renal cells to urine ($\rightarrow$ **A, u, v**).

1. Small molecules in plasma are filtered, but filtration may be hindered by protein binding since plasma proteins are too large to pass the glomerular filter ($\rightarrow$ **B**, T).

Proteins up to a molecular weight of about 80 000 are partly filtered. The extent of filtration depends on (a) molecular size and (b) on the charge of the proteins.

2. After being filtered into the tubular fluid, many substances are resorbed. These are chiefly electrolytes (Na^+, Cl^-, K^+, Ca^{2+}, HCO_3^-, phosphate, etc.), amino acids, uric acid, lactate, urea, peptides, proteins, ascorbic acid, and glucose ($\rightarrow$ **C**). **Driving forces for passive reabsorption** ($\rightarrow$ p. 16) may be (1) a *concentration gradient* established by water resorption (e.g. urea); (2) an *electrical gradient* (e.g. Cl^-); (3) a volume flux leading to *solvent drag* (Na^+, Cl^-, urea). In part Na^+ resorption is **primary active** ($\rightarrow$ p. 118). Glucose, amino acids, phosphate, lactate, Cl^- ($\rightarrow$ p. 118) etc. are actively *co-transported with Na^+* (**secondary active transport**). HCO_3^- resorption is active because it is driven by actively secreted H^+ ($\rightarrow$ p. 130). Active and carrier mediated passive transport processes have a *maximal rate* ("T_m") and can be *saturated*. Normally, resorption of glucose is nearly 100% and of amino acids 95–99.9% of the filtered ones. At elevated plasma levels (glucose: 10 mmol/l) excretion increases more or less sharply depending on the affinity of these solvents for their respective resorption carriers.

Substances that are resorbed by **simple passive diffusion** have no T_m; their resorption depends upon membrane permeability. The cell membrane functions as though it were a lipid barrier to diffusion. The greater the lipid solubility of a solute, the greater its resorption. For weak electrolytes, the ionic form is generally much less lipid soluble than the undissociated form (**non-ionic diffusion**). By changing urinary pH, the non-ionic fraction of molecules that are passively resorbed can be altered ($\rightarrow$ **A**, $y^- \rightleftharpoons y^0$). In alkaline urine, a weak acid becomes more ionized, diffuses less readily back to the blood, and is therefore excreted in greater amounts. In acid urine, a weak acid dissociates less; the undissociated lipid-soluble form predominates, diffuses more readily back to plasma, and its excretion is reduced. To respond to such influences, the pK_a of the weak electrolyte must fall within the range of urine pH (pH 4 to 8).

Maltose and oligopeptides (e.g. glutathione, angiotensin) are *hydrolyzed* by *brushborder enzymes* (maltase, aminopeptidases, γ-glutamyltransferase) within the tubular lumen and are reabsorbed subsequently i.e. glucose or amino acids. Proteins which can pass the glomerular filter are reabsorbed by *pinocytosis* ($\rightarrow$ p. 12).

3. **Active tubular excretion** (transcellular) ($\rightarrow$ **A, z**) occurs chiefly by two distinct mechanisms, one for weak acids and one for weak bases. Endogenous substrates (glucuronides, hippurates, sulfates) or drugs (penicillin, diuretics) or drug metabolites are excreted by these processes ($\rightarrow$ **C**). The clearance ratio for such compounds is greater than 1.0 ($\rightarrow$ p. 114). Some compounds (p-aminohippurate, PAH) are removed nearly completely from plasma as fast as they reach the tubules ($\rightarrow$ p. 110).

4. **Tubular cell secretion**. The nephron cells contain a variety of enzymes for synthesis of substances that are transferred to the tubular fluid ($\rightarrow$ **A, u, v**). These substances include NH_3 (glutaminase) ($\rightarrow$ p. 132), and H^+ (carbonic anhydrase) ($\rightarrow$ p. 130). Secretion of NH_3 is passive ($\rightarrow$ **A, v**), of H^+ active ($\rightarrow$ **A, u**).

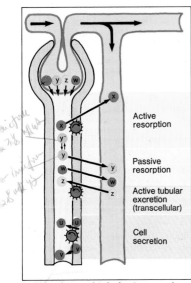

A. Filtration and tubular transport

Active resorption

Passive resorption

Active tubular excretion (transcellular)

Cell secretion

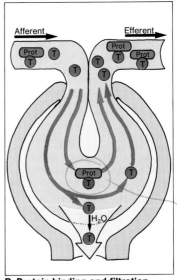

Afferent

Efferent

B. Protein binding and filtration

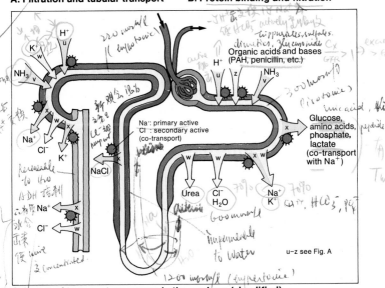

Organic acids and bases
H^+ (PAH, penicillin, etc.)

Na^+: primary active
Cl^-: secondary active (co-transport)

Glucose, amino acids, phosphate, lactate (co-transport with Na^+)

Urea

Cl^-
H_2O

Na^+
K^+

NaCl

Na^+

Cl^-

H^+

K^+

NH_3

Na^+

Cl^-

K^+

u–z see Fig. A

C. Locus of transport processes in the nephron (simplified)

Renal Function: Clearances

Glomerular filtration rate (GFR).

For the measurement of GFR, an indicator substance with specific properties must be present in the blood. The indicator must enter the tubule only by filtration and must not undergo resorption, tubular excretion, or metabolism. Furthermore, it should be inert and without influence on renal function. Suitable indicators are the carbohydrates mannitol and inulin, which can be infused into the blood to measure GFR. Under certain circumstances, creatinine, which is already present in the blood, may also be used.

The GFR may be determined ($\rightarrow$ **A**) if urine flow rate ($\dot{V}_u$) and plasma and urine concentrations of the indicator (P_{in} and U_{in}) are known. The rate at which the indicator is filtered is GFR (ml/min) $\times P_{in}$ (mg/ml). If GFR = 120 ml/min and P = 2 mg/ml, the indicator is filtered at 240 mg/min. Since all of the indicator must appear in the final urine, the rate of filtration = rate of excretion. If the urine concentration = 60 mg/ml and the urine flow = 4 ml/min, excretion = $U_{in} \times \dot{V}_u$ = = 240 mg/ml. GFR $\times P_{in} = U_{in} \times \dot{V}_u$; then GFR = $U_{in} \times \dot{V}_u / P_{in}$.

Aspects of GFR. About $^1/_5$ or 20% of the renal plasma flow (RPF, $\rightarrow$ p.110) is filtered at the glomerulum. This ratio, GFR/RPF, is called **filtration fraction (FF)**.

At a GFR = 120 ml/min, c. 180 l of fluid are filtered daily. Since a 70 kg man has 20% of his body weight, or 14 l, as ECF ($\rightarrow$ p. 124), his exchangeable water is reworked in the kidney 13 times/day and his plasma 50 times/day, or once every 30 min. Of the 180 l filtered per day, c. 99% is returned to the ECF by resorption out of the tubule ($\rightarrow$ p. 122); only 1 l is lost as urine. Large variations in GFR have no influence on urine volume.

GFR varies with capillary hydrostatic pressure (P_{cap}) ($\rightarrow$ p. 144), Bowman's capsular pressure (P_{BOW}), plasma oncotic pressure (π_{cap}) ($\rightarrow$ p. 144), glomerular surface area, and permeability. P_{cap} ($\rightarrow$ p. 111, **B**) at the afferent end of the glomerular capillary $\approx$ 6 kPa (45 mm Hg), at the efferent end $\approx$ 5.7 kPa (43 mm Hg). In contrast to other tissues where the filtration fraction is only 0.5% ($\rightarrow$ p. 144), filtration fraction in the glomerulus is 20%. π_{cap}, therefore increases from c. 2.7 kPa (20 mm Hg) at the afferent end to c. 4.4 kPa (33 mm Hg) at the efferent end. P_{BOW} is c. 1.6 kPa (12 mm Hg). The effective filtration pressure (= P_{cap} − P_{BOW} − π_{cap}) is, therefore, initially 1.7 kPa (13 mm Hg) and drops to 0 probably before the end of the glomerular capillary. Beyond this point filtration cannot occur. By altering prearteriolar and arteriolar resistance, filtration pressure and GFR can be varied. If the mean systemic blood pressure falls below 8 kPa (= 60 Torr), filtration ceases.

The expression $U_x \times \dot{V}_u / P$ is the **clearance**. The clearance of inulin (C_{in}) is the same as the GFR. The clearance of any other substance X is compared to the GFR by its **clearance ratio** (C_x/GFR) ($\rightarrow$ p.112). A substance X, which is removed from the tubule by **resorption** ($\rightarrow$ **B**, **1**) has a clearance less than the GFR and a clearance ratio less than 1.0. Other substances are removed from the blood by tubular cells and are added to the urine by **tubular excretion** ($\rightarrow$ **B**, **2**). Still others are added by cellular synthesis and **secretion**. In both of these cases, the amount appearing in the urine is greater than the amount filtered, and the clearance ratio is greater than 1.0.

The rate of tubular resorption or of tubular excretion can be calculated as the difference between the rate of filtration ($P_x \times$ GFR) and the rate of urinary excretion ($U_x \times \dot{V}_u$) ($\rightarrow$ **B**). In many cases, these tubular processes are active and can be saturated.

	Male	Female
GFR	124	109 ml/min*
RPF	654	592 ml/min*
FF	19.2%	19.4%

*) per 1.73 m² body surface.

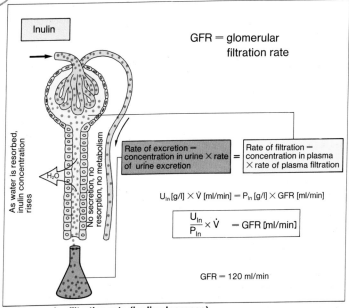

Inulin

GFR = glomerular filtration rate

As water is resorbed, inulin concentration rises

No secretion, no resorption, no metabolism

H_2O

Rate of excretion = concentration in urine × rate of urine excretion

=

Rate of filtration = concentration in plasma × rate of plasma filtration

$$U_{In} \, [g/l] \times \dot{V} \, [ml/min] = P_{In} \, [g/l] \times GFR \, [ml/min]$$

$$\frac{U_{In}}{P_{In}} \times \dot{V} = GFR \, [ml/min]$$

GFR = 120 ml/min

A. Glomerular filtration rate (inulin clearance)

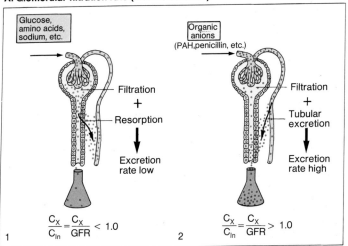

Glucose, amino acids, sodium, etc.

Filtration + Resorption

Excretion rate low

$$\frac{C_X}{C_{In}} = \frac{C_X}{GFR} < 1.0$$

1

Organic anions (PAH, penicillin, etc.)

Filtration + Tubular excretion

Excretion rate high

$$\frac{C_X}{C_{In}} = \frac{C_X}{GFR} > 1.0$$

2

B. Clearance ratios relative to filtration

Excretion

The daily diet furnishes a variety of organic anions, cations, and neutral substances for energy and growth. Substances that are physiologically inert must be rejected or eliminated, but the elimination process must distinguish between inert and biologically active substances. Processing begins in the **GI tract** (→ **A**, **1**). The intestinal tube resembles the proximal renal tubule in its function: it has a brush border and possesses active transport mechanisms for many physiologically useful materials. Like the kidney tubule, it is impermeable to many organic anions. In general, substances that the kidney conserves by tubular resorption (e.g., sugars) are also absorbed by the intestine and those that the kidney actively excretes (e.g., dyestuffs) are largely rejected by the GI tract.

From the intestine, all substances absorbed into the blood must pass through the **liver** (→ **A**, **2**), where as much as 95% of a substance may be extracted during the first passage (**first pass effect**). In the liver (as in other organs), microsomes perform a series of enzymatic reactions that include adding −OH or −COOH groups to the molecule (→ **A**, **2**). These groups are then **conjugated** (→ **A**, **3**) with glucuronic acid (on OH and COOH), sulfate (on OH), and amino acids (on COOH). Amino groups are acetylated. The resulting molecule, in most cases, is a more strongly ionized anion of the form $R-C$: $O-N-C-COOH$ or $R-OSO_2OH$ and is actively excreted into the **bile** (→ **A**, **4**). Many toxins and other xenobiotics are conjugated with the SH-group of hepatic *glutathione*. This *detoxication* process is catalyzed by glutathione-S-transferases. After further metabolization the conjugation product is excreted into the bile or into urine. By these conjugation reactions, a useless

compound is distinguished from a useful compound by being converted to one that is rapidly excreted. In the bile ducts, a fraction of these substances escapes into the venous and arterial blood by countercurrent exchange and is returned to the liver; the larger fraction enters the intestine. These compounds are poorly absorbed in the conjugated form in which they are excreted, but they may be deconjugated by intestinal or bacterial enzymes, and a fraction may be resorbed once again (**enterohepatic circulation** (→ **A**). Compounds that pass through the liver next enter the **lungs** (→ **A**, **5**), which contain a lipid matrix capable of absorbing highly lipid-soluble substances, chiefly organic cations (e.g., serotonin, methadone). Since many organic cations are active in the CNS, the lungs act as a filter to protect the brain from large concentrations of these substances. They are inactivated in the lung and either are returned to the blood or are excreted into the bronchi, where they are washed out with the mucus secretions.

Compounds that escape the filtering processes of the liver and lung are distributed to the tissues for oxidative or synthetic reactions. Inert metabolites of these compounds either are returned to liver and kidney for conjugation or are conjugated locally and are then pumped out of the tissue. In the **kidney**, the conjugated anions (e.g., PAH, some sulfa drugs) are eliminated by active tubular excretion (→ **A**, **6**). Organic cations (e.g., epinephrine, N-methylnicotinamide) are actively excreted by a second and distinct process. Glutathione conjugates are excreted by the kidney in the form of mercapturic acids. In the brain (CNS), the choroid plexus of the cerebral ventricles and the uveal tract of the eye can transport the same conjugated anions out of the tissue and into the blood as the liver and kidney transport out of the blood and into bile or urine.

Intake:
1. Absorption from intestinal tract to hepatic portal system

Metabolism (preparation for active excretion):
2. Microsomal oxidation to −OH and −COOH
3. Conjugation reactions with glucuronic acid, amino acids, glutathione, etc.

Elimination
4. Active hepatobiliary transport
5. Pulmonary extraction
6. Active renal tubular excretion

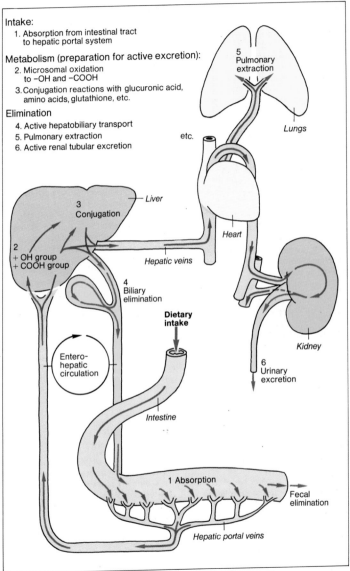

A. Absorption-metabolism-excretion of organic substances

Salt: Na⁺ and Cl⁻

NaCl is taken in the diet in amounts that vary, according to personal taste, from 8 to 15 g/day. Many hypertensive subjects consume NaCl at the upper end of this range. In the 180 l of plasma water that are filtered daily at the glomerulus ($\rightarrow$ p. 114), c. 1.2 kg NaCl are presented each day to the renal tubules. This is an absolute quantity approximately five times greater than the total body content. Of this amount, c. 99% is resorbed and less than 1% reaches the final urine. Tubular resorption of Na⁺ is adjusted to intake so that the concentration of Na⁺ in the ECF ($\rightarrow$ p. 124) and the amount of Na⁺ in the body remain constant. Energy sources of the kidney are expended chiefly on transport of Na⁺.

Proximal tubule. Throughout this segment, water and Na⁺ are resorbed in equivalent amounts. The resorbed fluid has the same osmolality (300 mOsm/l) as the plasma and the tubular fluid (**TF**), which remains in the tubular lumen (**isosmolar resorption**; $\rightarrow$ p. 122). Osmolality of the TF remains constant throughout the proximal tubule. By the end of the tubule, c. 70% of filtered water and Na⁺ have been resorbed ($\rightarrow$ **B**).

Mechanisms for Proximal Resorption of Na⁺ and Cl⁻

The cell membrane at the basal pole forms many infoldings. In this **basal labyrinth**, neighboured cells interdigitate. They are separated by the intercellular space. In the adjacent cell membranes, Na-K-ATPase actively transports Na⁺ into the intercellular space establishing a concentration gradient that permits water to follow by diffusion. From the intercellular space, water and Na⁺ reach the basal blood capillaries ($\rightarrow$ pp. 133, 209). The pumping by ATPase reduces intracellular Na⁺ allowing more Na⁺ to diffuse from the lumen into the cell. Passive transfer of Na⁺ across the cell occurs also (a) by exchange for actively secreted H⁺ and (b) by solvent drag when water is driven out of the tubular lumen by osmosis. Passive resorption of Na⁺ and Cl⁻ also occurs *between* tubule cells (*paracellular shunts*). **The loop of Henle** dips into the medulla where the osmolality of the ECF increases progressively toward the papillary tip ($\rightarrow$ pp. 120, 122). The medullary osmolar gradient determines how concentrated the final urine may become. In man, the limit is 1,200 to 1,400 mOsm/l ($\rightarrow$ p. 120). To achieve the medullary osmolar gradient, NaCl is actively pumped out of the thick ascending limb of the loop ($\rightarrow$ p. 122). Since the cell membrane of the thick limb is impermeable both to NaCl and to water, the TF cannot equilibrate with the medulla by water diffusing out of the TF or NaCl diffusing back in. In this way, the medullary ECF becomes **hyperosmolar** and the TF **hyposmolar**. These changes account for removal of 20% of filtered Na⁺ from the TF as it passes through the medulla. The TF leaving the loop is hyposmotic.

Distal tubule. Resorption of NaCl and water continues in this segment. Osmolality of the TF increases along the tubular length. Osmolality increases more during states of water deprivation than during hydration. At the end of the distal tubule, osmolality of the TF approaches 300 mOsm/l; less than 3% of filtered NaCl remains in the TF ($\rightarrow$ **B**). The chief action of aldosterone ($\rightarrow$ p. 137) in the kidney takes place in this segment where Na⁺ resorption is under hormonal influence.

Collecting duct. Final adjustment of salt concentration and excretion take place here. Final excretion of NaCl may vary from c. 5% to less than 1% of the filtered amount, depending on intake, fluid requirements, and hormonal state.

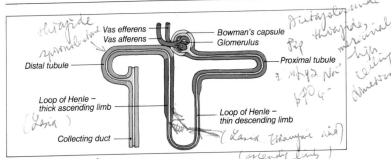

A. Nephron segments

Vas efferens
Vas afferens
Bowman's capsule
Glomerulus
Distal tubule
Proximal tubule
Loop of Henle –
thick ascending limb
Loop of Henle –
thin descending limb
Collecting duct

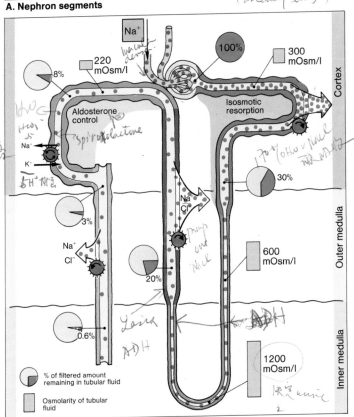

B. Fluxes of Na⁺ in the nephron

Na^+

100%

300 mOsm/l

Cortex

220 mOsm/l

8%

Isosmotic resorption

Aldosterone control

Na^+

K^+

30%

3%

Na^+

Cl^-

Na Cl

600 mOsm/l

Outer medulla

20%

0.6%

Inner medulla

1200 mOsm/l

% of filtered amount remaining in tubular fluid

Osmolarity of tubular fluid

Countercurrent Systems

Countercurrent systems are widely distributed in animals and serve a variety of functions.

1. **Simple exchanger.** Consider a heat exchanger in which two streams of water, one cold, the other hot, flow in parallel at the same rate ($\rightarrow$ **A, 1**). At the input end, the heat gradient is large and heat exchange is rapid. The gradient decreases along the length of the system until, at the output end, temperature has equilibrated; 50° is the limiting temperature for both the cold and the hot streams.

2. **Counter current exchanger.** If the direction of flow in one stream is reversed, there will be a temperature gradient at all points, allowing heat to be exchanged along the lengths of the two streams ($\rightarrow$ **A, 2**). In the same way, water may be exchanged instead of heat. Changes in rate of flow may alter the magnitude of exchange and of the gradient, but the general principle remains. Instead of heat, solutes may also be exchanged along a concentration gradient. Such an exchange takes place, for example, in the portal triads where bile and blood flow oppose each other.

The same conditions apply when a single stream reverses itself as in a hairpin loop ($\rightarrow$ **A, 3**). If the tip of the loops is in contact with a heat sink (ice), heat will be lost. A temperature gradient will be established, however, which will allow countercurrent heat exchange, and the returning stream will be only slightly cooler than the entering stream. This mechanism enables ducks, storks, and other birds to stand on ice without losing heat. The flippers of whales and seals function similarly. In the kidney, the vasa recta ($\rightarrow$ p. 110) in the medulla participate in a passive countercurrent exchange of water and solutes ($\rightarrow$ **A, 4**). They equilibrate osmotically with the ECF and loops of Henle. When the blood flow increases, they carry away more solute and reduce the medullary osmotic gradient.

3. A **countercurrent multiplier** is a system in which an osmotic gradient is established across the two limbs by expenditure of internal energy ($\rightarrow$ **A, 5**). The maximum gradient that can be achieved by a multiplier system is directly proportional to the length of the hairpin loop and to the concentration difference between the two limbs. It is inversely proportional to the square of the flow rate. In the renal tubule ($\rightarrow$ **A, 5, 6**), the thick ascending limb of the loop of Henle transports NaCl into the ECF ($\rightarrow$ **A, 5**). This action creates a concentration difference between the two limbs and simultaneously creates an osmotic gradient in the renal medulla; the longer the loops of Henle, the greater the medullary gradient that can be achieved. The tubular fluid leaving the loops is hyposmolar ($\rightarrow$ p. 122).

The final urine concentration is achieved in the collecting ducts (CD), which also pass through the osmotic gradient of the medulla ($\rightarrow$ **A, 6**, yellow). Urine concentration, therefore, is determined by permeability of the CD. Under the influence of ADH ($\rightarrow$ pp. 126, 214), the collecting duct is maximally permeable to water; the CD equilibrates with the high osmolarity in the medullary tip, and the urine becomes maximally concentrated. Its concentration, however, cannot exceed the concentration in the papillary tip ($\rightarrow$ **A, 5**, bottom). In the absence of ADH, the CD is relatively impermeable to water; the hyposmolar tubular fluid of the distal tubule does not equilibrate with the medullary ECF, and the final urine is dilute.

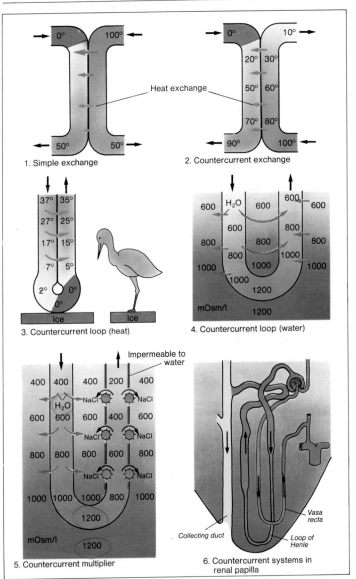

1. Simple exchange

2. Countercurrent exchange

3. Countercurrent loop (heat)

4. Countercurrent loop (water)

5. Countercurrent multiplier

6. Countercurrent systems in renal papilla

A. Countercurrent systems

Water: Concentration and Dilution in the Kidney

Plasma water is filtered in the kidney at 120 ml/min or c. 180 l/day ($\rightarrow$ p. 114). At the glomerulus, filtrate is isosmolar with plasma (300 mOsm/l). In contrast, in the terminal urine, the volume is c. 1 l and osmolality may vary from 40 to 1,200 mOsm/l according to water intake; urine may be hyposmolar with a flow up to 18 ml/min or hyperosmolar with a flow of only a few tenths of a ml/min.

Proximal tubule. Approximately two thirds of the tubular fluid (**TF**) is resorbed between the glomerulus and the end of this segment ($\rightarrow$ **A**). Resorption of Na^+ is the primary driving force for water resorption; it establishes a concentration gradient along which an osmotically equivalent volume of water diffuses passively (**isosmotic resorption**) ($\rightarrow$ p. 118). Oncotic pressure (π_{cap}) ($\rightarrow$ pp. 114, 144) in peritubular capillaries provides an additional force for water movement. When water is filtered out of the glomerular capillaries, it concentrates plasma proteins ($\rightarrow$ p. 112); the oncotic drive for water resorption is therefore especially high in the early efferent capillaries. The greater the amount of water filtered, the higher the oncotic pressure.

Loop of Henle. The mammalian kidney has two nephron types ($\rightarrow$ pp. 108, 110): (1) cortical nephrons (c. 80%) have short loops of Henle that do not penetrate beyond the inner stripe of the outer medulla; their efferent arterioles form a network close to the nephron of origin; (2) juxtamedullary nephrons (c. 20%) have loops that penetrate deep into the inner medulla; their efferent arterioles form long loops, **vasa recta**, which penetrate the medulla as far as the papillary tip. The thick ascending limb transports salt into the medullary ECF ($\rightarrow$ p. 118). Since cells of the thick limb are relatively impermeable to water, the fluid that remains in the tubule is made hyposmolar. NaCl, which is delivered into the medullary ECF, establishes an osmolar gradient that is greater toward the papillary tip. Urea ($\rightarrow$ **B**) and other solutes also contribute to the gradient. The **thin** loop of Henle dips into this gradient. Its TF equilibrates with the ECF; in the thin descending limb, net efflux of water occurs, but the water is returned to the thin ascending limb by net influx ($\rightarrow$ p. 120). The overall effect is to short-circuit water across the thin loop. Only animals with loops of Henle are able to concentrate urine above the osmolality of the blood. The **distal tubule** receives hyposmolar urine from the loops of Henle. In this tubular segment, although some resorption of water occurs, the osmolality of the TF probably remains less than that of the blood. Thus, hyposmotic fluid is delivered to the **collecting ducts**, where the final adjustment of urinary volume and concentration takes place ($\rightarrow$ p. 118). In the absence of **antidiuretic hormone (ADH)** ($\rightarrow$ pp. 126), NaCl is resorbed more than water and the TF is diluted further. However, when the need for water conservation develops, ADH is secreted and increases permeability of the collecting ducts to water and urea. Water ($\rightarrow$ **A**) and urea ($\rightarrow$ **B**) diffuse out of the collecting ducts into the higher osmotic environment of the medullary ECF. The absorbed water in the medullary ECF equilibrates with blood in the vasa recta and is carried away. By this means, urine in the collecting ducts is concentrated, but the concentration cannot exceed that of the medullary osmolality ($\rightarrow$ p. 120). Medullary osmolality can itself be influenced by a number of factors: increased medullary blood flow can wash out solute (NaCl and urea) from the medulla; osmotic diuresis abolishes the medullary osmotic gradient; water diuresis reduces the medullary gradient, limiting it to the inner medulla.

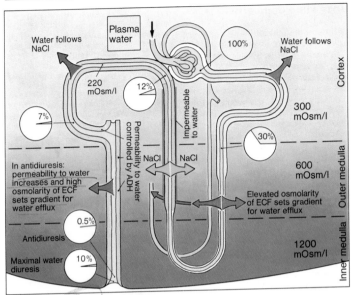

Plasma water

100%

Water follows NaCl

Water follows NaCl

220 mOsm/l

12%

7%

Impermeable to water

Permeability to water controlled by ADH

300 mOsm/l

30%

NaCl   NaCl

In antidiuresis: permeability to water increases and high osmolarity of ECF sets gradient for water efflux

600 mOsm/l

Elevated osmolarity of ECF sets gradient for water efflux

0.5%

Antidiuresis

Maximal water diuresis

10%

1200 mOsm/l

Cortex

Outer medulla

Inner medulla

A. Fluxes of water in the nephron

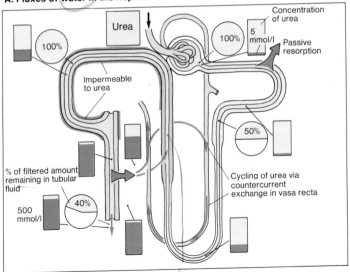

Concentration of urea

Urea

100%

100%

5 mmol/l

Passive resorption

Impermeable to urea

50%

% of filtered amount remaining in tubular fluid

500 mmol/l

40%

Cycling of urea via countercurrent exchange in vasa recta

B. Fluxes of urea in the nephron

Body Fluid Compartments

Without water, life on earth could not exist. The cells of all multicellular animals are surrounded by fluid, the **internal environment**, which approximates the composition of the primordial sea in which life first took form. This fluid environment, the **extracellular fluid** (ECF), is the connecting link between the external world and the inner space of the cell; it carries nutrients, disposes of wastes, and carries humoral messages.

The ECF comprises 27% of the lean body weight (→ **C**). It has two major compartments, the **plasma volume** within the blood vessels and the **interstitial fluid** (inulin space) around and about the cells. These make up 4.5% and 21% of body weight, respectively. In the mean, a 70 kg adult has 19 l ECF (= 0.27 × 70 kg) with 3.1 l plasma volume, 14.9 l interstitial volume and 1.0 l transcellular volume (gut lumina, cerebrospinal fluid; c. 1.5% fo body weight). Since plasma/red blood cell ratio amounts to c. 57/43 (→ p. 58), the whole blood volume approximates 7.8% (range 6 to 8%) of body weight. The rest of the body water is in the cells as intracellular fluid (ICF), c. 35% of body weight. The ECF and ICF are in osmolar equilibrium.

Measurement of fluid volumes. When a known amount of a test substance is dissolved in an unknown volume of fluid, the volume can be estimated by determining the concentration of the substance after thorough mixing. If the test dose is 50 g and the concentration after mixing is 5 g/l, the distribution volume of the substance (V_D) is 10 l (= dose/conc). Volumes of fluid compartments in the body are similarly estimated by determining the V_D of appropriate substances after intravenous injection and mixing (→ **C**). **Total body water** approximates the V_D for antipyrine or D_2O which penetrate all the cells. There is no perfect indicator for total ECF. It can be estimated as the V_D for inulin (c. 16% of body weight) or as the V_D for chloride isotopes (c. 30% of body weight). The dis-

crepancy arises mainly from the fact that inulin can neither enter the ECF of bones and connective tissue, nor the transcellular space. V_D for Cl^- is a slight overestimation of ECF. The ICF cannot be measured directly; it is estimated as the difference between total body water and ECF. **Blood volume** is calculated from the hematocrit and either the plasma volume or the red cell mass. The V_D of Evans blue dye (T-1824) measures plasma volume because the dye binds to plasma proteins and does not diffuse out of the blood vessels. The red cell mass is determined by injecting radioactively labeled cells (chromium) and observing their V_D.

$$Blood\ volume = \\ = plasma\ volume \times \frac{100}{100 - Hct\%}.$$

The fraction of body water (→ **B**) varies with body fat content. Most cells contain c. 70% water, but fat cells are relatively free of water. Therefore, the more fat, the less the fraction of body water. Total body water is 64% in young men, 53% in young females, and decreases with age. Control of ECF volume resides in the kidney; control of ICF volume is a function of each cell. Protein content is high in the ICF, low in the ECF; this generates an oncotic pressure that draws water into the ICF. The cell does not swell, however, because the extracellular position of Na^+ counterbalances the oncotic effect.

Depletion of body water results in thirst (→ **A**; p. 270). A general loss of ICF stimulates thirst receptors in the hypothalamus: signal, increase in osmolality; response, ADH release. A loss of ECF stimulates volume receptors in the thorax and elsewhere: signal, atrial pressure; response, angiotensin and ADH production. In both cases, the kidney retains water and an appropriate amount of salt (→ p. 126).

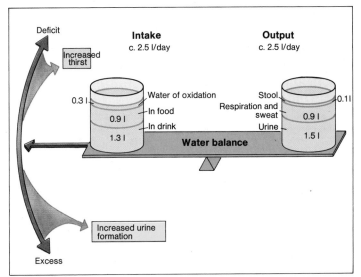

A. Water balance

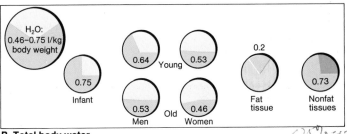

B. Total body water

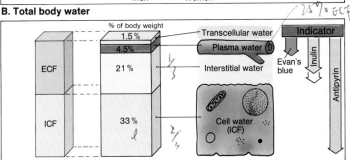

C. Fluid compartments

Salt and Water: Hormonal Control

Internal fluids of animal organisms, with rare exceptions, have an osmolar concentration of c. 290 mOsm/l. Intake of food and water can change osmolarity acutely: increased water and/or decreased salt intake lead to hypo-osmolarity; converse changes to hyperosmolarity. Since extreme changes in volume and concentration can be lethal, internal fluids are normally maintained within a narrow range. In mammals, such **osmolar and volume homeostasis** is under hormonal control. Water and salt concentrations in the body respond to two separate but interdependent mechanisms: antidiuretic hormone (**ADH**) ($\rightarrow$ p. 122) controls water, **aldosterone** influences salt ($\rightarrow$ p. 139, B). Signals are recognized promptly to resist extreme changes. Although daily intake of salt and water may vary greatly, these mechanisms respond so delicately that only as much salt and water are retained as is necessary to match body losses; all surplus is eliminated.

Water deficit ($\rightarrow$ **A, 1**). When water intake is insufficient to replace water losses (sweat, urine, lungs, GI tract) ($\rightarrow$ p. 125, A), plasma becomes hyperosmolar. A rise of only 3 mOsm/l is sufficient to stimulate increased water retention in the kidney. Neurosecretory cells of the hypothalamus respond to hyperosmolarity (**osmoreceptors**) and to an decrease in atrial pressure (Henry-Gauer reflex) to release ADH from nerve terminalis in the posterior pituitary ($\rightarrow$ **A, 1**). ADH is carried to the kidney where it reduces urine volume and increases urine concentration ($\rightarrow$ p. 122).

Water excess ($\rightarrow$ **A, 2**). Water intake dilutes the blood. A decrease in plasma osmolarity of only 3 mOsm/l decreases ADH release from the posterior pituitary. The kidney receives less ADH and excretes a larger volume of urine with lower osmolarity ($\rightarrow$ p. 122). Response is rapid; urine dilution begins within a few minutes of water intake, and the total water load can be excreted within an hour. A rapid intake of a large volume of water, as in severe thirst or after exertion, may produce extreme dilution of the blood. At levels below 250 mOsm/l, *water intoxication* can occur with prostration, nausea, vomiting, and shock.

Salt deficit ($\rightarrow$ **A, 4**). Reduced salt intake without a corresponding change in water intake leads to hypo-osmolarity. ADH secretion decreases and water excretion increases. The result is to decrease plasma volume and total body water. The reduced plasma volume influences juxtaglomerular cells in the kidney to release **renin** ($\rightarrow$ p. 138), angiotensin II is generated, and the adrenal cortex is stimulated to release **aldosterone** ($\rightarrow$ p. 136). Aldosterone acts on salt-transporting cells, such as those of the distal renal tubule, to diminish salt excretion. Other responsive cells include intestinal, sweat, and salivary glands. Aldosterone does not directly affect water losses, but by stimulating salt retention it allows a secondary effect on water retention via ADH, which restores fluid volume.

Salt excess ($\rightarrow$ **A, 3**). A salt load increases plasma osmolarity. Water is then retained via ADH to restore normal osmolarity, but in the process plasma volume and total body water increase. This increase is a signal to reduce renin and angiotensin II generation ($\rightarrow$ p. 138). Less aldosterone is released. The renal tubules, under a lesser steroidal influence, excrete the salt excess. Hypo-osmolarity develops, and water excretion follows as a secondary effect until the normal plasma volume is restored. Some investigators believe that a *natriuretic hormone* ("*third factor*") is also involved.

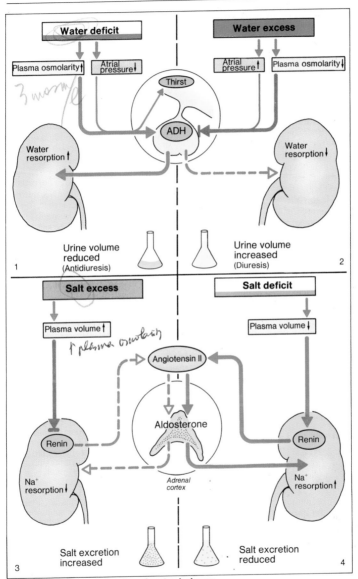

A. Hormonal control of salt and water balance

Disturbances in Salt and Water Homeostasis

Disturbances in salt and water homeostasis (→ **A**) occur because of a change in **balance**, **distribution**, and **hormonal regulation**.

1. **Isosmotic loss** (→ **A, 1**). ECF vol–dec; osm – no change. Ex: vomiting, diarrhea, diuretic therapy, blood loss, burns, drainage of ascites.
2. **Water deficit** (→ **A, 2**). ECF vol–dec; osm – inc; fluid shifts from ICF. Ex: sweating, hyperventilation, osmotic diuresis, chronic renal disease, diabetes insipidus.
3. **Salt deficit** (→ **A, 3**). Osm – dec; fluid shifts to ICF → ECF vol–dec; Ex: vomiting, diarrhea, sweating, adrenal insufficiency, hypokalemia, CNS lesions, salt-losing nephritis.
4. **Isosmotic excess** (→ **A, 4**). ECF vol–inc; osm – no change. Ex: heart failure, nephrosis, acute glomerulonephritis, decompensated cirrhosis.
5. **Water excess** (→ **A, 5**). ECF vol–inc; osm – dec; fluid shifts to ICF. Ex: water drinking, inappropriate ADH secretion, intensive gastric lavage, infusion of glucose solutions during oliguria.
6. **Salt excess** (→ **A, 6**). Osm – inc; fluid shifts from ICF → ECF vol–inc; Ex: infusion of hypertonic saline, adrenal hyperactivity, steroid therapy, drinking sea water, CNS lesion.

Diuresis and Diuretics

The upper limit of urine osmolarity (c. 1,400 mOsm/l in man) determines the volume of water that must accompany excreted solutes; excretion of 1,400 mOsm requires a minimum of 1 l of water. In the proximal tubule, the tubular fluid is reduced in volume by isosmolar water resorption (→ p. 118), but in distal segments more water may be resorbed than is necessary for isosmotic resorption. This excess is **free water**. When resorption of free water increases, or when its clearance decreases, urine becomes more concentrated.

Water diuresis. Water drinking dilutes plasma and reduces ADH secretion (→ p. 126). Hyposmolar urine (min = 40 mOsm/l) is excreted. The same events occur when there is a failure of ADH secretion, as in diabetes insipidus. **Osmotic diuresis**. When a nonresorbable solute is excreted, it must be accompanied by a corresponding volume of water. Mannitol, glucose (as in diabetes mellitus), and even excesses of salt may act in this manner to drag fluid into the urine. **Pressure diuresis**. When the blood pressure rises, autoregulation prevents an increase in RPF in the cortex (→ p. 110). In the medulla, however, autoregulation is less effective; medullary blood flow increases and washes out the concentration gradient in the medulla (→ p. 122). This action reduces the maximum urine osmolality and allows a diuresis.

Diuretic agents (→ **B**) act on the renal tubule to suppress resorption of solutes. Because an osmotically equivalent volume of solvent (water) must accompany the solute, urine volume increases; the extra urinary water is derived from the ECF. A reduction of ECF volume is a therapeutic objective in edema or hypertension. However, depletion of ECF activates the renin aldosterone system, which resists salt and water loss (*secondary hyperaldosteronism*). Some diuretic agents inhibit carbonic anhydrase (CA) (→ pp. 130–131) and produce a modest diuresis (high pH, increased $NaHCO_3$, and decreased NH_4^+ in urine). Agents that inhibit NaCl transport in the ascending thick limb of the loop of Henle (→ p. 118) produce a copious diuresis and decrease free water clearance as well as the medullary osmotic gradient. Thiazide diuretics act as inhibitors of CA and of Na^+ transport; they act in the proximal and distal tubules, and possibly also in the loop of Henle. Aldosterone antagonists (→ p. 138) produce a modest loss of Na^+, but conserve K^+; all other diuretics may increase K^+ excretion.

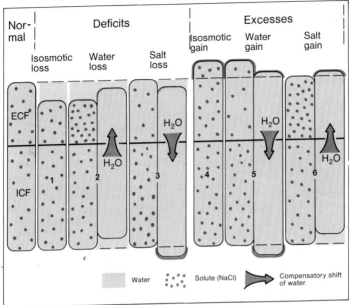

A. Disturbances in salt and water balance

Normal | Deficits (Isosmotic loss, Water loss, Salt loss) | Excesses (Isosmotic gain, Water gain, Salt gain)

ECF / ICF

H_2O

1 2 3 4 5 6

Water | Solute (NaCl) | Compensatory shift of water

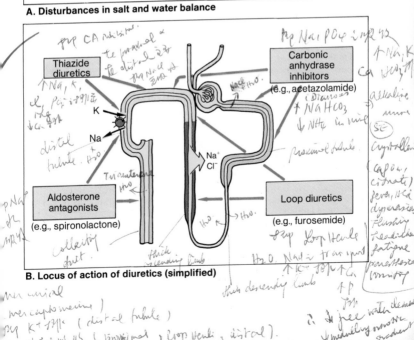

B. Locus of action of diuretics (simplified)

Thiazide diuretics

Carbonic anhydrase inhibitors (e.g., acetazolamide)

K

Na

Na^+ Cl^-

Aldosterone antagonists (e.g., spironolactone)

Loop diuretics (e.g., furosemide)

pH Homeostasis – Nitrogen Excretion

The chief function of the kidneys is to maintain both the volume and the concentration of body water constant. Control of volume is achieved by varying water excretion, of osmolality by varying solute excretion. The pH is kept within a narrow range (pH 7.40 → p. 102) by varying excretion of H^+ and HCO_3^-. The enzyme carbonic anhydrase (CA) is central to pH homeostasis (→ **A**). CA catalyzes the splitting of H_2CO_3 or, alternatively, the splitting of HOH; the OH^- then reacts with CO_2 in plasma to form HCO_3^-. Whichever mechanism operates, the products of the reaction are identical.

Wherever CA is found in polar cells, it is involved in pumping H^+ in an acid fluid across one pole of the cell and OH^- or HCO_3^- in an alkaline fluid out the other. This mechanism accounts for the ability of the stomach to produce acid gastric juice and of the intestines to produce alkaline secretions. The erythrocyte is unusual since CA reactions take place wholly within the cell.

In the kidney, the process is polarized to secrete H^+ into the tubular lumen. The H^+ thus generated serves three functions: (1) acidification of urine, generation of titratable acid (→ p. 104) and resorption of Na^+ (→ **B**), (2) absorption of HCO_3^- from tubular fluid (→ **C, 1**), and (3) generation of NH_4^+ (→ **C, 2**).

1. Acidification of urine (→ **B**): A normal diet generates 40 to 80 mEq H^+/day of which H_2SO_4 and H_3PO_4 are the principal acids. Only a small fraction of H^+ can be eliminated as such because the kidney cannot form a urine that is more acidic than pH 4 (limiting H^+ gradient). The remaining H^+ must be excreted as buffer salts. Losses of cations (Na^+ and K^+) by this route are reduced by two facts: (1) For each H^+ secreted one Na^+ is reabsorbed; (2) urinary buffer salts carrying H^+ are anions ($H_2PO_4^-$) and cations (NH_4^+) in similar amounts.

The kidney normally excretes 8–34 mmol/d H^+ in the form of titratable acid (→ p. 104) and 30–50 mmol/d in the form of NH_4^+; in extreme acidosis, as may occur in diabetes mellitus, NH_4^+ excretion may be as much as 300 to 500 mmol/day.

In the *proximal tubule*, CA is distributed on the tubular cell membrane. The glomerular filtrate, which initially has the same pH as the blood, is gradually acidified as it moves along the tubule. The drop in pH is not great, having a minimum value of pH 6.4 to 6.7, but the absolute amount of H^+ secreted is large. In the *distal tubule*, the limiting H^+ gradient is greater (1,000) and allows a lower pH (4) to be achieved, but because the buffering capacity of the distal tubule is small, the absolute amount of H^+ secreted is also small. In both tubular segments, secretion of H^+ allows resorption of a corresponding amount of Na^+ and K^+. Inhibition of CA by a diuretic would reduce secretion of H^+, allow the development of systemic acidosis, and enhance loss of Na^+ and K^+ in the urine.

2. Absorption of HCO_3^- (→ **C, 1**). All buffers in the body fluids are in equilibrium with each other; changes in one buffer influence all other buffers (→ pp. 98–107). The most common buffer pair is CO_2/HCO_3^- (pK_a 6.1). Respiration controls pCO_2 and the kidneys control HCO_3^-. Since kidneys filter 4,500 mmol HCO_3^-/day, an efficient resorptive mechanism is essential to prevent serious disturbances in acid-base balance: 99.9% of filtered HCO_3^- is normally resorbed by action of CA, ca. 90% in the proximal tubule. In absence of CA (inhibition by acetazolamide, → p. 128) c. 50% of filtered HCO_3^- can still be reabsorbed by passive diffusion (HCO_3^- gradient is established by volume resorption).

Since cell membranes are relatively im-

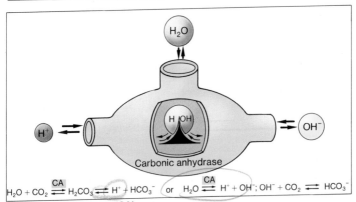

$$H_2O + CO_2 \underset{CA}{\rightleftharpoons} H_2CO_3 \rightleftharpoons H^+ + HCO_3^- \quad \text{or} \quad H_2O \underset{CA}{\rightleftharpoons} H^+ + OH^-;\ OH^- + CO_2 \rightleftharpoons HCO_3^-$$

A. Carbonic anhydrase (CA)

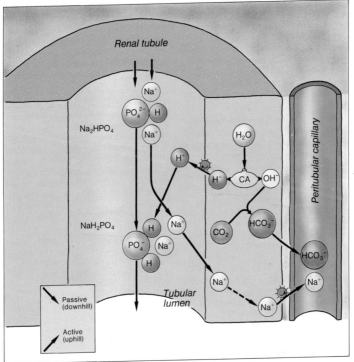

B. H⁺ excretion as H₂PO₄⁻

permeable to ions and are more permeable to undissociated molecules ($\to$ p. 112), CO_2 would be expected to diffuse readily through tubular cells, whereas HCO_3^- would only to a much smaller extent. HCO_3^- remaining in the glomerular filtrate would be lost by excretion. Carbonic anhydrase (CA) and a H^+ pump mechanism prevents loss of HCO_3^- by secreting H^+ into the lumen, where it reacts to generate CO_2 and water. Both of these diffuse readily into the cell along their respective concentration gradients. In the cell, CO_2 is rehydrated by CA to form H^+ and HCO_3^-. By these reactions, H^+ secreted into the tubular lumen is buffered, and the H^+ gradient is kept low. At the same time, H^+ replaces Na^+, which is liberated for resorption with HCO_3^-, both of which are returned to the blood. CA is essential for this process since the rate of H^+ necessary for HCO_3^- resorption is too great to be generated by the uncatalyzed reaction. pCO_2 also has an influence: when elevated, it stimulates increased secretion of H^+ and resorption of HCO_3^- ($\to$ p. 104).

3. **Nitrogen metabolism and excretion** ($\to$ **C, 2**). Metabolic degradation of proteins, amino acids, and nucleotides generates nitrogen, which must be excreted. In man nucleotide-N is excreted as urate. Amino acid-N (c. 0.8 mol/d) is excreted after conversion in the liver to urea (90%), as NH_4^+ or creatinine (each c. 5%). The buffer pair NH_4^+/NH_3 (pK$_a$ 9.1) is like the pair CO_2/HCO_3^-; it can accept H^+ without lowering the urinary pH. Since NH_3 is freely diffusible ($\to$ p. 102), it is uniformly distributed in tissue water. However, when H^+ is secreted into the tubular lumen, it reacts with NH_3 to form NH_4^+, which cannot easily diffuse out of the tubule. In effect, NH_3 is trapped by being converted to NH_4^+. Thus excretion of ammonia is increased, change in urine pH is resisted, and Na^+ and K^+ are liberated for resorption.

Origin of renal NH_3. Although 15% to 35% of NH_3 is preformed in blood, the majority is generated in the kidney mainly from **glutamine**. Several glutaminases in the kidney generate ammonia: the mitochondrial, phosphate dependent, glutaminase, the most important, splits the amido-N of glutamine and forms NH_3 and glutamate. A further mitochondrial enzyme (glutamate dehydrogenase) subsequently splits the amino-N forming a second NH_3 (and 2-oxyglutarate). A less important glutaminase II pathway in the cytosol first splits the amino-N (glutamine ketoacid amino transferase) and, in a second step (ω-amidase), splits the amido-N forming again 2 mol NH_3 per 1 mol glutamine. Also the γ-glutamyltransferase of the brush border could serve as (phosphate independent) glutaminase. The importance of this pathway is debatable. Mainly the mitochondrial mechanism is stimulated by acidosis, which can increase ammonia production from glutamine up to ten times.

3. **Nitrogen metabolism and excretion** ($\to$ **C, 2**). Metabolic degradation of proteins, amino acids, and nucleotides generates nitrogen, which must be excreted. In man nucleotide-N is excreted as urate. Amino acid-N (c. 0.8 mol/d) is excreted after conversion in the liver to urea (90%), as NH_4^+ or creatinine (each c. 5%). The buffer pair NH_4^+/NH_3 (pK$_a$ 9.1) is like the pair CO_2/HCO_3^-; it can accept H^+ without lowering the urinary pH. Since NH_3 is freely diffusible ($\to$ p. 102), it is uniformly distributed in tissue water. However, when H^+ is secreted into the tubular lumen, it reacts with NH_3 to form NH_4^+, which cannot easily diffuse out of the tubule. In effect, NH_3 is trapped by being converted to NH_4^+. Thus excretion of ammonia is increased, change in urine pH is resisted, and Na^+ and K^+ are liberated for resorption.

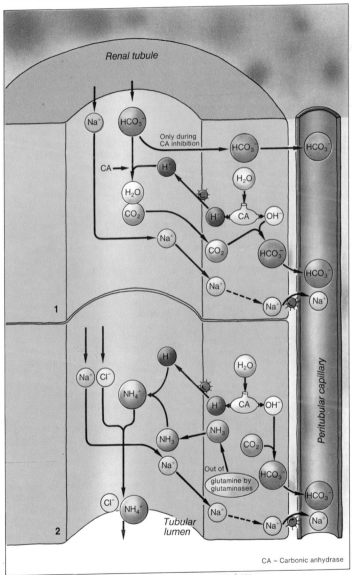

C. Bicarbonate resorption (1); H$^+$ excretion as NH$_4^+$ (2)

Potassium

Potassium (K^+) is the most abundant **intra**cellular cation. Of the total body K^+, c. 4,500 mmol/70 kg, only 1 % to 2 % is **extra**cellular (plasma = 3.4—5.2 mmol/l). K^+ intake is 50 to 150 mmol/ day; minimum requirement = 25 mmol/ day. Excretion is predominantly lower than filtration because K^+ undergoes **net** resorption, although under certain circumstances (e.g., high K^+ intake) the amount excreted may exceed the filtered (**net** tubular secretion). The kidney is better developed to prevent hyperkalemia with the result that K^+ depletion may occur more readily than K^+ excess.

Proximal tubule ($\rightarrow$ **A**). By the end of the proximal tubule, a constant fraction of the filtered K^+ (c. 60% to 70%) is resorbed. This fraction is independent of the final amount excreted, i.e., this segment does not control K^+ excretion. A probably small fraction of the reabsorption represents active transport.

Loop of Henle. The descending limb is poorly permeable to K^+ so that as water is resorbed, the K^+ concentration rises. In the ascending limb, K^+ is resorbed in large amounts along with Cl^- and Na^+. Only 10% of the filtered K^+ reaches the distal tubule ($\rightarrow$ **A**). At the end of the ascending limb, the concentration of K^+ in tubular fluid (TF) is less than one half of that in plasma (TF/P = 0.4).

Distal tubule ($\rightarrow$ **B**). K^+ moves into the TF along the electrochemical gradient. Some water is resorbed also so that at the end of this segment TF/P may exceed 4.0. The movement is probably passive since it never exceeds the amount predicted by the Nernst equation ($\rightarrow$ p. 22). Under certain circumstances, some active K^+ resorption also occurs.

Collecting duct ($\rightarrow$ **B**). Active resorption in this segment reduces the K^+ in the TF leaving the distal tubule.

Variations in the extent of this terminal resorption account for final control of K^+ excretion.

Summary. A relatively constant fraction of filtered K^+ is resorbed in the proximal tubule and loop of Henle. In the distal tubule, electronegativity of the tubular lumen and intracellular concentration determine movement of K^+ into TF; as each increases, more K^+ enters TF. Final control of K^+ excretion occurs in collecting ducts where active resorption and Na^+-dependent secretion of K^+ may operate.

Influences on K^+ Excretion

1. **Intracellular K^+ concentration** establishes the chemical gradient for passive K^+ diffusion ($\rightarrow$ **B**). When K^+ in the cell is high, urinary excretion is increased. The K^+ content of the cell is influenced by the acid-base state ($\rightarrow$ **B**; p. 102). Intracellular K^+ and H^+ have a reciprocal relationship: in acidosis, when intracellular H^+ is increased, there is less K^+; in alkalosis, the converse occurs. Similarly, a high dietary intake of K^+ increases intracellular K^+ and lowers H^+ leading to cellular alkalosis. Cellular changes are reflected in urinary excretion of K^+, which is elevated in alkalosis and decreased in acute acidosis. (For unknown reasons, K^+ excretion is elevated in chronic acidosis). **Low K^+ excretion** (and acid urine) occurs in K^+ depletion, aldosterone deficiency, and increased pCO_2. Excretion of H^+ and NH_4^+ is increased; HCO_3^- is decreased. **High K^+ excretion** and usually increased HCO_3^- excretion occurs in high K^+ intake, high HCO_3^- intake, excess aldosterone, dehydration and use of carbonic anhydrase inhibitors.

2. **Flow rate of tubular urine.** The concentration of K^+ in TF is in equilibrium with the cellular K^+. As flow increases, K^+ concentration in the TF remains constant as more K^+ diffuses into the TF. Increased tubular flow is

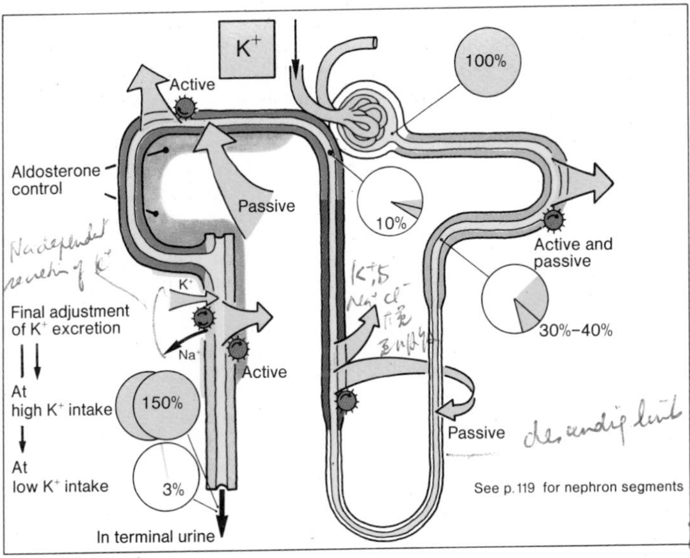
K+
Active
Aldosterone control
Passive
100%
10%
Active and passive
30%–40%
K+
Final adjustment of K+ excretion
Na+
Active
At high K+ intake
150%
At low K+ intake
3%
Passive
See p. 119 for nephron segments
In terminal urine

A. Fluxes of K+ in the nephron

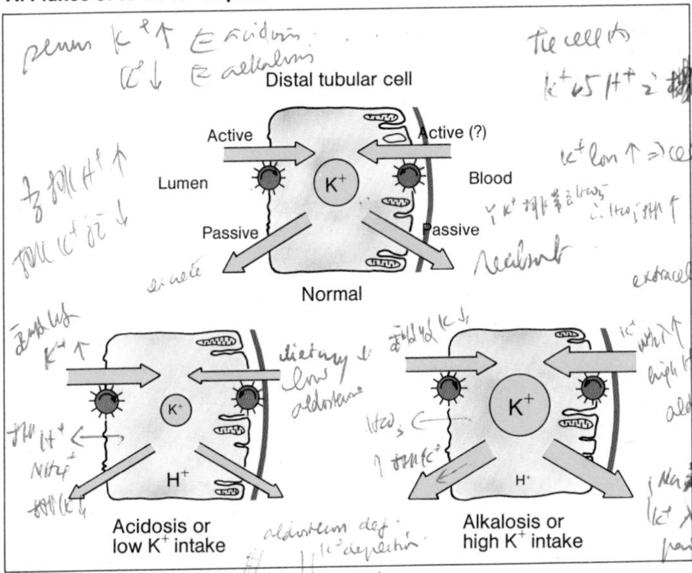
Distal tubular cell
Active
Active (?)
Lumen
K+
Blood
Passive
Passive
Normal
K+
H+
Acidosis or low K+ intake
K+
H+
Alkalosis or high K+ intake

B. Variations in distal tubular fluxes of K+

thus one of the factors involved in increased K^+ excretion in response to diuretic therapy ($\rightarrow$ p. 128).

3. **Sodium.** Na^+ activates Na^+-K^+-ATPase in the collecting duct. There follows increased resorption of Na^+ and active secretion of K^+ into the TF. Normally, only small amounts of Na^+ reach the collecting ducts, and only net resorption of K^+ occurs. When delivery of Na^+ to this segment is increased, as with diuretic therapy, secretion of K^+ may be observed.

4. **Anions** that are poorly resorbed in the distal tubule, such as HCO_3^-, increase electronegativity of the lumen, change the gradient for K^+, and allow greater K^+ excretion since more K^+ moves passively into the TF along the steeper electrochemical gradient.

5. **Aldosterone** increases K^+ excretion. It increases effective cell permeability for K^+, tubular membrane potential, and concentration of K^+ in the cells of the distal tubule. Effects of aldosterone on Na^+ and K^+ may be dissociated under certain conditions ($\rightarrow$ p. 137).

6. **Other hormones.** Insulin increases entry both of glucose and of K^+ into cells; the AV difference for K^+ in muscles is increased by insulin. Adrenergic activity also lowers plasma K^+ by a mechanism independent of insulin. This action depends on β_2-adrenergic receptors.

Mineralocorticoids

Mineralocorticoids stimulate **salt retention** ($\rightarrow$ p. 126) in all tissues that transport salt (e.g., gall bladder, intestinal tract, salivary and sweat glands, etc.). Of these tissues, the kidney is quantitatively the most important. Aldosterone (ALD) is the most potent mineralocorticoid and accounts for 95% of the salt-retaining activity of the adrenal cortex. Other mineralocorticoids (corticosterone and desoxycorticosterone) and glucocorticoids ($\rightarrow$ p. 240)

also have significant salt retaining effects.

Chemistry. ALD is a C_{21} steroid ($\rightarrow$ p. 239). It is synthesized in the zona glomerulosa ($\rightarrow$ p. 241) of the adrenal cortex. Although the adrenal cortex is capable of synthesizing sufficient cholesterol, it accumulates cholesterol from the plasma as a starting product. ACTH ($\rightarrow$ p. 241) stimulates the rate-limiting reactions that split the side chain of cholesterol and hydroxylate C_{20}. The zona glomerulosa lacks the C_{17}-hydroxylase necessary for glucocorticoid synthesis ($\rightarrow$ p. 238) and therefore is capable of producing only mineralocorticoids. The secretion rate for ALD = 80 to 240 µg/day; plasma concentration = 0.1 to 0.15 µg/l (varies with salt intake and posture). Secretion is highest in the early morning and is least in the late evening. ALD is inactivated in the liver and is excreted in bile and urine chiefly as a glucuronide ($\rightarrow$ p. 200).

Control. ALD stimulates a generalized salt retention throughout the body. Secondarily, water is also retained to resist an increase in plasma osmolality. The end effect of ALD secretion is to increase blood volume. It is natural, therefore, that physiologic changes that are associated with reduction of blood volume should stimulate secretion of ALD. These include hyponatremia and production of angiotensin II and ACTH. It is not clear which of these represents the primary stimulus for release of ALD from cells of the adrenal cortex. cAMP ($\rightarrow$ p. 222) and prostaglandins are involved in the release.

Effects. ALD influences the renal tubular cell (and other salt-transporting cells) to resorb Na^+ and to lose K^+. In the absence of ALD, Na^+ is lost and K^+ is retained. The effects of ALD are first observed after a latent period of 1 h and reach their peak in c. 4 h. The latent period reflects the time required for the intracellular sequences responsible for steroid hormonal effects ($\rightarrow$ p. 224). ALD stimulates intracellular synthesis of an **aldosterone-induced protein** (AIP), the operant substance in achieving ALD effects. AIP has several possible mechanisms of action: (1) it stimulates ATP synthesis to provide energy for the Na^+ pump; (2) it activates the Na^+ pump to transport Na^+ at a faster rate; (3) it increases the apparent passive permeability of the cell membrane for Na^+.

Hyperaldosteronism occurs when the adrenal secretes excess amounts of ALD. *Primary hyperaldosteronism* does not respond to normal feedback controls ($\rightarrow$ p. 218). Na^+ retention leads to an increase in fluid volume and hypertension; K^+ loss results in a hypokalemia with acidosis intracellularly and hypokalemic alkalosis of the ECF. *Secondary hyperaldosteronism* is much more common and occurs when there is a reduction of the **effective** plasma volume, as occurs in heart failure, chronic diuretic therapy, dietary salt restriction, cirrhosis of the liver with ascites, nephrosis, and pregnancy. In each of these conditions, renin ($\rightarrow$ p. 138) is released, angiotensin II is produced, and ALD secretion is enhanced. In many cases, this sequence worsens the original disease. In *chronic aldosterone administration*, at first, Na^+ is retained and K^+ is excreted. After 2 weeks, the tubule "escapes" from aldosterone influence on Na^+ excretion but not on K^+ (dissociation).

Excretion of Ca^{2+} and Phosphate

The kidney is most important for the Ca^{2+} balance because Ca^{2+} can be excreted in the urine ($\rightarrow$ pp. 234–237).

Total Ca^{2+} plasma concentration amounts to 2.3 to 2.7 mmol/l (1 mmol = 2 mEq = 40 mg Ca^{2+}). In plasma Ca^{2+} is present in **three forms**: (1) *free*, ionized Ca^{2+} (1.3 mmol/l); (2) Ca^{2+} *complexed* by phosphate, citrate, bicarbonate etc. (0.2 mmol/l); (3) Ca^{2+} *bound to plasma proteins* (0.8–1.2 mmol/l). The last fraction cannot be filtered at the glomerulus. The daily filtered load of Ca^{2+}, therefore, amounts to c. 270 mmol (180 l/d × 1.5 mmol/l). Only about **0.5 to 3**% of this load are **excreted** in the urine.

Ca^{2+} is reabsorbed along the whole nephron except the collecting ducts. Reabsorption of Ca^{2+} and Na^+ ($\rightarrow$ p. 118) often go in parallel. This is true for the effects of diuretic drugs ($\rightarrow$ p. 128) and the site of *final adjustment of urinary excretion* which takes place *in the distal tubule*. Parathyrin (PTH, $\rightarrow$ p. 234) and, to a smaller extent, the *D-hormone* (1,25-$(OH)_2$-calciferol, $\rightarrow$ p. 236) *decrease* Ca^{2+} excretion, whereas *calcitonin* ($\rightarrow$ p. 236) *increases* Ca^{2+} excretion.

80–95% of the **phosphate** filtered at the glomerulus are resorbed, mainly in the *proximal tubule*. According to the plasma pH and the pK_a of 6.8 the ratio of $HPO_4^{2-}/H_2PO_4^-$ in the glomerular filtrate is c. 4/1. Because of the low urinary pH the excreted species is mainly $H_2PO_4^-$. *Intratubular titration of phosphate* ($HPO_4^{2-} + H^+ \rightarrow H_2PO_4^-$) serves for H^+ excretion in the form of *titratable acid* ($\rightarrow$ p. 104). Phosphate excretion is *increased* by *parathyrin* and *calcitonin*, and *decreased* by *D-hormone* ($\rightarrow$ p. 236).

Renin − Angiotensin

The juxtaglomerular apparatus (JGA) (→ **A**) lies at the transition point where the ascending thick limb of the loop of Henle contracts the afferent arteriole of the glomerulus and becomes the distal tubule. The JGA is ideally situated to sense differences in volume and composition of tubular fluid in the renal tubule and in pressure and flow rate of blood in the glomerulus. Its responses to these signals are both local, with modulation of glomerular resistance, and systemic, with modulation of blood pressure and volume.

Anatomy of the JGA. Components are the afferent and efferent arterioles, the renal tubule, and three classes of specialized cells: (1) the *macula densa* is a group of renal tubular cells at the point of contact between tubule and glomerulus (→ p. 108), which respond to Na^+ concentration in distal tubular fluid (TF) and influence renin release; (2) *granular juxtaglomerular cells* are myoepithelial cells in the arteriolar media (more in the afferent than in the efferent): one pole of the cell is in contact with arteriolar intima and the other with macula densa cells; the granules contain **renin**; (3) *agranular juxtaglomerular cells* (lacis cells; Pollkissen) are interstitial mesangial cells that seem to be precursors of the granular cells.

Chemistry (→ **B**): **Renin** (molecular weight c. 40,000) is a proteolytic enzyme produced by the kidney. Renin-like activity is found in blood and other tissues (uterus, placenta, brain, etc). Its plasma half-life is 20 min. It acts on **renin substrate** (angiotensinogen) (molecular weight c. 60,000), which is produced in the liver and occurs in blood and lymph. Renin splits the substrate at a Leu-Leu bond, which joins it to the N-terminal of angiotensin I (**A-I**). A-I is an inactive decapeptide but is activated to **angiotensin II** (**A-II**), an octapeptide, by the action of **converting enzyme** (CE), a dipepti-

dase. (The half-life of A-II is c. 1 min, inactivation occurring in kidney and liver.) CE exists in many tissues, but the lung is the major site. CE is non-specific since it also hydrolyzes brady-kinin.

Control. Mechanisms for control of renin and A-II have not been clearly defined. The JGA is ideally situated to maintain blood volume by local intra-renal control of Na^+ excretion and to maintain blood pressure by systemic control of vascular resistance. Thus, a high concentration of Na^+ in TF reaching the distal tubule stimulates renin release; reduced GFR follows and less Na^+ is excreted; or, reduction of blood volume reduces glomerular perfusion pressure and releases renin, which raises systemic blood pressure. Renin release is mediated, at least in part, by β_2-adrenergic receptors (→ p. 54) and responds to circulating epinephrine. Some β-blockers, but not all, reduce renin release. Both A-II and aldo-sterone (ALD) have a negative feed-back effect, which reduces renin release.

Actions of Angiotensin II

1. **Cardiovascular.** A-II is the body's most potent vasoconstrictor. It achieves its effect by direct action on vessels, chiefly on arterioles. A-II has a direct action on Ca^{2+} availability and an indirect action, which prolongs the constrictor response to catecholamines.

2. **CNS.** A-II stimulates central sympathetic discharge for secondary support of its direct vasocontrictor influence.

3. **Renal.** A-II initiates intrarenal vasoconstriction and reduces GFR and RPF; FF increases (→ p. 114).

4. **Adrenal.** A-II is a direct stimulant of ALD secretion (→ p. 136). ALD stimulates the renal tubule to retain Na^+ and supports the direct effect of A-II on GFR, which also results in Na^+ retention.

5. **Biochemical** effects include activation of phosphodiesterase, which reduces local concentration of cAMP (→ p. 222).

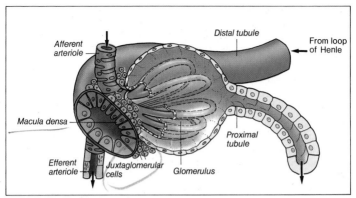

A. Juxtaglomerular apparatus

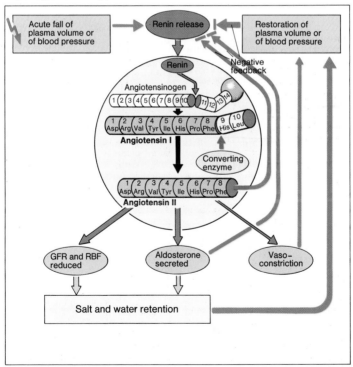

B. Renin-angiotensin system

Cardiovascular System

The cardiovascular system (CVS) has two components: **central** (the heart) and **peripheral** (the blood vessels). Its function is (1) to supply O_2, nutrients, and humoral signals to tissues according to need and (2) to carry metabolic waste products from the tissues. The CVS also controls body temperature. **Cardiovascular failure** occurs when the CVS no longer performs these functions optimally.

The normal **heart rate** is c. 70–80 strokes/min, the normal **stroke volume** amounts to about 70 ml. Heart rate x stroke volume is the **cardiac output** and amounts to c. 5.5 l/min (at a body weight of 70 kg or a body surface of 1.73 m^2).

Cardiac output can be standardized by dividing it by body surface. The result, the *cardiac index*, varies with age: Rising rapidly to 4 l/min per m^2 at 10 years of age it declines to c. 2.4 l/min per m^2 at the age of 80.

Cardiac output can be determined according to Fick's principle: Cardiac output = oxygen utilization ($\to$ p. 82) divided by arterio-venous O_2 concentration difference ($\to$ pp. 82, 83, 89 A).

Delivery of blood to tissues depends upon metabolic priorities ($\to$ **A**):

1. Interruption of blood flow to **brain** for only a few seconds results in loss of consciousness; for minutes, in permanent damage. Blood flow to the brain is maintained constant even if flow to other organs must be sacrificed (as in heart failure).

2. The **heart** muscle receives blood chiefly through the coronary arteries. Interruption of blood flow causes cardiac ischemia and may cause permanent damage.

3. The **lungs** receive blood from two sources: (1) 100% of blood from the right ventricle (RV) via pulmonary arteries is delivered at **low pressure** for gas exchange ($\to$ p. 68); (2) 5% of blood from the left ventricle (LV) via bronchial arteries is delivered at **high pressure** for metabolic needs. All blood leaves the lungs via pulmonary veins for delivery to the left atrium (LA). Pulmonary blood volume varies with respiration (0.6 to 1.2 l). The lungs are a reservoir that changes its volume to meet varying needs of the systemic circulation.

4. **Kidneys** receive c. one fourth of cardiac output, although only a small fraction of the renal blood flow (RBF) is needed for metabolism. The majority is processed for excretion of waste products. The RBF can be decreased for prolonged periods without irreversible tissue damage. Thus, during exercise ($\to$ p. 47, A) or insufficient circulation (heart failure, shock), a large volume of blood can be shunted from the kidney (and other organs) to high priority organs.

5. **Muscles, GI tract**. Blood flow in these two organ systems varies according to metabolic activity. During exercise, muscles can receive more than two thirds of cardiac output for short periods; after a meal, the GI tract may receive equally large flows. Clearly, the cardiac output is not sufficient to supply both systems with large flows at the same time. After a meal and during digestion, the GI tracts has priority and strenuous exercise cannot be maintained; muscle metabolism soon exceeds its blood supply causing anoxia and cramps.

6. Blood flow to the **skin** varies with systemic metabolic activity. It functions to regulate body temperature ($\to$ p. 180).

$$CO = HR \times SV = \frac{O_2\,utilization}{pO_{2A} - pO_2}\;(l/min)$$

$$Cardiac\;Index = \frac{CO\;(l/min)}{Body\;surface\;(m^2)}$$

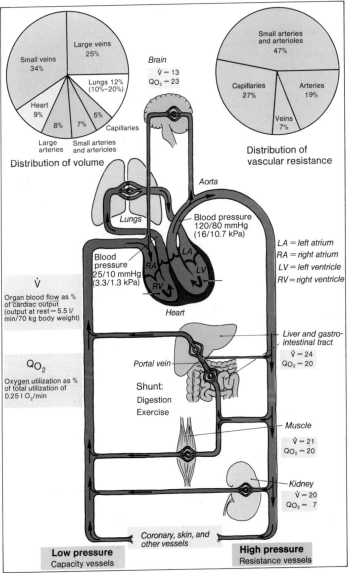

Distribution of volume

Small veins 34%

Large veins 25%

Lungs 12% (10%–20%)

Heart 9%

Large arteries 8%

Small arteries and arterioles 7%

Capillaries 5%

Brain
V̇ = 13
QO₂ = 23

Distribution of vascular resistance

Small arteries and arterioles 47%

Capillaries 27%

Arteries 19%

Veins 7%

Aorta

Blood pressure 120/80 mmHg (16/10.7 kPa)

Lungs

Blood pressure 25/10 mmHg (3.3/1.3 kPa)

RA   LA
RV   LV

Heart

LA = left atrium
RA = right atrium
LV = left ventricle
RV = right ventricle

V̇

Organ blood flow as % of cardiac output (output at rest = 5.5 l/min/70 kg body weight)

QO₂

Oxygen utilization as % of total utilization of 0.25 l O₂/min

Portal vein

Shunt:
Digestion
Exercise

Liver and gastro-intestinal tract
V̇ = 24
QO₂ = 20

Muscle
V̇ = 21
QO₂ = 20

Kidney
V̇ = 20
QO₂ = 7

Coronary, skin, and other vessels

Low pressure
Capacity vessels

High pressure
Resistance vessels

A. Cardiovascular system

Blood Vessels

The blood vessels and the heart comprise a closed circulating system. There are three categories of blood vessels, each with different characteristics:

1. **Exchange vessels**. The **capillaries** are small vessels with walls of only one cell thickness. They are semipermeable and allow exchange of plasma constituents (except larger proteins) between blood and interstitial space ($\rightarrow$ p. 144). The exchange is favoured by the low flow velocity in capillaries. Their common cross-sectional area ($\rightarrow$ **A, 2**) is c. 700 times larger than that of the aorta. Flow velocity, consequently, is 700 times lower in the capillaries (c. 0.4 mm/s versus c. 300 mm/s in the aorta).

2. **Resistance vessels**. The **arteries** possess smooth muscle walls that are under neurohumoral influence ($\rightarrow$ pp. 162–166) and contribute to regulation of blood pressure (BP). At precapillary sites, they also possess sphincters that open and close to control how much of the capillary surface will be exposed to blood ($\rightarrow$ p. 144). BP in the arterial circuit is necessarily high to maintain adequate tissue perfusion.

In analogy to **Ohm's law**, in the circulation *flow rate × resistance = pressure difference* (ΔP) between the two ends of the vessel(s). This interrelationship is valid for a single vessel as well as for the whole vascular system. Here ΔP equals the aortic blood pressure because central venous pressure ≈ 0:

mean blood pressure = cardiac output × total peripheral resistance.

Mean systemic blood pressure ($\rightarrow$ p. 146) is c. 13.3 kPa (100 mm Hg); if cardiac output ($\rightarrow$ p. 140) is 5.5 l/min or 92 cm³/s, total peripheral resistance (TPR) is 145 MPa · m^{-3} · s ($\rightarrow$ pp. 5–6). TPR is the sum of several resistances (R) *in series* (R of aorta + R of all arteries + R of all arterioles etc., $\rightarrow$ p. 141). Total R of resistances *in parallel* (e.g. of the single systemic arterioles) is calculated from $1/R_{total} = 1/R_1 + 1/R_2 \dots$ etc.

Windkessel effect ($\rightarrow$ p. 149): The arterial wall is elastic and exhibits **compliance** (= change in volume as a response to change in pressure). As the heart contracts and BP increases, the arteries stretch and store potential energy. When the heart relaxes, BP decreases and the stretched arteries rebound.

3. **Capacitance vessels**. The **veins** and the pulmonary circulation comprise a low-pressure circuit that has high capacity ($\rightarrow$ **A, 3**, p. 141) and high compliance. Because of these characteristics, the low pressure vessels function as a blood reservoir. They have up to 200 times the compliance of arteries. Consequently, if the circulation receives an extra 1 l of blood ca. 99.5% of it distributes into the low pressure vessels and only 0.5% is found in the arteries. In both vessels changes in volume are followed by changes in pressure. Arterial pressure, however, is strictly regulated ($\rightarrow$ pp. 146, 162–173). *Central venous pressure*, therefore, is a much better indicator for blood volume than arterial pressure.

Poiseuille's law. Flow through a vessel varies directly with the fourth power of the vessel radius (r⁴). The resistance varies inversely with r⁴. Only a small change in vessel radius elicits a large change in flow and resistance. Blood viscosity and vessel length also influence flow. Blood viscosity increases with an increasing hematocrit ($\rightarrow$ p. 58) and with a decreasing flow velocity. The reason for the latter effect is a lowered shear stress which is followed by an increased aggregation of red blood cells.

Laplace relationship ($\rightarrow$ also p. 146). Wall tension in a vessel = (BP-interstitial pressure) × vessel radius ($\rightarrow$ **A, 1**). A small radius vessel (capillary) is protected against bursting even with a high BP because its wall tension remains low. When it dilates and the diameter increases, however, the wall tension must increase to retain the pressure.

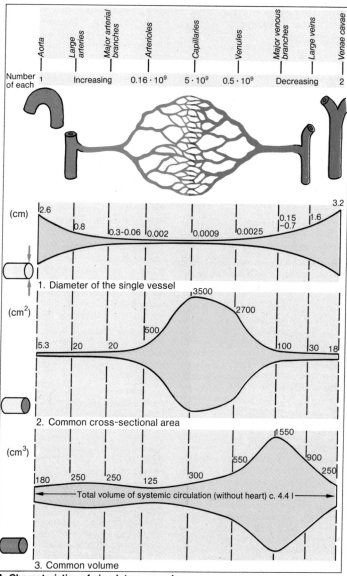

	Aorta	Large arteries	Major arterial branches	Arterioles	Capillaries	Venules	Major venous branches	Large veins	Venae cavae
Number of each	1	Increasing		0.16·10⁹	5·10⁹	0.5·10⁹	Decreasing		2

(cm) 2.6 — 0.8 — 0.3–0.06 — 0.002 — 0.0009 — 0.0025 — 0.15–0.7 — 1.6 — 3.2

1. Diameter of the single vessel

(cm²) 5.3 — 20 — 20 — 500 — 3500 — 2700 — 100 — 30 — 18

2. Common cross-sectional area

(cm³) 180 — 250 — 250 — 125 — 300 — 550 — 1550 — 900 — 250

Total volume of systemic circulation (without heart) c. 4.4 l

3. Common volume

A. Characteristics of circulatory vessels

Fluid Exchange Across Capillaries (Starling Hypothesis)

Cell nutrition takes place across the capillary wall, which is only one cell thick ($\rightarrow$ p. 142). Its large effective pore size allows free filtration and diffusion of all substances in the blood except cells and large proteins. Forces that influence loss and restoration of fluid in the capillaries are **hydrostatic** (P) and **oncotic** (π) pressures as well as **permeability** and **surface area** of the capillaries. It is assumed that these functions are uniform along the capillary length; actually, they increase toward the venous end. At the arterial end, filtration out of the capillary is favored; at the venous end, resorption back into the capillary is favored.

Filtration ($\rightarrow$ **A**). Unimpeded filtration can occur faster than blood can be delivered to the capillaries and would quickly empty the blood vessels. Normal rate is c. 20 l/day. Of this amount, 18 l/day are resorbed, and the remainder is returned to the blood via the lymphatics ($\rightarrow$ **A**). The principal filtering force is the hydrostatic pressure difference (ΔP) between the capillary lumen (P_c) and the interstitium (P_t). $\Delta P = $ c. 4 kPa ($= 30$ Torr) at the arterial and 2 kPa ($= 15$ Torr) at the venous end.

Resorption ($\rightarrow$ **A**) is the primary force opposing filtration and depends chiefly on the plasma oncotic pressure (π_c), which is that part of the osmotic pressure contributed by protein: 65% of π_c comes from plasma albumin. When plasma water is lost from the capillary, the protein concentration increase is small, but the π_c increases disproportionately more. For 20% loss of water, π_c increases 10% instead of the expected 2%. This property of protein solutions amplifies the restraining effect of π_c and protects against extreme fluid loss.

The Starling hypothesis defines events in an idealized vessel: fluid movement $= k \times [(P_c + \pi_t) - (P_t + \pi_c)]$ or $= k \times (\Delta P - \Delta \pi)$. The equation indicates that filtration out of the capillary is assisted by tissue oncotic pressure and is resisted by tissue hydrostatic pressure, but these effects normally are small and become significant only in disease states. When ΔP exceeds $\Delta \pi$, filtration is favored; when it is smaller, resorption is favored. The balance between filtration and resorption is influenced by several factors:

Venous pressure. An elevation ($\rightarrow$ **B, 1**) as in heart failure, raises P_c at the venous end and more fluid is filtered. If lymph flow is insufficient to carry away the excess, **edema** fluid accumulates and P_t rises to oppose the excess filtration.

Arterial pressure. The capillaries are protected against elevations by the precapillary sphincters; arterial pressure changes have little influence on fluid exchange. However, changes in the resistance vessels control how much of the capillary network will be perfused.

Oncotic pressure. A decrease in plasma albumin ($\rightarrow$ **B, 2**), as in hepatic cirrhosis or nephrosis, reduces π_c and allows greater filtration, which can also lead to **edema**.

In the **lungs**, P_c is low, c. 1 kPa ($= 7$ Torr), so that resorption always exceeds filtration. Absorption of fluid is so rapid that aerosol inhalation of drugs is one of the fastest ways of administering medicines. When P_c in the lungs exceeds π_c (c. 3.2 kPa $= 24$ Torr), filtration will exceed absorption and the lungs will accumulate fluid ($=$ pulmonary edema).

Edema may also develop if lymphatic flow is hindered or if the permeability of the capillaries is increased by bacterial toxins or by histamine ($\rightarrow$ p. 60). This tissue hormone ($\rightarrow$ p. 215) also increases P_c by its vasodilating effect ($\rightarrow$ p. 162).

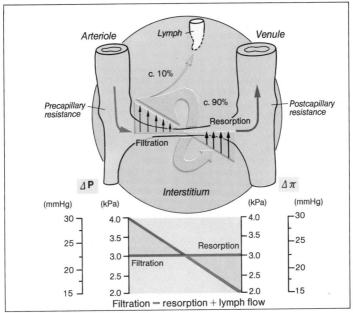

A. Capillary fluid exchange

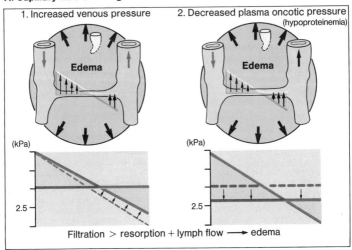

B. Causes of edema

Blood Pressure

Arterial blood pressure (BP) must be maintained to supply blood to the tissues. If the BP falls too far, nutrient supply to the tissues is inadequate, anoxia develops, and shock may ensue. If the BP becomes chronically elevated, it contributes to degeneration of vessels in the heart, kidney, and brain, among others. Measurement of the BP may be **direct**, by a needle in a vessel, or **indirect**, by means of an occlusive cuff. **Systolic pressure** ($\rightarrow$ **A**) is the peak pressure produced when the ventricles contract; it propels blood through the arteries. **Diastolic pressure** is the pressure when the ventricles relax; it is influenced chiefly by arteriolar resistance. **Pulse pressure** is the difference between systolic and diastolic pressures. **Mean pressure** is the geometric mean of systolic and diastolic pressures ($\rightarrow$ **A**).

BP is a function of blood flow and vascular resistance. Blood flows from high to low pressure areas. Rate of flow is determined by the magnitude of the pressure gradient. In the **systemic** circulation, BP averages 16/10.7 kPa (= 120/80 Torr; systolic pressure is 120 Torr and diastolic is 80 Torr). As the vessels become more distant from the heart, their resistance increases ($\rightarrow$ p. 141). Sphincters in the arterioles ($\rightarrow$ p. 144) contribute to resistance and control rate of runoff of blood from the systemic circulation.

The **pulmonary** circulation has several unusual features: (1) The lungs are the only organs through which 100% of the cardiac output flows. (2) Pulmonary pressure is 3.3/1.3 kPa (= 25/10 Torr) ($\rightarrow$ **B, 2**) and is much lower than the systemic BP of 16/10.7 kPa (=120/80 Torr) ($\rightarrow$ **B, 1**). (3) Blood takes just as long to flow through the pulmonary as through the systemic circulation even though the linear distance in the lungs is shorter.

This circumstance reflects the enormous cross-sectional area of the lung capillaries. (4) Blood flow is a function of the pressure gradient between the pulmonary artery and the left atrium. It is further affected by cyclic changes in intrathoracic pressure during respiration.

The **venous** circulation has low pressure and high compliance ($\rightarrow$ **B, 3**; pp. 142, 170). Venous pressure (VP) rises when blood volume increases. A high VP alters capillary equilibrium and edema may develop ($\rightarrow$ p. 144). A high VP can increase return of blood to the heart and affect cardiac output. The mean rate of venous return to the heart is greater during quiet respiration than during breath holding because respiration furnishes a pumping action on the great veins. On inspiration, intrathoracic and left atrial pressures fall ($\rightarrow$ **B, 3, 4**) and increase the gradient from extrathoracic vessels; on expiration, the reverse process occurs. During inspiration, venous return to the heart increases. Consequently also the cardiac output of the right ventricle rises ($\rightarrow$ **B, 6, 7**).

Posture: In the supine posture arterial and venous pressures are not significantly influenced by the hydrostatic pressure of the blood. On standing, the pressures in the head drop but those in the lower parts of the body rise. In the vessels of the feet a blood column of c. 1 m increases the vascular pressures by c. 10 kPa. Because of their high compliance mainly the veins dilate. Thus return of blood to the heart and, as a consequence, cardiac output and BP decrease. If the compensatory mechanisms ($\rightarrow$ p. 166) are sluggish *orthostatic hypotension* with dizziness and even fainting develops.

Laplace relationship.

Wall tension in a spherical organ like the ventricles of the heart or like the alveoli of the lungs ($\rightarrow$ pp. 80,81) = $^1/_2$ x transmural pressure difference x radius. If the radius of the ventricle decreases (ejection of blood) at constant wall tension the ventricular pressure rises ($\rightarrow$ p. 149, phase II). On the other hand, if the heart is enlarged the myocardium must develop an increased wall tension for maintaining the normal systolic pressure.

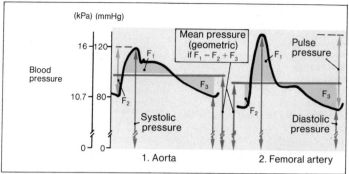

(kPa) (mmHg)

Blood
pressure

16 — 120
10.7 — 80
0 — 0

F_1
F_2
F_3

Mean pressure
(geometric)
if $F_1 = F_2 + F_3$

Systolic
pressure

Pulse
pressure

Diastolic
pressure

F_1
F_2
F_3

1. Aorta

2. Femoral artery

A. Arterial pulse waves

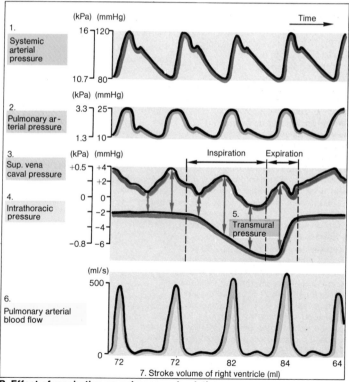

1.
Systemic
arterial
pressure

(kPa) (mmHg)
16 — 120
10.7 — 80

Time

2.
Pulmonary ar-
terial pressure

(kPa) (mmHg)
3.3 — 25
1.3 — 10

3.
Sup. vena
caval pressure

4.
Intrathoracic
pressure

(kPa) (mmHg)
+0.5 — +4
— +2
0 — 0
— -2
— -4
-0.8 — -6

Inspiration   Expiration

5.
Transmural
pressure

6.
Pulmonary arterial
blood flow

(ml/s)
500

0

72   72   82   84   64

7. Stroke volume of right ventricle (ml)

**B. Effect of respiration on pulmonary circulation
and venous pressure (schematic)**

Cardiac Cycle

The cardiac cycle concerns mechanical, electric and acoustic events that recur with each heartbeat. There are two phases: **systole**, or ventricular contraction and ejection of blood, and **diastole**, or ventricular relaxation and filling with blood.

Events of the Cardiac Cycle (numbers in parentheses refer to figure).

Atrial systole (IVc). The sinoatrial (SA) node depolarizes to initiate atrial contraction and to generate the P wave (→ **A, 1**). As long as atrial pressure (→ **A, 4**) exceeds ventricular pressure (→ **A, 3**), blood flows into ventricles. Ventricular diastole ends with closure of atrio-ventricular (AV) valves (→ **A, 3, 4: Z**). The ventricular end diastolic volume (EDV) (→ **A, 6**) normally is 125 ml but may be as large as 250 ml. If EDV at the end of one contraction is large, the force of the following ventricular contraction becomes greater (Frank-Starling's law) (→ p. 164).

Isovolumetric contraction (I) occurs when all valves are shut. Tension develops rapidly until ventricular pressure (→ **A, 3**) exceeds arterial pressure (→ **A, 2**); the semilunar valves open and blood is ejected into the great vessels (→ **A, 7**). Rate of change of ventricular pressure (dP/dt) is maximal in this phase.

Rapid ventricular ejection (early systole, **IIa**). Ventricular pressure (→ **A, 3**) exceeds aortic (→ **A, 2**). 50 % of stroke output is achieved. Ventricular volume (→ **A, 6**) and aortic blood flow (→ **A, 7**) change rapidly. Maximal aortic flow (→ **A, 7**) occurs before peak aortic pressure (→ **A, 2**) is achieved.

Reduced ventricular ejection (late systole, **IIb**) begins when aortic flow (→ **A, 7**) starts to decline. Aortic pressure falls (→ **A, 2**) but remains above ventricular pressure (→ **A, 3**). Semilunar valves close at the end of systole because blood flow in aorta (→ **A, 7**) becomes retrograde when the aortic

pressure exceeds ventricular pressure (→ **A, 2, 3**).

The **dicrotic notch** (→ **A, 2**) reflects vibrations in blood when the aortic valve snaps shut.

Windkessel effect: when the aorta stretches during ventricular ejection (→ **A, II**), it creates potential energy; as pressure falls in ventricular relaxation (→ **A, III**), the aorta rebounds, converts its potential energy to kinetic energy, and maintains the forward movement of blood. This action also damps pulsatile flow in the peripheral small vessels.

Isovolumetric relaxation (III): all cardiac valves are closed. As heart muscle relaxes, ventricular pressure (→ **A, 3**) falls below atrial pressure (→ **A, 4**) and the AV valves open for ventricular filling.

Rapid ventricular filling (early diastole, **IVa**). Filling is passive, according to the pressure gradient (→ **A, 3, 4**).

Diastasis (IVb): slow ventricular filling from the great veins (→ p. 146). At normal heart rate, **IVa** and **IVb** account for 70% to 80% of filling and **IVc** for only 20%. At rates of 125/min, **IVb** does not occur; above 150/min, **IVa** becomes extremely short and **III** is followed almost immediately by **IVc**.

Coronary flow (→ **A, 8**) occurs principally in diastole when ventricles relax. During systole, the high tension in the ventricular wall creates resistance to flow by squeezing the coronaries. In the left coronary (→ **A, 8**), blood may actually be squeezed out of the vessel.

The **venous pulse** (→ **A, 5**) has three components that can be observed in the right atrium. The **a wave** of atrial contraction pressure is transmitted retrograde to the jugular vein. The **c wave**: ventricular contraction at **I** bulges the AV valve into the atrium and raises pressure. The **v wave**: blood filling the atria raises pressure as long as the AV valves remain closed (**III**). At the X point, ventricular contraction pulls the AV ring toward the apex, enlarges the atrium, and lowers atrial pressure (important for atrial filling). The **y wave** is the drop in atrial pressure just after the AV valves open.

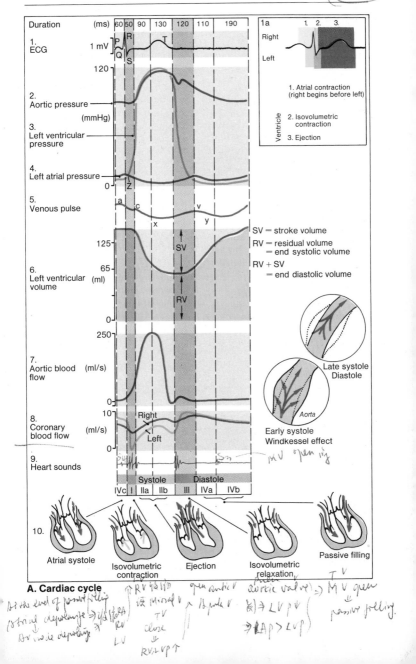

A. Cardiac cycle

| Duration | (ms) | 60 | 50 | 90 | 130 | 120 | 110 | 190 |

1. ECG
1 mV
P R Q S T

1a
Right
Left

1. Atrial contraction (right begins before left)
Ventricle
2. Isovolumetric contraction
3. Ejection

2. Aortic pressure (mmHg)
120

3. Left ventricular pressure

4. Left atrial pressure
0
Z

5. Venous pulse
a c v
x y

SV = stroke volume
RV = residual volume = end systolic volume
RV + SV = end diastolic volume

6. Left ventricular volume (ml)
125
65
0
SV
RV

7. Aortic blood flow (ml/s)
250
0

Late systole
Diastole

8. Coronary blood flow (ml/s)
10
0
Right
Left

Aorta

Early systole
Windkessel effect

9. Heart sounds

Systole — Diastole
IVc | I | IIa | IIb | III | IVa | IVb

10.
Atrial systole | Isovolumetric contraction | Ejection | Isovolumetric relaxation | Passive filling

Electric Conduction in the Heart

The heart contains two types of cells: (1) conducting cells that initiate and propagate impulses and (2) muscle cells that respond to stimuli by contracting. These cells have combinations of the following four inherent properties, which permit the heart to perform as a pump: (1) **automaticity**, or **chronotropism**, determines heart rate and is characterized by dV/dt of phase 4 (diastolic depolarization or DD ($\rightarrow$ pp. 24, 153, 154 A); (2) **conductivity**, or **dromotropism**, refers to the rate at which stimuli are propagated along a conduction pathway; (3) **excitability** is determined by threshold potential (TP); and (4) **contractility**, or **inotropism** (q.v.). The heart is a functional syncytium; although the cells are not actually contiguous, a stimulus originating anywhere in the heart produces contraction of the entire organ. The impulse for contraction is generated within the heart cells; nerves do not initiate contraction, they only influence its rate and force ($\rightarrow$ p. 49–51). The situation is quite different from that in striated muscle where the stimulus for contraction is initiated by nerves and must be transmitted to the muscle cells ($\rightarrow$ p. 42).

The conduction pathway ($\rightarrow$ **A**). The impulse for a heart beat normally starts in the **sinoatrial (SA) node,** the **pacemaker**, which lies in the right atrium (RA) near the superior vena cava. The impulse spreads through both atria to the **atrioventricular (AV) node,** which lies at the septal border of the tricuspid valve. Conduction through the AV node is slower than in the atria. The septum is activated from left to right. The **bundle of His** in the neck of the AV node extends to the **bundle branches** (R and L), which are the main connecting trunks to the corresponding ven-

tricles. The **Purkinje network** consists of terminal strands of conducting tissue, which carry the impulse from the bundle branches to all areas of the ventricular muscle. Muscle is activated from endo to epicardium. The impulse travels faster through conduction tissue than through muscle.

Propagation of a pacemaker impulse depends upon amplitude of the AP, the slope (dV/dt) of phase 0, and the instability of phase 4 (diastolic depolarization or DD ($\rightarrow$ p. 24). The rate of impulse formation (**rhythmicity** or frequency) is influenced by several variables:

1. The refractory period (ref pd), or the time needed for the cell to recover from a previous stimulus. A distinction must be made between **absolute ref pd** ($\rightarrow$ p. 28), when the cell cannot respond to any stimulus, and **relative ref pd,** when the cell responds incompletely to a normal stimulus.

2. The threshold potential (TP) ($\rightarrow$ p. 153, **A**, **1**). As the TP falls toward 0 mV, the time for DD to reach TP is increased and the rate of firing decreases.

3. Rate of DD (dV/dt of phase 4) ($\rightarrow$ p. 153, **A**, **2**). As the rate of DD falls, it takes longer to reach TP and the rate of firing decreases. Of all the automatic cardiac cells, the SA node has the fastest DD (steepest slope). This is the reason that the SA node has a pacemaker function since its firing rhythm is propagated to more distal cells and causes them to depolarize before they would have done so spontaneously. If the SA node becomes relatively slowed, the next fastest area may take over as pacemaker.

4. Maximum diastolic potential (MDP) ($\rightarrow$ pp. 24, 153, **A**, **3**). As the MDP becomes larger (more negative), the cell membrane becomes hyperpolarized. The time for DD to reach TP is increased and the rate of firing decreases.

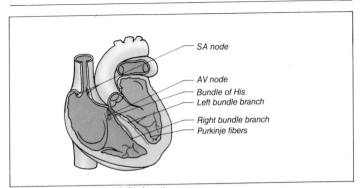

A. Conducting system of the heart

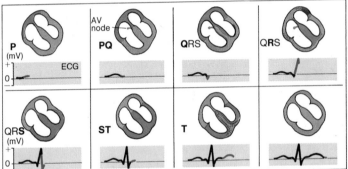

B. Correlation of ECG with depolarization wave

Event		Time (ms)	ECG	Conduction velocity (m/s)	Intrinsic auto-maticity (min⁻¹)
SA node Impulse generated		0	P wave	0.05	70–80
Atrial depolarization	right	50		} 0.8–1.0	
	left	85			
AV node Arrival of impulse		50	} P–Q interval		40–60
Transmission of impulse		125		0.05	
Bundle of His activated		130		1.0–1.5	
Bundle branches activated		145		1.0–1.5	
Purkinje fibers activated		150		3.0–3.5	
Endocardium depolarized	right ventricle	175	} QRS complex		30–40
	left ventricle	190			
Myocardium depolarized	right ventricle	205		} 1.0 in myocardium	
	left ventricle	225			

C. Depolarization sequence

Physiologic Basis of Arrhythmias

Arrhythmias result from disturbances in impulse **formation** (automaticity) or **propagation** (conduction). They are influenced by electric inhomogeneity of the heart. Cardiac cells exhibit two types of electric activity ($\rightarrow$ p. 24).
1. **Fast response** occurs in contracting and conducting cells. Depolarization is by Na^+ flux through fast channels of the cell membrane. dV/dt of phase $0 = c.$ 800 V/s, RMP $= -90$ mV, AP $= 100$ to 300 ms, and conduction is rapid at 0.5 to 5 m/s.

2. **Slow response** occurs chiefly in SA and AV nodes. Depolarization is by Na^+ and Ca^{2+} flux through slow channels. dV/dt of phase $0 = c.$ 1 V/s, RMP $= -70$ to -60 mV, AP $= 35$ to 75 ms, and conduction is slow at 0.01 to 0.1 m/s.

Disturbances in automaticity. Pacemaker cells in the SA node generate an impulse that initiates each heart beat. The rate of impulse formation depends upon dV/dt of phase 4 (diastolic depolarization or DD), threshold potential (TP), and maximum diastolic potential (MDP) ($\rightarrow$ **A**, **B**; pp. 24, 150). Cells of the SA node have the shortest refractory period and the steepest DD of any automatic cells in the heart. If the TP, MDP, or DD change so that the pacemaker cells **slow** their firing rate or so that ectopic cells **increase** their rate, an ectopic pacemaker may take control of the heart ($\rightarrow$ p. 160). **Disturbances in conduction** are attributable to **circus movement** or **reentry** phenomena, which are associated with the development of **decremental conduction.** One model for reentry ($\rightarrow$ **C**) requires at least two conduction paths to a muscle fiber; one path is normal ($\rightarrow$ **C**, **a**), the other ($\rightarrow$ **C**, **b**) has a unidirectional block that may result from local tissue damage or from altered metabolism locally. The impulse cannot be conducted in the forward direction ($\rightarrow$ **C**, **1**), but it can in the reverse direction.

Decremental conduction requires long conduction paths (as in an enlarged heart), short refractory periods, and inhomogeneous repolarization to develop. Cell injury, as by ischemia, may partially depolarize the membrane enough for inactivation of the fast Na^+ channels but not enough for the slow channels ($\rightarrow$ p. 24). Such cells are capable of a slow response, but the AP that they develop is smaller and has a slower rate of rise. The AP is therefore less effective in propagating an impulse to the next cell. As a result, adjacent cells in the conduction path respond with still smaller and slower AP. Transmission of the impulse becomes progressively less effective until it no longer elicits a response (decremental conduction with unidirectional block ($\rightarrow$ **C**, **1**). However, the tissue beyond the damaged site can respond normally to a sufficient stimulus. Thus, although the impulse cannot be conducted forward through path B, the impulse from path A may reenter at path B and travel in reverse ($\rightarrow$ **C**, **2**). If it then reaches path A when the cells are no longer refractory, path A may be reexcited. In this way, a circus movement may be established, which can become the focus of a self-sustaining impulse and arrhythmia ($\rightarrow$ p. 160).

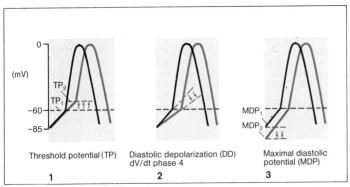

A. Influences on pacemaker rate (see also p. 150)

1 — Threshold potential (TP)
2 — Diastolic depolarization (DD) dV/dt phase 4
3 — Maximal diastolic potential (MDP)

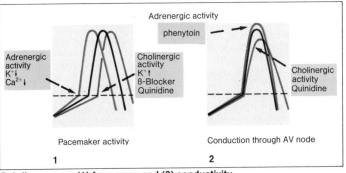

B. Influences on (1) frequency and (2) conductivity

1 — Pacemaker activity
2 — Conduction through AV node

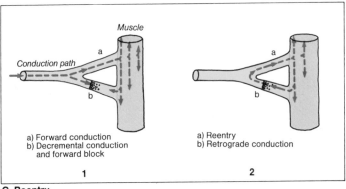

C. Reentry

1
a) Forward conduction
b) Decremental conduction and forward block

2
a) Reentry
b) Retrograde conduction

Electric Properties of the Heart — ECG

The electrocardiogram (ECG) is a means of measuring the electric forces generated by the heart. The **standard electric leads I, II, and III are bipolar leads** ($\rightarrow$ **D**); they record the difference in voltage (potential) between two points electrically equidistant from the heart, which is a two-dimensional representation of cardiac potentials in the frontal plane of the body ($\rightarrow$ **C**). Additional leads are necessary to obtain a three-dimensional representation. These are the **unipolar** V leads ($\rightarrow$ **F**); they record the potential difference between the $\oplus$ electrode and electric ground zero. They are not equidistant from the heart, and their magnitudes are not comparable; those that are closer to the heart will record larger deflections. The Goldberger variations in the frontal plane ($\rightarrow$ **E**) are amplified V leads, which allow a more precise measurement of cardiac forces as vectors (aVR, aVL, aVF).

Each portion of the heart generates electric forces when it polarizes and repolarizes. However, since leads are placed at a distance from the heart rather than directly on it, only the **average** forces for the entire heart are recorded. The components of the ECG are ($\rightarrow$ **B**):

P wave: atrial depolarization. The wave for atrial repolarization is rarely seen because it is overlapped by other waves.

QRS complex: ventricular depolarization. Q wave is the first downward $\ominus$ deflection before the R wave; R wave is the first upward $\oplus$ deflection; S wave is the downward $\ominus$ deflection after the R wave. It is always referred to as the QRS complex even if all component waves are not present.

T wave: ventricular repolarization; a slower process than depolarization.

PQ interval: from beginning of P to the beginning of QRS. It is the time taken by the impulse from the SA node to pass through the AV node.

QRS duration: time required for ventricular depolarization.

QT interval: from beginning of QRS to end of T wave. It represents the time required for ventricular repolarization. It varies with heart rate.

Significance of the ECG. The ECG can give two kinds of information: (1) it shows disturbances of rate, rhythm, and conduction ($\rightarrow$ p. 160) and (2) it shows disturbances in the ventricular gradient i.e., the phase relationship between depolarization and repolarization; these disturbances may be metabolic, hemodynamic, anatomic, or physical.

Vector analysis, the ventricular ECG. Since the electric forces of the heart have direction and amplitude, they can be represented as vectors ($\rightarrow$ **A**). Usually, only the vectors for the **average** QRS ($\rightarrow$ **D**) and T forces are analyzed. Little information is obtained from the P vectors. For precise analysis, vectors may be derived for selected time segments; for example, the initial QRS vector is significant in the diagnosis of myocardial infarct, and the terminal QRS vector sometimes gives information on intraventricular conduction disturbances. When the ends of all the instantaneous vectors are connected serially, the vector loop is obtained ($\rightarrow$ **C**). To derive the **mean** QRS vector in the frontal plane, leads I, II, and III are examined ($\rightarrow$ **G**). For precise analysis, the **areas** under each QRS should be determined; in practice, the heights (amplitudes) generally suffice. Downward $\ominus$ deflections are subtracted from upward $\oplus$, and an average value is determined. The direction of the mean QRS vector relative to the three standard lead axes will be either (1) parallel to the lead that shows the largest deflection or (2) perpendicular to the lead that shows the smallest deflection. By convention, the vector

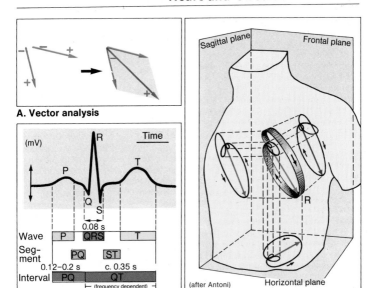

A. Vector analysis

B. ECG

Wave | P | QRS | T
Segment | | PQ | ST
Interval | PQ | QT

0.08 s

0.12–0.2 s   c. 0.35 s

(frequency dependent)

C. Spatial vector loop

Sagittal plane   Frontal plane

R

(after Antoni)   Horizontal plane

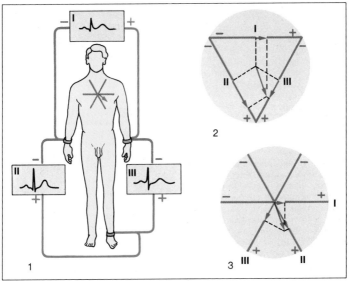

D. Bipolar leads (I, II, III) Einthoven triangle

points toward the ⊕ pole of the lead. The T vector is derived by the same means. In Fig. **G, 1**, QRS is isoelectric (net zero) on lead I; it is therefore perpendicular to I. On II and III, QRS is ⊕ and of equal magnitude. It is therefore equidistant from II and III. It points to the ⊕ poles of all three leads. Such an ECG is characteristic of people with long, narrow chests. As the QRS vector shifts toward the left shoulder, its representation becomes smaller on III and larger on I and II (→ **G, 2**). When the vector is parallel to I (→ **G, 3**), the QRS is largest on I and equal on II and III; however, it is ⊕ on II and ⊖ on III.

Chest leads. The V leads when used with the standard leads allow a three-dimensional description of QRS and T vectors (→ **C**). Since they are not equidistant from the heart (→ **F**), they are not analyzed in the same way as the standard leads (I, II, III). The chest is assumed to be a cylinder, which is divided by a plane into two electric halves (→ **F, 4**). Over one half, the recorded QRS is ⊕, over the other, it is ⊖. At the transition between the two halves, the recorded QRS is isoelectric (net zero). The dividing plane is perpendicular to the vector, and the vector points to the ⊕ half of the chest cylinder. There will be a different position for the plane of the QRS and of the T vector. In the normal heart, the mean QRS vector points to the left and somewhat posteriorly; therefore, it will be ⊖ in V1 and V2; at V3 it will be either isoelectric or ⊕ and at V4, V5, and V6 it will be ⊕. The dividing plane passes through the position where the QRS is isoelectric (V3). The mean T vector is normally in the frontal plane and points to the left; it will therefore be ⊖ on V1 and ⊕ on V2 to V6.

Axis (→ **H**). The mean QRS vector normally points to the left, downward and posteriorly (normal limits, → **F, H**). When it points outside of the normal limits, right or left axis deviation (RAD or LAD) is noted. Body build influences the axis; stocky, squat people have a horizontal axis, sometimes with LAD; thin, elongated people have a vertical axis. Infants have a relative right ventricular hypertrophy that is seen on the ECG as RAD. Dextrocardia shows RAD. Increased pulmonary artery resistance introduces increased work on the right ventricle and causes. RV hypertrophy and RAD. As the individual ages, the axis shifts toward the left. In left ventricular hypertrophy or in hypertension, there is commonly LAD. Large myocardial infarcts also may shift the axis.

QRST angle (→ **J**). **Small angle:** In the ECG, the normal angle between QRS and T vectors is c. 60° regardless of the axis. The angle reflects the phasic relationship between ventricular depolarization and repolarization (ventricular gradient). **Right angle:** From middle age on, the T vector tends to move anteriorly and the QRST angle to widen. It often is a normal finding, but it may also reflect ventricular ischemia. It is commonly a nonspecific change associated with advancing age. **Large angle:** When the QRST angle approaches 180°, one of three diagnoses is likely:

1. Intraventricular hypertension as occurs in essential hypertension, as well as in aortic stenosis or insufficiency. The QRST angle may, but need not, widen in any of these conditions. Widening reflects altered phasic relationships in depolarization and repolarization.

2. Bundle-branch block. When the conduction path is damaged, the impulse must be conducted through muscle and takes longer. In this case, the QRS wave is prolonged. Since the damaged ventricle will depolarize late, the terminal forces will arise from that side. The terminal QRS vector (rather than the average) will point to the

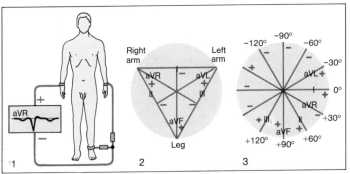

E. Unipolar leads in frontal plane (Goldberger)

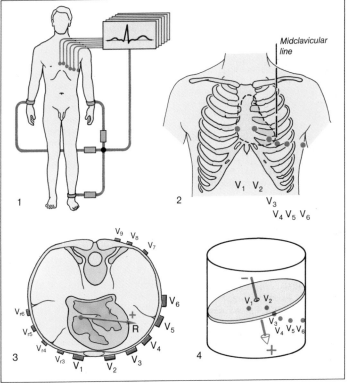

F. Unipolar chest leads (Wilson)

damaged side. The delay also alters the repolarization sequence and changes the direction of the T vector.

3. Digitalis effect. Changes in electrolyte fluxes shorten repolarization (short QT interval) and alter the phasic sequence. Digitalis may also slow conduction through the AV node, which would lengthen the PQ interval.

Myocardial infarction (→ **K**). Three events that are characteristic of transmural infarction are represented in the ECG:

1. **Death of muscle.** Dead muscle in the infarcted zone generates no electric potential. Absence of this potential is particularly obvious in the first 0.04 s, when the depolarization impulse sweeps through the endocardium. The average of the forces that remain is therefore deviated away from the infarcted region, accounting for an initial or 04 vector pointing **away** from dead muscle. Infarcts are almost always found in the left ventricle so that the abnormal 04 vector will point away from the LV. Since this direction is usually different from the mean QRS vector, initial forces and mean forces will not be parallel. Thus, in many leads, initial forces will be negative when the mean forces are positive and a Q wave will be written. The converse is also true; in leads where the mean QRS is negative, the 04 vector will be positive. In either case, the abnormal initial force must have a duration of 0.04 s.

2. **Ischemia.** Surrounding the dead muscle, there is an ischemic area where metabolic processes of the cell continue to function, but at altered rates. As a result, the sequence of repolarization in this area will be out of phase with the rest of the heart. The mean T vector will be displaced away from the infarct.

3. **Injury.** The infarcted area will be surrounded by a zone of injured tissue, where permeability of the cellular

membrane to ions has been altered. The ionic leaks generate an "injury current," which can be seen on the ECG as a displacement of the ST segment above or below the baseline of the recording. A separate vector can be derived for this ST force. Since the injury forces are generated only in the vicinity of the infarct, the ST vector will naturally point in the direction of the injury.

The three zones, dead tissue, ischemia, and injury, are not necessarily concentric so it should not be expected that the three vectors will be strictly parallel. These considerations apply of course only to single infarcts. When more than one infarct has occurred, the vectors will necessarily represent an average of the existing forces and may be of little value in localizing the infarct. As the infarct heals, the abnormal forces slowly regress. The injury current (ST vector) is the first to subside if healing takes place. The abnormal T wave may last for months (→ **K, 2**) and the abnormal 0.04 vector for years (→ K, **3**). Nevertheless, an infarct may heal so thoroughly with time that it can be very difficult to find evidence for its presence on the ECG.

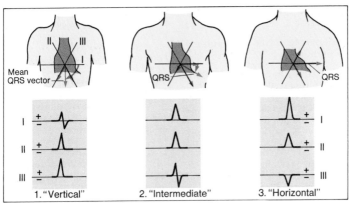

G. Normal mean QRS vectors

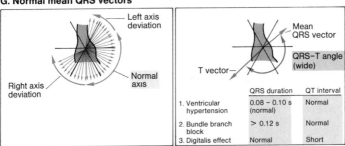

H. Electric axis of the heart

J. Interpretation of a wide QRS–T angle

	QRS duration	QT interval
1. Ventricular hypertension	0.08 – 0.10 s (normal)	Normal
2. Bundle branch block	> 0.12 s	Normal
3. Digitalis effect	Normal	Short

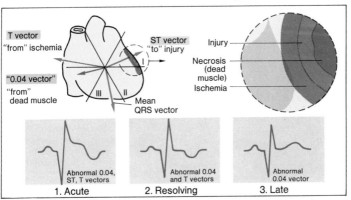

K. Myocardial infarction

Arrhythmias and Conduction Defects

Arrhythmias Based on Impulse formation (→ p. 152).

1. **Variations in SA node.** (a) **Sinus tachycardia** (rate above 100/min) and **sinus bradycardia** (rate below 50 to 60/min). Impulse arises in SA node. Rhythm is regular. (b) **Sinus arrhythmia.** Vagal activity alternates in phase with respiration: heart rate increases on inspiration, decreases on expiration, a normal event in adolescents. (c) **Sinus arrest.** Excess acute vagal activity suppresses automaticity of the SA node; it may even stop the node and the heart. The blood pressure falls and elicits reflex sympathetic discharge and escape from vagal control.

2. **Supraventricular arrhythmias with ectopic focus.** Includes atrial and nodal arrhythmias, which are often clinically indistinguishable. (a) **Premature atrial contraction** (PAC) or extrasystole (→ **A**, **1**). On the ECG, the QRS and T are normal because conduction through the AV node is not changed. (b) **Atrial tachycardia** (→ **A**, **2**). Atrial rate up to 200/min. Rhythm regular. Essentially a series of ectopic atrial contractions that may be transient (paroxysmal) or continuous. It may stop spontaneously or require treatment. Ventricles are capable of responding to atrial impulses up to a maximum rate of 200/min (1 : 1 response). Above this rate, the atrial impulse reaches the ventricle during its absolute refractory period and only every second impulse (2 : 1 response) (→ **A**, **2**) or rarely every fourth (4 : 1 response) produces ventricular contraction. (c) **Atrial flutter** (→ **A**, **3**) (may involve reentry). Atrial rates above 300/min. P waves are coarse with a sawtooth pattern. Ventricular response is regular or irregular depending on atrial rate and ventricular ref. pd. (d) **Atrial fibrillation** (→ **A**, **4**) (may involve reentry). Atrial rate c. 500/min. Ventricular response is totally irregular. Only those impulses that arrive at the AV node at its relative refractory pd. are transmitted to the ventricles. Ventricular response, size of QRS, cardiac output, and systolic blood pressure vary from beat to beat.

3. **Ventricular arrhythmias with ectopic focus** (a) **Premature ventricular contraction** (→ **A**, **7**). Impulse originates in ventricle and produces bizarre QRS on the ECG. (b) **Ventricular tachycardia** (→ **A**, **8**). An extended sequence of ectopic ventricular contractions. (c) **Ventricular fibrillation.** Ventricular contractions are asynchronous and ineffective; cardiac output and blood pressure fall precipitously.

Conduction Disturbances (→ p. 152).

4. **AV block.** (a) **First-degree AV block.** Decremental conduction (→ p. 152) prolongs the PQ interval. (b) **Second-degree AV block.** The AV node is intermittently nonconductive and an occasional ventricular beat drops out. (c) **Third-degree AV block.** AV node is completely unresponsive. Usually, a ventricular pacemaker with slow idioventricular rhythm takes over. The block may be transient or permanent. Epinephrine sometimes may improve responsiveness of the AV node. An artificial pacemaker may be necessary.

5. **Intraventricular block.** Right or left bundle branch block. The corresponding bundle fails to transmit the impulse that is compelled to take an aberrant path through the ventricle. Conduction through muscle is slow and the depolarization process is prolonged. On the ECG, the QRS complex is longer than 0.12 s and bizarre in form. The uninvolved bundle conducts normally.

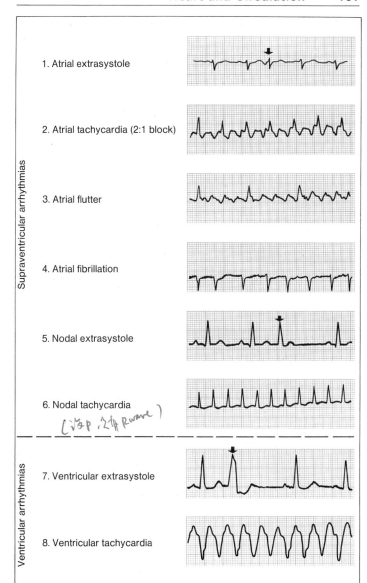

Supraventricular arrhythmias

1. Atrial extrasystole

2. Atrial tachycardia (2:1 block)

3. Atrial flutter

4. Atrial fibrillation

5. Nodal extrasystole

6. Nodal tachycardia

Ventricular arrhythmias

7. Ventricular extrasystole

8. Ventricular tachycardia

A. ECG expression of arrhythmias

Control of the Circulation – Blood Vessels

Intrinsic Control or Autoregulation

This is the intrinsic capacity of the microcirculation to vary its resistance to maintain blood flow. Although it is independent of neural control, it may be overridden by reflex and hormonal influences. It is most prominent in organs where neural control of the vessels is minimal (e.g., coronaries and brain). It has two functions: when local metabolism is constant, an increase in blood pressure (BP) does not increase local flow because the vessels constrict to hold flow constant; when local metabolism increases, the vessels dilate and increase local flow independently of the BP. Autoregulation ceases to function when the BP falls below a critical level. Several factors play a role: (1) **Myogenic factors.** Vascular smooth muscle contracts in response to stretch when the BP rises. This effect increases resistance and reduces flow. (2) **Metabolites** (CO_2, H^+, ADP, K^+) accumulate in the ECF and relax vascular smooth muscle fibers. Blood flow increases to wash away the metabolites and to reduce their influence. (a) **Oxygen** is a vasoconstrictor in most tissues except in the lung where it is a vasodilator. In the tissues, anoxia causes vasodilation and increases local O_2 supply; in the lungs, anoxia causes vasoconstriction and slows blood flow to allow more time for gas exchange. Furthermore, vasoconstriction in regions of low pO_2 reroutes the blood to alveoli supplied with more O_2. (b) **Reactive hyperemia.** If blood flow to an area is occluded, as with a tourniquet, metabolites accumulate and anoxia develops. Releases of occlusion is followed by up to fivefold increase in blood flow above normal. Duration of the elevated flow is equivalent to the duration of occlusion. Prostaglandins seem to play an important role in reactive

hyperemia. (3) **Tissue pressure.** When BP rises, capillary filtration increases ($\rightarrow$ p. 144). More fluid enters the ECF and tissue pressure rises. It increases vascular resistance and flow is reduced.

Autonomic Control

All blood vessels contain both α- and β-receptors ($\rightarrow$ pp. 50–57) which vary in number and in proportion in different tissues. α-receptors elicit vasoconstriction; β-receptors elicit vasodilation. The predominant receptor in vessels of skin and kidney is α; of skeletal muscle is β; in coronaries and viscera both α and β are active. Some receptors receive sympathetic fibers in which norepinephrine is the transmitter; other receptors have no innervation and respond only to circulating catecholamines (e.g., vessels of muscle). Some vessels, as in the brain and genitals, have dual sympathetic and parasympathetic innervation. The functional role of the latter is not yet defined. Regulation of **venous tone** influences return of blood to the heart; regulation of **arterial tone** ($\rightarrow$ **C**) influences peripheral resistance and tissue perfusion. Sympathectomy or spinal anesthesia may interfere with such control and cause hypotension.

In an area where high metabolic demand requires an increased blood supply, sensitivity of the α-receptors decreases (= functional sympathectomy). Vasoconstriction can still take place systemically, but it does not occur in the local area. In this way, blood can be shunted from an area of low metabolic activity (and high receptor sensitivity) to metabolically active areas.

Humoral Control

In addition to catecholamines, a number of agents – kinins, cortisol, histamine, angiotensin II ($\rightarrow$ p. 138), serotonin – influence the microcirculation.

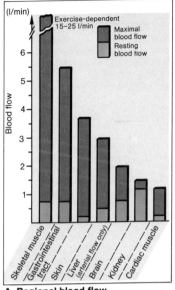

(l/min)

Exercise-dependent
15–25 l/min

■ Maximal
blood flow

Resting
blood flow

Blood flow

6

5

4

3

2

1

Skeletal muscle
Gastrointestinal tract
Skin
Liver (arterial flow only)
Brain
Kidney
Cardiac muscle

A. Regional blood flow

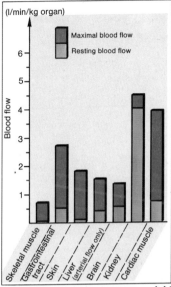

(l/min/kg organ)

■ Maximal blood flow

Resting blood flow

Blood flow

6

5

4

3

2

1

Skeletal muscle
Gastrointestinal tract
Skin
Liver (arterial flow only)
Brain
Kidney
Cardiac muscle

B. Blood flow relative to organ weight

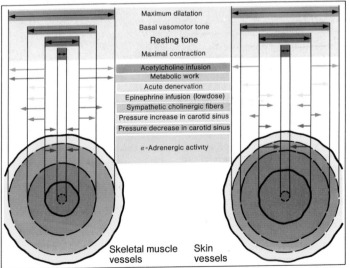

Maximum dilatation

Basal vasomotor tone

Resting tone

Maximal contraction

Acetylcholine infusion

Metabolic work

Acute denervation

Epinephrine infusion (lowdose)

Sympathetic cholinergic fibers

Pressure increase in carotid sinus

Pressure decrease in carotid sinus

α-Adrenergic activity

Skeletal muscle vessels

Skin vessels

C. Vasomotor influences

(after Koepchen)

Control of the Circulation – Blood Vessels and Heart

Vasomotor control by CNS (→ D)

In the **medullopontine** area (→ p. 252), near the respiratory control area, there is a concentration of afferent and efferent cardiovascular nerves closely associated with fibers from stretch and pressure receptors in arteries and veins (→ **D, 2**). Together, these comprise the vasomotor control area. Laterally, there is an area of continuous sympathetic discharge for vasoconstriction and cardioacceleration. Medially, an area modulates intensity of sympathetic discharge and adjusts sympathetic tone. Increased activity of the medial area reduces activity of the lateral area. Vagal parasympathetic fibers for cardiodeceleration are close by. This control area is intrinsically automatic and supervises cardiovascular reflexes for maintaining blood pressure (BP).

The **hypothalamus** allows communication between the vasomotor area and higher centers. It controls cardiac acceleration, heat loss, and sympathetic inhibition.

Stimulation of the **motor cortex** (→ p. 268) elevates BP, dilates arteries in muscle, and constricts them in skin, splanchnic, and renal vessels.

Heart

Intrinsic. (1) **Heterometric autoregulation** alters the force of ventricular contraction by changing the length of the muscle fiber (→ pp. 38–39) according to the **Frank-Starling law of the heart**: the *stroke work of the heart* (*tension*) *is a function of the end diastolic volume* (*EDV*) (*length*). In essence, what goes into the heart must come out. (a) **Increased venous return** to the heart enlarges end diastolic volume (EDV, → p. 149, A 6) and stretches muscle fibers; cardiac output increases to compensate for the greater return and the original EDV is restored. (b) **Increased peripheral resistance**

reduces the ejection volume and increases end systolic volume (ESV). The ESV adds to the EDV, stretches muscle fibers, and increases cardiac output. Both BP and ESV are elevated. (c) Output of the two ventricles is equalized by this mechanism. If right ventricle (RV) output exceeds left ventricle (LV) output, more blood is delivered to the LV. The LV fibers stretch and output increases to compensate (→ pp. 169, 171).

The length-tension curve (→ pp. 38–41) for the RV is normally to the left of that for the LV. For the same EDV, RV stroke output is greater than LV, or, the RV generates the same stroke work as LV for a smaller initial fiber length. As a consequence, LA pressure normally is greater than RA; in atrial septal defect, this circumstance prevents unsaturated blood from the right atrium from entering the arterial circulation.

(2) **Homeometric autoregulation** allows force of ventricular contraction to change even though muscle fiber length remains constant. Heart differs from striated muscle because the heart can vary the position of its length-tension diagram. For example, increased peripheral resistance or venous return stretches muscle fibers. If the increase is of long duration, the fibers return to their original length while retaining the increased force of contraction. In essence, the length-tension diagram has been shifted to the left. Other influences on this property include autonomic nerves and cations (Ca^{2+}, K^+).

Contractility (inotropism) of the heart is defined by the force-velocity curve (→ pp. 38–41) for a given fiber length. Homeometric regulation alters contractility. A positive inotropic response follows stimulation of cardiac β-receptors. These responses are the result of a shift in the force-velocity diagram (→ p. 169, B 2).

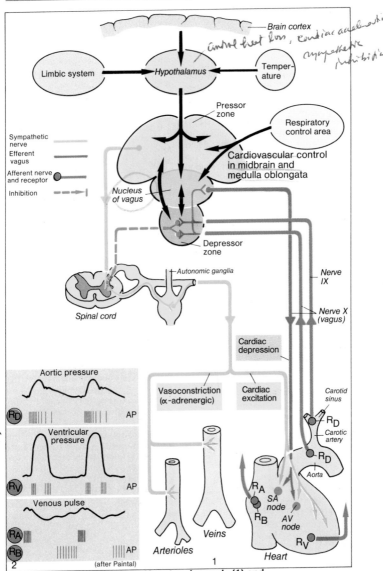

Brain cortex

control heart loss, cardiac acceleration sympathetic inhibition

Limbic system

Hypothalamus

Temperature

Pressor zone

Respiratory control area

Cardiovascular control in midbrain and medulla oblongata

Sympathetic nerve
Efferent vagus
Afferent nerve and receptor
Inhibition

Nucleus of vagus

Depressor zone

Nerve IX

Autonomic ganglia

Spinal cord

Nerve X (vagus)

Cardiac depression

Aortic pressure

R_D AP

Vasoconstriction (α-adrenergic)

Cardiac excitation

Carotid sinus

R_D

Carotic artery

R_D

Aorta

Ventricular pressure

R_V AP

Venous pulse

R_A
R_B AP

Arterioles

Veins

R_A
R_B
SA node
AV node

R_V

Heart

2   (after Paintal)   1

D. Neurohumoral regulation of heart and vessels (1) and action potentials (AP) in detectors (2)

Control of the Circulation – Heart

Neurohumoral Regulation

Autonomic nerves. (1) **Parasympathetic.** The SA node receives fibers chiefly from the right vagus ($\to$ p. 48), the AV node from the left vagus. Vagal activity, by releasing AcCh ($\to$ p. 52), suppresses the SA node and slows the heart rate ($\to$ p. 152) by (a) increasing MDP ($\to$ pp. 24, 153, A3) by as much as -30 mV and (b) decreasing the slope of diastolic depolarization ($\to$ pp. 24, 153, A2) (dV/dt of phase 4). With stronger vagal activity, hyperpolarization may become so great as to suppress automaticity completely. Excitability (dV/dt of phase 0, $\to$ p. 24) is reduced and decremental conduction ($\to$ p. 152) enhanced, especially in the AV node. The His bundles are not greatly influenced and are capable of taking over the pacemaker function (vagal escape rhythm). There is a small indirect negative inotropic response to the vagus caused by the lowered heart rate ($\to$ p. 169, B2). (2) **Sympathetic** ($\to$ pp. 48–57) activity and catecholamines increase dV/dt of phase 4 ($\to$ p. 152) and cause tachycardia. Since this effect is exerted on all areas of the heart, catecholamines can increase automaticity, excite ectopic pacemakers, and initiate arrhythmias. They have little effect on TP and MDP ($\to$ pp. 24, 153 A). Excitability is increased, and there is a strong positive inotropic response (p. 169, B2).

Reflexes ($\to$ E)

The cardiovascular system contains special receptors, which respond to changes in hemodynamics and in chemical composition of the blood. The **afferent** paths are predominantly in the vagus. There is also the carotid sinus nerve (Hering's nerve), which runs in the hypoglossal nerve (IX). (1) The **carotid and aortic sinus – depressant – reflex** ($\to$ **D, E**): an acute rise in BP in the range of 6.7 to 26.7 kPa ($=$ 50 to 200 Torr) stimulates stretch receptors in the arterial circuit. These fire during systole when the blood pressure (BP) is highest ($\to$ **D, 2, R_D**). They depress the vasomotor control area in the medulla and stimulate the vagus. The response is vasodilation, slowing of the heart rate, decreased contractility, and lower BP. The fall in BP is mainly the result of vasodilation in splanchnic vessels, but also in muscle and skin. The reflex can adapt: if BP is sustained at a higher level, activity of the stretch receptors declines and the reflex is deactivitated. The receptors respond to mean pressure, pulse pressure, and the rate of rise of systolic pressure. For a given mean pressure, the greater the pulse pressure or the rate of rise, the greater is the reflex depressant response. (2) The **atrial (Bainbridge) – excitatory – reflex**: increased atrial filling and an elevated central venous pressure produce a reflex tachycardia ($\to$ **E**). The reflex is sensitive to a 10% change in volume. The receptors fire during the v wave ($\to$ **D, 2**; p. 148) of atrial filling. The response is influenced by the original heart rate: if slow, a tachycardia is produced; if fast, the rate may be slowed. Significance of this reflex has not been established. (3) **Chemoreceptors** in the arterial limb of the vascular system ($\to$ pp. 92, 93) respond principally to anoxia (low pO_2), but also to elevated arterial CO_2 and decreased pH. They occur in the aorta and carotid sinuses and are most sensitive at lower BP (5.3 to 10.7 kPa $=$ 40 to 80 Torr), as might occur in hemorrhage or shock. The reflex involves excitement of the respiration and vasomotor control areas and intense vasoconstriction. This response is supported by the intrinsic response of the vasomotor area, which elicits a general vasoconstriction and elevation of BP when the local concentration of CO_2 in the medulla rises.

peripheral chemoreceptors – carotid, and
                                    aortic bodies
Central " – The medulla of p conc.
                                    pH.

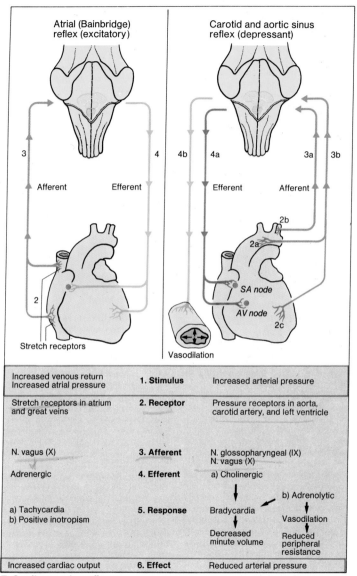

Atrial (Bainbridge) reflex (excitatory)		Carotid and aortic sinus reflex (depressant)

Increased venous return Increased atrial pressure	**1. Stimulus**	Increased arterial pressure
Stretch receptors in atrium and great veins	**2. Receptor**	Pressure receptors in aorta, carotid artery, and left ventricle
N. vagus (X)	**3. Afferent**	N. glossopharyngeal (IX) N. vagus (X)
Adrenergic	**4. Efferent**	a) Cholinergic
a) Tachycardia b) Positive inotropism	**5. Response**	Bradycardia → Decreased minute volume b) Adrenolytic → Vasodilation → Reduced peripheral resistance
Increased cardiac output	**6. Effect**	Reduced arterial pressure

E. Cardiovascular reflexes

Myocardial Function I

Myocardial function is best expressed in terms of developed pressure and cardiac ejection; in clinical terms, these refer to ventricular end diastolic pressure (EDP) or volume (EDV) (right or left), arterial pressure (BP), and cardiac output (CO). Myocardial function is influenced by:

1. The **preload** (EDV) which determines the length of the heart fiber before contraction.

2. The **afterload** or aortic output impedance (BP).

3. **Contractility.**

A reduced preload (orthostasis, hypovolemia, pericardial tamponade) may reduce CO by limiting fiber length. An increased preload (hypervolemia) may lead to pulmonary edema by raising LEDP. An increased afterload (at constant EDV) requires more energy to raise the pressure to the point of opening the aortic valve (isometric) ($\rightarrow$ p. 148) so that less energy is available for ejection of blood (isotonic) and the stroke volume (CO) falls.

In Fig. **A**, pressure/volume curves for active (green and purple curves) and resting (blue curve) tension are shown. When the heart contracts against a closed aortic valve, its volume is constant (isometric); the maximum pressure achieved is a function of fiber length (EDV) ($\rightarrow$ **A, 1** vertical arrows, green curves). When the heart contracts against an open aortic valve, its tension may be held constant (isotonic), but its fibers shorten; the maximum blood volume that can be ejected is shown by the horizontal arrows and the purple curve. A single contraction of the left ventricle starts at A on the passive (resting) tension curve. EDV = 125 ml, filling pressure c. 1 kPa (= 7.5 Torr). Systolic contraction is isometric until the aortic valve opens (D). Beyond D, BP continues to rise and blood is ejected until S is reached. Stroke volume (SV) is the difference in volume between D and S or 125 – 60

= 65 ml. The maximum ventricular contraction pressure for a given SV is established by the line joining M, the isotonic maximum for A, with T, the isometric maximum. BP falls during diastole. The aortic valve closes at K. The end systolic volume (ESV) is shown at V. Diastolic filling follows the passive tension curve (V-A). At A, the next contraction begins. The enclosed space (yellow area) represents systolic work; the triangular orange area under the diagram represents diastolic work. The whole sequence requires 0.9 s at a heart rate of 66 beats/min.

Increased ventricular filling ($\rightarrow$ **C, 1**). When venous return to the heart increases, EDV is greater (A_1) (180 ml). If afterload remains constant, the aortic valve opens at the same pressure (D_1), but the maximum ventricular pressure is determined by new isotonic (T_1) and isometric (M_1) maxima. Stroke volume almost doubles ($D_1 - S_1$) but ESV hardly changes. More work has been performed.

Increased peripheral resistance ($\rightarrow$ **C, 2**). A sudden increase in afterload reduces the CO and initiates a number of adaptive changes to restore the original CO. (In Fig. **C, 2**, one intermediate stage is shown marked with D', S', V', SV', ESV', and a striped orange area). Because of the increased BP, the intraventricular pressure must rise to a higher level (D') before the aortic valve opens. Stroke work is thus dissipated in raising pressure at the expense of CO. The stroke volume falls (SV'), and the ESV increases (ESV'). The ESV is added to the normal diastolic filling and the EDV enlarges ($\rightarrow$ **A, 2**). The next contraction can therefore generate more tension, and the stroke volume increases. This process continues, the stroke volume increasing with each contraction until the original SV is again achieved (SV = SV_2), but at an elevated ESV (ESV_2) and EDV ($ESV_2 + SV_2$).

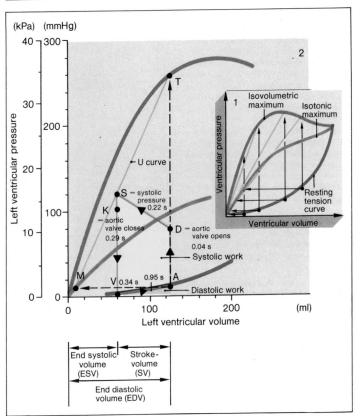

A. Cardiac work diagram (left ventricle)

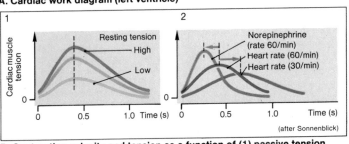

(after Sonnenblick)

B. Contraction velocity and tension as a function of (1) passive tension and (2) heart rate

Myocardial Function II

Inotropism ($\to$ **C, 3**). Unlike striated muscle, the heart is not fully activated during contraction ($\to$ p. 34); its Ca^{2+} stores are smaller, and its activation is influenced by variations in the transmembrane availability of Ca^{2+}. Agents such as norepinephrine or digitalis increase Ca^{2+} inward current and augment contractile force without major changes in the AP. There is no effect in the **resting** length-tension relationship, but a greater **active** tension is produced for any given fiber length. The curves for isotonic and isometric maxima are shifted to the left. The fibrils shorten faster, duration of systole is less, rate of rise to peak tension is faster, and peak tension is greater ($\to$ **B2**). Figure **C, 3** shows two separate inotropic responses, but in reality they overlap to produce increased SV, systolic BP, and rate.

1. If the SV does not change, systolic ejection pressure (D_3/S_3) and pulse pressure rise.

2. If systolic BP does not change, SV increases (D-S_4).

Venous Pressure and Flow

The **venous circulation** is responsible for returning blood to the heart. It is characterized by:

1. **Low venous pressure** (VP). VP falls from c. 2 kPa ($=$ 15 Torr) at the capillaries to c. 0.8 kPa ($=$ 6 Torr) in the extrathoracic large veins. VP is lowest in the right atrium (RA); the pressure must be raised by the right ventricle to enable the blood to flow through the lungs.

2. **High compliance** ($\to$ p. 142).

3. **Large blood volume** (BV) (c. 60% of total, $\to$ p. 141). Venous blood flow occurs along small pressure gradients so that even the smallest variation in resistance affects return flow. The upright posture makes great demands on energy for venous return. The main drive comes from the arteriovenous pressure gradient (P_{art}-P_{ven}) across the resistance vessels ($=$ vis a tergo). The effect of gravity retards venous return; the large compliance of the veins allows marked distension without large increases in VP. This effect can result in venous pooling of a large fraction of the total blood volume ($\to$ p. 146). Venous return is assisted by skeletal muscles and venous valves and by respiration that produces cyclic variations in intrapleural pressure ($\to$ p. 146). Venous return is greatest on inspiration when the intrapleural pressure falls from -0.35 kPa ($= -2.6$ Torr) to -0.8 kPa ($= -6$ Torr). Simultaneously, descent of the diaphragm into the abdomen raises intra-abdominal pressure and increases the gradient to the thorax. Both of these influences further venous return to the right atrium.

If the circulatory system were at complete rest with zero blood flow, the pressure in the system ($=$ mean circulatory pressure $=$ MCP $=$ 1 kPa [$=$ 7.5 Torr]) would be determined by blood volume and vascular tone. From this condition, when the heart starts beating again, the cardiac output adds blood to the arterial volume by removing blood from the venous volume; P_{art} increases and P_{ven} decreases. The arteriovenous gradient rises until it is sufficient to drive the cardiac output across the resistance vessels. Thus, cardiac output $=$ venous return (on average). In a number of conditions, an increase in MCP allows an increase in venous return and in cardiac output. An increased MCP can be achieved in the short term by an increase in venous tone or in the long term by an increase in absolute blood volume. Short-term increases in *effective* blood volume may follow vasoconstriction in the body's blood reservoirs such as the skin, liver, lungs, and spleen ($\to$ p. 142).

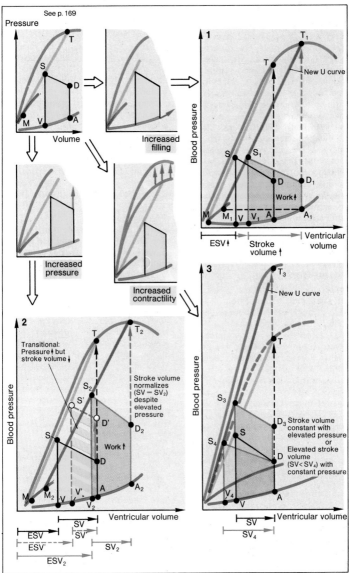

See p. 169

Pressure

Volume

Increased filling

Increased pressure

Increased contractility

1

Blood pressure

Ventricular volume

ESV ↑ Stroke volume ↑

New U curve

Work ↑

3

Blood pressure

New U curve

Stroke volume constant with elevated pressure
or
Elevated stroke volume (SV< SV$_4$) with constant pressure

Ventricular volume

SV

SV$_4$

2

Blood pressure

Transitional: Pressure ↑ but stroke volume ↓

Stroke volume normalizes (SV — SV$_2$) despite elevated pressure

Work ↑

Ventricular volume

SV
SV'
ESV
ESV'
SV$_2$
ESV$_2$

C. Cardiac work curve: (1) increased filling (preload), (2) increased pressure (afterload), and (3) increased contractility

Hypovolemic Shock

Shock is a syndrome in which the **effective** blood volume is reduced until perfusion of tissues becomes insufficient. It is characterized by low blood pressure (BP), tachycardia, weak pulse, oliguria, and vasoconstriction. Shock is therapeutically reversible but it may progress to an irreversible stage if therapy is inadequate. There are many causes of shock, including central causes affecting the heart (terminal heart failure, pulmonary embolus, heart infarct, arrhythmia) and peripheral causes affecting blood vessels (hemorrhage, diarrhea, anaphylaxis, heat stroke, or exhaustion). In the former, stroke output of the heart is affected even though venous return is usually sufficient. In the latter, circulating blood volume is reduced, and venous return is insufficient. Compensatory responses are stimulated to restore (a) perfusion **pressure** and (b) perfusion **volume**.

Pressure compensations. The fall in BP stimulates baroreceptors ($\rightarrow$ pp. 162–167) and activates both neural and humoral sympathetic discharge ($\rightarrow$ p. 162), which exerts positive intotropic (force of contraction) and chronotropic (tachycardia) actions. When circulating volume and venous return are low, stroke output cannot increase materially. In this case, minute output is increased by tachycardia; the pulse is weak. Vasoconstriction in skin, muscles, kidney, and splanchnic circulations shunts blood to brain and coronary vessels. Vasoconstriction can be so intense as to cause acute renal failure and intestinal sloughing. When there is also volume depletion, warming the patient before restoring fluid will reduce vasoconstriction and may worsen the patient's state.

When BP falls to low levels, chemoreceptors are activated. Cerebral ischemia and elevated P_{CO_2} increase adrenergic discharge. In the kidney, renin release generates angiotensin II ($\rightarrow$ p. 138). This vasoconstricting substance has both local intrarenal and systemic effects, although their full significance has not been clarified.

Volume compensations. To increase volume, salt and water must be retained ($\rightarrow$ p. 118). Reduced GFR and RPF in the kidney contribute to such retention. In addition, renal release of renin and subsequent generation of angiotensin II stimulates aldosterone release from the adrenal cortex. The stress of shock releases ACTH from the anterior pituitary which supports stimulation of aldosterone synthesis and release. The posterior pituitary is stimulated to release ADH, which enhances water retention ($\rightarrow$ p. 126). Oliguria or even anuria occurs.

Arteriolar constriction and reduced BP lower P_c in the capillaries to such an extent that net water resorption predominates ($\rightarrow$ p. 144). By this means, as much as 1 l of fluid can be restored to the vascular system in 1 h. The hemodilution that occurs lowers blood viscosity, hematocrit, and plasma oncotic pressure. Consequences of hemorrhage: 10% blood volume loss – increased pulse rate, decreased pulse pressure; 20% loss – in addition to above, cardiac output declines; 30% loss – in addition, peripheral resistance increases; 40% loss – in addition, blood is shunted from skin, muscle, renal and splanchnic circulations; 50% loss is fatal.

Irreversible shock may develop by the following mechanisms:

1. Blood volume $\downarrow$ $\rightarrow$ vasoconstriction $\rightarrow$ disturbance of tissue metabolism $\rightarrow$ vessel damage $\rightarrow$ vasodilation and filtration into interstitium $\uparrow$ $\rightarrow$ blood volume $\downarrow\downarrow$ ...

2. Vasoconstriction and low BP $\rightarrow$ blood flow velocity $\downarrow$ $\rightarrow$ blood viscosity $\uparrow$ $\rightarrow$ resistance $\uparrow$ $\rightarrow$ blood flow $\downarrow\downarrow$ ...

3. BP $\downarrow$ $\rightarrow$ O_2 deficiency and acidosis $\rightarrow$ damage of myocardium $\rightarrow$ cardiac output $\downarrow$ $\rightarrow$ BP $\downarrow\downarrow$ ...

4. BP $\downarrow$ $\rightarrow$ tissue metabolism $\downarrow$ $\rightarrow$ vessel damage $\rightarrow$ blood clotting $\rightarrow$ clotting factors $\downarrow$ $\rightarrow$ blood los $\rightarrow$ BP $\downarrow\downarrow$ ...

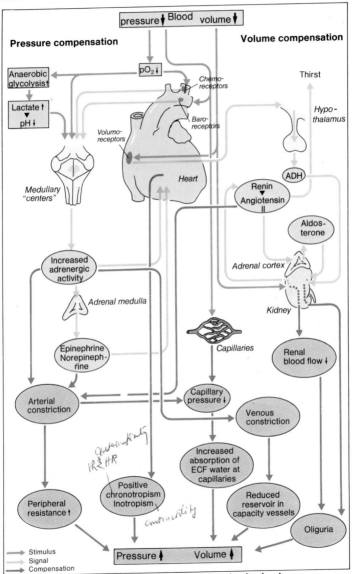

pressure↓ Blood volume↓

Pressure compensation

Volume compensation

pO₂↓

Chemo-receptors

Anaerobic glycolysis↑

Lactate↑
pH↓

Thirst

Hypo-thalamus

Baro-receptors

Volumo-receptors

Heart

ADH

Renin
Angiotensin II

Medullary "centers"

Aldosterone

Increased adrenergic activity

Adrenal cortex

Adrenal medulla

Kidney

Epinephrine Norepinephrine

Capillaries

Renal blood flow↓

Arterial constriction

Capillary pressure↓

Venous constriction

Positive chronotropism Inotropism

Increased absorption of ECF water at capillaries

Peripheral resistance↑

Reduced reservoir in capacity vessels

Oliguria

Pressure↑ Volume↑

Stimulus
Signal
Compensation

A. Compensatory responses in reversible hypovolemic shock

Coronary Blood Flow and Myocardial Metabolism

The coronary arteries are the major blood supply to the heart, carrying O_2 and nutrients. The primary driving force for coronary blood flow ($\rightarrow$ p. 148) is the aortic diastolic pressure. Modulation of flow is achieved by varying coronary artery resistance. Autoregulation ($\rightarrow$ p. 162) plays a role in this control. Increased metabolic activity of the heart initiates a reduction in coronary resistance, which allows increased blood flow and increased delivery of O_2 and nutrients to the heart muscle. Intraventricular pressure also has an important effect on coronary flow, especially in the left ventricle (LV) ($\rightarrow$ p. 148). In systole, pressure in the ventricle is high and impedes flow through the vessels in the ventricular wall. Since the pressure gradient is distributed across the wall, vessels at the endocardium are influenced more than those at the epicardium. Vessels in the right ventricle (RV) are not affected as greatly because the intraventricular pressure is lower. Maximum coronary flow takes place during diastole. In tachycardia, diastole is shortened and tends to reduce coronary flow, but this effect is compensated for by reduced coronary resistance, which follows the increased metabolic demands of the muscle. The coronary vessels contain predominantly β-adrenergic receptors ($\rightarrow$ pp. 54, 162) which respond to adrenergic stimulation with vasodilation. The larger coronary vessels also contain α-adrenergic receptors for vasoconstriction and show mixed responses to adrenergic stimulation. Coronary blood flow ($\rightarrow$ **A**) is c. 250 ml/min in the resting heart but may increase up to four times on exercise ($\rightarrow$ p. 46). The arteriovenous O_2 difference (AVDO$_2$) is c. 0.12 ml O_2/ml blood at rest and may increase to 0.15 on exercise. Since O_2 extraction is high and AVDO$_2$ cannot be greatly increased (venous O_2 in coronaries is low), O_2 needs of the heart

must be met by increasing O_2 delivery. coronary blood flow. O_2 utilization = (coronary flow) $\times$ (AVDO$_2$) = 30 ml O_2/min at rest and over 90 ml O_2/min in exercise.

The heart is versatile in its use of **energy sources** ($\rightarrow$ **A**); it can use fatty acids, glucose, and lactate. In contrast to skeletal muscle, which shows a net release of lactate ($\rightarrow$ p. 44), the heart can extract lactate from the blood.

Efficiency of the heart is a measure of work accomplished relative to the total energy used. The heart pump has an efficiency at rest of c. 15%. However, O_2 utilization does not correlate well with work done because there are two components to cardiac work: work/beat = stroke work = stroke volume (SV) $\times$ mean aortic pressure. O_2 requirement is much greater for the pressure component of work than for the volume component. An increase in SV at constant aortic pressure at a requires only a small increment of O_2 utilization. In contrast, an increase of aortic pressure at a constant SV requires a large increment. For the same work output, there can be several combinations of SV and aortic pressure and therefore several different values for O_2 utilization. Efficiency of the heart is greatest when the volume component makes up the greater fraction of total work as it does during exercise. A good correlation is found in most instances between O_2 utilization and the area under the systolic pressure curve of the LV (= tension-time index).

Blood flow may be measured by several techniques:

1. Plethysmography ($\rightarrow$ **B**): an extremity is placed in a chamber and the venous outflow is occluded. Blood flow can be measured as a rate of volume change.

2. Flow through a magnetic coil can be measured by the magnitude of the induced current ($\rightarrow$ **C**).

3. An **indicator gas** ($\rightarrow$ **D**) (argon, krypton) is inspired for a fixed time, and concentrations in artery and vein are determined repeatedly. From the arterio-venous indicator concentration difference (mean C_{art} – mean C_{ven}), the time required to reach equilibrium and the equilibrium concentration (concentrations in artery, vein, and tissue are equal: C_e), the blood flow V can be calculated ($\rightarrow$ **D**).

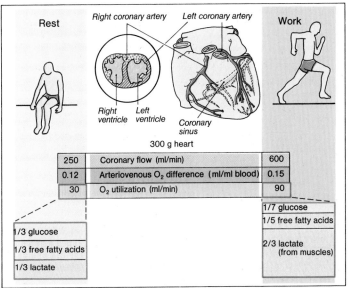

250	Coronary flow (ml/min)	600
0.12	Arteriovenous O_2 difference (ml/ml blood)	0.15
30	O_2 utilization (ml/min)	90

1/3 glucose

1/3 free fatty acids

1/3 lactate

1/7 glucose

1/5 free fatty acids

2/3 lactate (from muscles)

A. Blood flow, substrate, and O_2 utilization in heart muscle

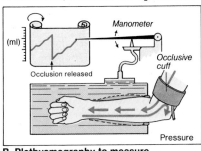

B. Plethysmography to measure blood flow

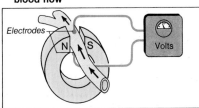

C. Magnetic flowmeter to measure blood flow

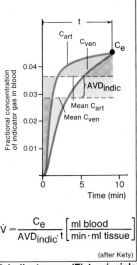

$$\dot{V} = \frac{C_e}{AVD_{indic} \cdot t} \left[\frac{ml\ blood}{min \cdot ml\ tissue} \right]$$

(after Kety)

D. Indicator gas (Fick principle) (Kety and Schmidt)

Fetal Circulation

The fetal circulation is designed to meet the needs of a rapidly growing organism existing in a state of relative hypoxia. Two factors allow a sufficient O_2 supply to the fetus even when placental blood is not fully oxygenated: (1) fetal hemoglobin (Hb) has a greater affinity for O_2 than adult Hb and can be more fully saturated at the same pO_2 ($\rightarrow$ p. 89, C); (2) fetal tissues are more resistant to hypoxia than adult tissues. Whatever O_2 is supplied to the fetus is optimally distributed so that tissues with the greater need receive the greater amounts. Nonfunctioning tissues such as lung and liver are largely bypassed. Four peculiarities of the fetal circulation must be noted: (1) the **placenta** functions as a fetal lung, GI tract and kidney; (2) the **ductus venosus** carries blood from the placenta directly to the right atrium (RA), largely bypassing the liver; it mixes with blood returning from the lower fetal circulation in the inferior vena cava (IVC); (3) the **foramen ovale** is an opening between the two atria, which allows blood entering the RA from the placenta to be largely diverted to the LA; and (4) the **ductus arteriosus** connects the pulmonary artery to the aorta. It allows mixing of right and left ventricle blood at a point beyond the origin of vessels that supply the brain.

Blood is delivered to the fetus from the placenta at 80% O_2 saturation. After mixing with blood from the legs it reaches the RA via the inferior vena cava, mixing with blood from the upper torso and brain (superior vena cava). Mixing, however, is incomplete; instead, cross-over of streams of blood with differing levels of O_2 saturation takes place. A fold in the vena cava is designed to separate these streams; the larger stream with higher O_2 saturation reaches the LA, the other the RV. Pressures within the atria are approximately equal and further mixing is minimal. From the RV, because of the high pulmonary arterial resistance in the fetus, only one third of the output reaches the lungs via the pulmonary artery. The blood has low saturation and is supplied for nutrient purposes, not for gaseous exchange. The remaining two thirds of the blood reaches the aorta via the ductus arteriosus. From the LV, blood with the greatest O_2 saturation is ejected. One third of the output reaches the brain. The remaining two thirds mixes with blood from the ductus arteriosus. Part of this volume supplies the lower torso, and the remainder returns via the umbilical artery to the placenta for exchange with maternal blood. The umbilical vein carries blood at higher O_2 saturation than the umbilical artery.

At **birth**, after separation from the placental circulation, peripheral resistance and aortic pressure rise. Gasping respiratory movements expand the lungs and pulmonary resistance falls so that pulmonary blood flow increases. The greater return of blood to the LA raises its pressure above that in the RA and closes the valve in the foramen ovale. Aortic pressure exceeds pulmonary pressure and flow in the ductus arteriosus is reversed. The higher O_2 tension is a stimulus for functional closure of the ductus arteriosus within 2 days after birth; anatomic closure may take several weeks. The adult state, in which the two ventricles pump in series rather than in parallel, is achieved by these closures.

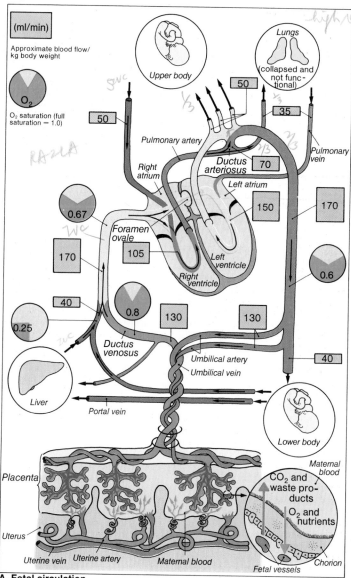

A. Fetal circulation

Heat Metabolism

The body temperature of lower animals varies with the temperature of the environment. The body temperature of cold-blooded (**poikilothermic**) animals varies over a wide range. Higher animals, including man, are **homeothermic** and maintain body temperature within a narrow range. Homeothermic animals that hibernate are poikilothermic during hibernation. Even in homeothermic animals, the range of temperature variation differs; the core temperature, which is measured deep within body cavities, is relatively constant at $37° \pm 0.5$ °C, but the skin and the extremities can vary their temperature over several degrees ($\to$ p. 181, A). Maintenance of a constant temperature, therefore, requires a steady state between heat production and heat loss.

Heat production ($\to$ **A**) is a function of energy metabolism ($\to$ pp. 182–185). At rest, the basal heat is generated by the inner organs (to more than 50% of total heat production) and muscles and skin (together up to 20%); during exertion, heat production increases absolutely, and muscular activity can account for 90% of the total. Shivering is a supplementary muscular mechanism for heat generation ($\to$ p. 181, D).

Heat loss ($\to$ **B**). "Core" heat produced by metabolic processes is transferred to the blood and is transported to the body surface when the skin temperature is below the core temperature. Control of body temperature is primarily dependent on circulation in the skin ($\to$ p. 180).

Heat transmission ($\to$ **B**).

(1) **Radiation** is the transfer of heat from one body to another. The amount of heat transferred by this means is a fourth-power function of the temperature of the radiating body. The temperature of the air through which heat radiates has little influence on heat transfer; heat input to a body from the sun or from a heat lamp takes place in spite of intervening cold air or vacuum.

(2) **Conduction** is heat exchange by contact between two bodies at different temperatures ($\to$ p. 181, B). Rate of heat transfer depends on the thermal gradient (temperature difference) between the two bodies. Heat can be given off to the surrounding air by conduction. **Convection** ($\to$ **B, 2**) moves masses of air or fluid by virtue of an environmental temperature gradient and aids conduction. A breeze is cooling because it replaces the mass of warm and wet air around the body with cooler and dryer air. Convection enhances heat transfer by conduction by maintaining a large thermal gradient.

(3) Radiation and conduction are insufficient to prevent warming of the body during heavy exertion. Under these circumstances, heat loss is enhanced by **evaporation of water** ($\to$ **B, 3; C**), as in perspiration. Fluid reaches the skin surface by (a) diffusion and (b) via the sweat glands (innervation $\to$ p. 51). Evaporation cools the skin below environmental temperatures. At an environmental temperature above 36 °C, heat loss can be effectively achieved only by water evaporation. At temperatures much above 37 °C ($\to$ **C**), the body gains heat from the environment by radiation and conduction; in this case, perspiration becomes profuse to maintain the balance between heat uptake and loss. One liter of perspiration can account for a heat loss of up to 2428 kJ ($= 580$ kcal).

By this mechanism *dry heat* (desert) can be tolerated at temperatures much higher than the body temperature as long as water and salt lost with the sweat are replaced. If the environmental air is highly water saturated (tropic jungle) air temperatures above c. 34 °C cannot be tolerated because evaporation is impossible.

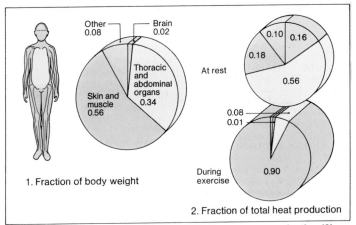

1. Fraction of body weight

At rest

During exercise

2. Fraction of total heat production

A. Body organs: Fractions of body weight (1) and of heat production (2)

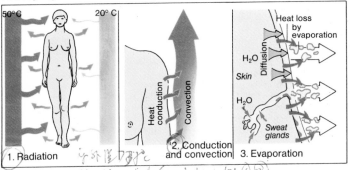

1. Radiation

2. Conduction and convection

Heat conduction   Convection

Heat loss by evaporation

H₂O

Skin

Diffusion

H₂O

Sweat glands

3. Evaporation

B. Mechanisms of heat loss

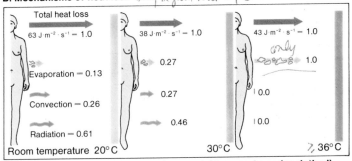

Total heat loss
63 J·m⁻²·s⁻¹ = 1.0

Evaporation = 0.13

Convection = 0.26

Radiation = 0.61

Room temperature 20°C

38 J·m⁻²·s⁻¹ = 1.0
0.27
0.27
0.46
30°C

43 J·m⁻²·s⁻¹ = 1.0
1.0
0.0
0.0
36°C

C. Heat loss at rest at different environmental temperatures (unclothed)

Thermoregulation

Thermoregulation maintains a constant body temperature in spite of variations in heat uptake, generation, and loss. A change in one of these processes must be compensated for by a change in the others. On average, body temperature is held at 37 °C with a diurnal variation of ±0.5 °C. For most people, the maximum body temperature is achieved in the evening (5 : 00 PM) and the minimum at 3 : 00 AM. This daily cycle is controlled by an internal "biologic clock", which regulates other diurnal biologic cycles (→ p. 272). A longer sequence corresponds with the menstrual cycle; the temperature is lower in the first 2 weeks and may rise by a full 1 °C in the second 2 weeks (→ p. 243).

Thermoregulation is centered in the **hypothalamus** (→ p. 270). Heat-sensitive thermoreceptors respond to changes in core temperature, which are transmitted by the blood stream. Additional information is received from thermoreceptors in the skin (→ p. 256) and spinal cord. The hypothalamus integrates these data and initiates a variety of responses to counteract deviations from normal body temperature:

1. If **body temperature rises,** heat loss (→ p. 178) is enhanced by (a) increased perspiration and (b) increased blood flow to the skin, especially to the extremities. Since the arms and legs have a large surface area, they are more effective in losing heat than other parts of the body. The increased flow allows a greater volume of blood and a greater quantity of heat per time unit to be carried to the skin for heat exchange. The increased blood flow has another effect: heat is normally transmitted between adjacent arteries and veins by countercurrent exchange (→ **B**; p. 116). At slow rates of blood flow, this process conserves heat. At high rates of flow, exchange is reduced, and more heat is carried to the periphery.

2. If **body temperature falls,** (a) heat loss by perspiration and from the skin is reduced and (b) heat production is stimulated. Production can be increased up to fourfold by voluntary muscle activity and by shivering (→ **D**). In infants, brown fat, located in the shoulders and back, is a further source of heat. Its high rate of metabolism generates heat; it is in effect an electric blanket. Some animals have brown fat, but it is not found in adult humans. These mechanisms permit body temperature to be maintained at a constant value in spite of environmental temperature variations from 0° to 50 °C and, in extremes, to over 65 °C, as in a sauna. Body temperature can be influenced by behavior, such as choice of appropriate dress, heating of living quarters, and seeking shade in summer. These actions assume considerable importance at temperatures beyond the extremes of 0° and 50 °C when they are the only means of maintaining body temperature (→ **C**).

Continuing exposure to high temperatures, as in the tropics, results eventually in **acclimatization**: thirst and water intake increase and sweat production increases, but salt content of the sweat decreases.

Fever is produced by circulating **pyrogens**, which disturb hypothalamic regulatory mechanisms. The body "thermostat" is set at a higher value, and processes for heat production and loss are adjusted to maintain a higher temperature. As the fever develops, the body is relatively cold and chills and shivering accompany the rising temperature. Conversely, as fever resolves and temperature declines, the body is relatively hot, and vasodilation and perspiration occur in spite of a falling temperature.

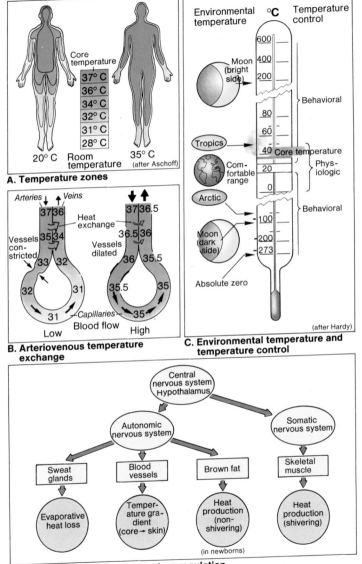

A. Temperature zones

Core temperature
37° C
36° C
34° C
32° C
31° C
28° C

20° C Room temperature   35° C

(after Aschoff)

B. Arteriovenous temperature exchange

Arteries ↓ ↑ Veins

Heat exchange

Vessels constricted

Vessels dilated

Capillaries

Blood flow

Low   High

C. Environmental temperature and temperature control

Environmental temperature   °C   Temperature control

600
400
200

Moon (bright side)

Behavioral

80
60
40   Core temperature
20   Comfortable range
0

Tropics

Comfortable range   Physiologic

Arctic

Behavioral

100
200
273

Moon (dark side)

Absolute zero

(after Hardy)

D. Neural influences on temperature regulation

Central nervous system Hypothalamus

Autonomic nervous system   Somatic nervous system

Sweat glands   Blood vessels   Brown fat   Skeletal muscle

Evaporative heat loss   Temperature gradient (core → skin)   Heat production (non-shivering)   Heat production (shivering)

(in newborns)

Nutrition

The need for energy arises because living matter is a thermodynamically unstable system that will run down unless energy is continuously added. Furthermore, living matter is constantly engaged in performing various kinds of work: motion, synthesis, active transport, heat production, each of which requires energy input. Energy for these functions is derived from foodstuffs. Proper nutrition, therefore, requires adequate energy sources, a minimum quantity of protein containing all of the essential amino acids, carbohydrates, essential fats, minerals, vitamins, and trace elements. In addition, water must be available.

Basic daily energy needs for a 70 kg man at rest are c. 8,400 kJ ($=$ 2,000 kcal) (1 cal $=$ 4.18 J [joule]) ($\rightarrow$ p. 184). Additional energy required for daily activity may range from 1,700 kJ ($=$ 400 kcal) for light office work to c. 10,500 kJ ($=$ 2,500 kcal) for heavy work, such as mining. Women at rest require less energy (6,700 kJ), but the requirement rises during pregnancy and in the postpartum period. Children require a lower **absolute** caloric intake because of their size, but a larger **relative** intake is needed to promote growth.

Energy needs are covered primarily by the three basic nutrients: proteins, carbohydrates, and fats ($\rightarrow$ **A**). The *minimum* **protein** intake is c. 0.5 g protein/kg body weight. This intake is necessary to balance the output of endogenous protein lost as a result of degradative reactions. However, for normal activity, about twice this value is needed (*functional minimum*), of which half should be in the form of animal proteins to furnish the essential amino acids. Many plant proteins are deficient in one or more of the essential amino acids, and their nutritive value is correspondingly less. (An **essential amino acid** is one that cannot be synthesized by the organism.)

Most of the energy requirements are satisfied by **carbohydrates** and **fats**, which are largely interchangeable. The energy contribution of carbohydrate, normally 60% of the total, can fall to 10% before metabolic disturbances appear. Fats are superfluous, provided there is a supply of essential fats (e.g., linoleic) and of fat-soluble vitamins (A, D, E, K). The fat intake, normally 25% of the total, can increase when energy needs mount. Fats are the most compact energy source because of their high yield ($\rightarrow$ **A**).

Minerals must be included in the diet to maintain health. Na^+ and K^+ occur in sufficient quantities in the diet; their renal excretion is adjusted to their intake. Dietary deficiency of minerals can have serious consequences, e.g., calcium (req.: c. 1 g/day), iron (req.: 10 mg/day for males, 15 mg/day for females), and iodine (req.: 0.2 mg/day). **Trace elements** are also needed for a balanced diet; Al, Br, Cr, Cu, Mn, Mg, Mo, Zn.

Vitamins are organic substances not synthesized de novo by the organism. They are essential for life and catalyze a number of biochemical reactions. They are present in the body in low concentrations. Characteristic disturbances are associated with vitamin deficiency, e.g., night blindness (vitamin A), scurvy (vitamin C), rickets (vitamin D), megaloblastic anemia (folic acid), pernicious anemia (vitamin B_{12}), beri beri (vitamin B_1), pellagra (nicotinic acid) and blood clotting disorders (vitamin K).

Intoxication with vitamin A results in diseases of the skeleton (painful periosteal proliferation); overdosage of vitamin D leads to disturbances of Ca^{2+} homeostasis (hypercalcemia, metastatic calcification, etc., $\rightarrow$ p. 236).

Intestinal absorption of the majority of the vitamins is a carrier-mediated (partly active) process and takes place mainly in jejunum and ileum ($\rightarrow$ also pp. 64, 194). Overdosage (mega-vitamin therapy) is often ineffective because absorption is saturated.

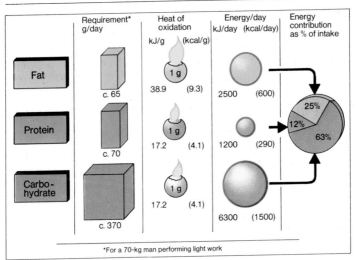

	Requirement* g/day	Heat of oxidation		Energy/day		Energy contribution as % of intake
		kJ/g	(kcal/g)	kJ/day	(kcal/day)	
Fat	c. 65	38.9	(9.3)	2500	(600)	25%
Protein	c. 70	17.2	(4.1)	1200	(290)	12%
Carbo-hydrate	c. 370	17.2	(4.1)	6300	(1500)	63%

*For a 70-kg man performing light work

A. Energy requirements and sources

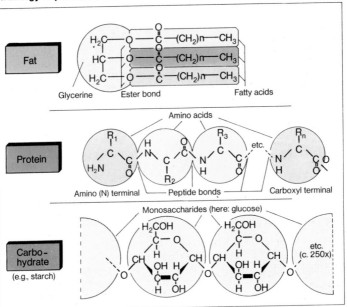

Fat

H_2C—O—$\overset{O}{\underset{||}{C}}$—$(CH_2)n$—$CH_3$

HC—O—$\overset{O}{\underset{||}{C}}$—$(CH_2)n$—$CH_3$

H_2C—O—$\overset{O}{\underset{||}{C}}$—$(CH_2)n$—$CH_3$

Glycerine   Ester bond   Fatty acids

Protein

Amino acids

H_2N ... R_1, C, C=O ... $\overset{H}{N}$, C, R_2, C=O ... R_3, C, C=O ... etc. ... $\overset{R_n}{N}$, C, C=O

Amino (N) terminal —— Peptide bonds —— Carboxyl terminal

Carbo-hydrate (e.g., starch)

Monosaccharides (here: glucose)

H_2COH ... H_2COH ... etc. (c. 250x)

B. Chemical structure of basic nutrients

184   Nutrition and Digestion

Energy Metabolism

Metabolism of foodstuffs results in physical work and in heat production. At rest and under standardized conditions, heat production predominates; muscular activity is limited chiefly to the heart and respiratory muscles. At constant body temperature, heat production corresponds to energy metabolism, which determines the **basal metabolic rate** (BMR). The BMR as measured, however, is not truly basal; a lower metabolic rate can be measured during sleep.

Calorimetry. A gram **calorie** is the amount of heat required to warm 1 g water from 15 °C to 16 °C ($\Delta 1$ °C). A kcal is 1,000 cal or 4.185 kJ ($\rightarrow$ p. 182). When food is metabolized in the body, it liberates the same energy as when food is burned outside the body. This energy can be measured by **direct calorimetry** in the bomb calorimeter ($\rightarrow$ **A**), a metal vessel contained in an envelope of water. When substances are burned in the calorimeter, the change in temperature of the water is a measure of the calories produced (the physical caloric value, C_{PHY}). In the organism, fats and carbohydrates are **completely** oxidized to CO_2 and H_2O so that the available caloric value (C_{AV}) and C_{PHY} are identical. Proteins are incompletely oxidized to CO_2 and H_2O; they form urea and other residual molecules as end products; for proteins, C_{AV} is less than C_{PHY}. The caloric value of carbohydrates = 17.2 kJ/g (= 4.1 kcal/g) and of fat = 38.9 kJ/g (= 9.3 kcal/g). For protein, C_{PHY} = 24 kJ/g (= 5.7 kcal/g) and C_{AV} = 17.2 kJ/g (= 4.1 kcal/g) ($\rightarrow$ p. 182, A).

Energy production may also be determined by **indirect calorimetry**. In this case, measurement is made of all the metabolic end products (CO_2, O_2, urea, etc.) or of the utilization of O_2 ($\dot{V}_{O_2}$), which is proportional to energy liberation ($\rightarrow$ p. 82). For exact measurements by this technique, it is

necessary to establish the **caloric equivalent** (CE) for each foodstuff being oxidized. The average CE = 20.2 kJ/l O_2 used (= 4.82 kcal/l O_2). For 1 mol of glucose (180 g) ($\rightarrow$ **C**), 6 mol of O_2 (6×22.4 l) are needed; this yields 15.7 kJ/g (= 3.75 kcal/g) or 2,827 kJ (= 675 kcal) from 180 g of glucose. The CE for carbohydrates is therefore 21.2 kJ/l O_2 (= 5.05 kcal/l O_2); for fats, it is 19.6 kJ/l O_2 (= 4.84 kcal/l O_2); for proteins, it is 19.65 kJ/l O_2 (= 4.69 kcal/l O_2).

Energy metabolism can be calculated from the CE, provided the nutrient that is being oxidized is known. Energy metabolism can be approximated by determining the **respiratory quotient** (RQ) ($\rightarrow$ **C**, **D**, p. 82). The RQ is the ratio of CO_2 produced to O_2 utilized ($RQ = \dot{V}_{CO_2}/\dot{V}_{O_2}$). For carbohydrate, RQ = 1.00, as can be seen from the reaction: $C_6H_{12}O_6 + 6 O_2 \rightarrow 6 CO_2 + 6 H_2O$. For the fat tripalmitin, the reaction is $2 C_{51}H_{98}O_6 + 145 O_2 \rightarrow 102 CO_2 + 98 H_2O$. The RQ = 102/145 = 0.7. Energy production (E) can then be calculated from $E = CE \times \dot{V}_{O_2}$.

When food is ingested and digested, the BMR rises. Energy is required to break down the nutrients, absorb them from the intestinal tract, and assimilate them into the body. This capacity to raise the BMR is the **specific dynamic action** (SDA). For 419 kJ (= 100 kcal) of nutrient, the SDA is 25 kJ (= 6 kcal) for carbohydrate and fat but is 126 kJ (= 30 kcal) for protein; only 70 % of the protein intake is useful for energy production; the remainder is used for the assimilation processes, a fact that is useful in designing reducing diets. Energy metabolism is influenced by **hormones**, mainly by thyroid hormones ($\rightarrow$ p. 232).

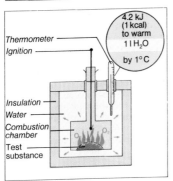

A. Bomb calorimeter

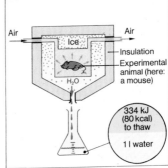

B. Direct calorimetry (Lavoisier)

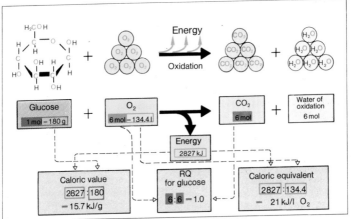

C. Oxidation of glucose

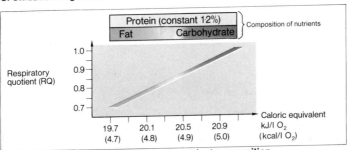

D. Caloric equivalent and RQ relative to nutrient composition

Digestive Organs – Transit Time

The gastrointestinal (GI) system is the boundary between the external and the internal environments in which nutrients are prepared for passage to the internal environment. Nutrients are broken down into smaller units (**digestion**) that are absorbed through the intestinal tract mucosa (**absorption**) into the lymph or portal blood. The absorption process takes place by diffusion, carrier transport, or pinocytosis.

Digestion begins in the **mouth**, where the large food particles are reduced in size, mixed with saliva ($\rightarrow$ p. 188) and converted to a more or less homogeneous semifluid mass. Swallowing ($\rightarrow$ p. 190) is a reflex that transfers the chewed food to the **esophagus**. The food passes through the esophagus to the **stomach**, where it is mixed with gastric juice ($\rightarrow$ p. 194). Gastric juice is strongly acid. Proteins are denatured in the stomach. The stomach contents pass through the pyloric sphincter into the **duodenum**. The duodenum receives bile and pancreatic secretions ($\rightarrow$ pp. 198–205) and produces digestive tract hormones. At the duodenum, secretions of the intestinal cells and digestive juices from the pancreas and gallbladder are added to the intestinal contents, **the chyme**. These secretions are alkaline and neutralize the gastric acid. A number of tissue hormones that contribute to digestion are produced in the region of the duodenum and lower stomach. The small intestine (**jejunum** and **ileum**) has three muscular layers in its walls that churn, mix, and propel the chyme. Most of the absorption of organic molecules, water and electrolytes takes place in this segment. There is a two-way flux across the mucosal surface, an exchange of electrolytes between the intestinal contents and the blood. In severe diarrhea, the equilibrium between these fluxes may be disturbed and may lead to large losses of Na^+, K^+, HCO_3^-, and water. In the **large intestine**, final absorption of electrolytes and water takes place ($\rightarrow$ p. 208). The contents of the end of the large intestine, **feces**, consist chiefly of unabsorbable vegetable matter, desquamated cells and bacteria, and a minimum of water. They are stored in the **rectum** until they are voluntarily eliminated (**defecation**). The transit time through the gastrointestinal tract, especially the lower portion, varies widely and depends in part on the fraction of undigestible constituents in the diet. (See Fig. **A** for average values.)

Transit time from mouth to cecum averages 8 h. It is influenced by the gastric emptying time. In the interdigestive period, the stomach empties quickly, but during digestion, it may delay emptying for up to 6 h. From the stomach to the cecum generally requires c. 2 h. From the cecum to rectal emptying, transit time ranges widely, normally from 16 to 40 h, but transit time may be up to 270 h in constipation and much less in diarrhea.

Blood supply to the GI tract is derived from three branches of the aorta. Blood flow is assisted by peristalsis and is increased by local reflexes; it is largely independent of the systemic blood pressure (**autoregulation**) ($\rightarrow$ p. 162). **Ileus**, reduced intestinal motility, leads to reduced blood flow and local accumulation of gases (CO_2, methane, H_2S, etc.); the increase in intraluminal pressure that results further reduces blood flow.

Nutrients in the blood from the intestinal tract are drained to the liver in the hepatic portal system before they reach the general circulation. Some nutrients, chiefly fats, are drained in the intestinal lymphatics and thus bypass the liver. Water and absorbable substances introduced into the rectum enter the systemic venous drainage and also bypass the liver.

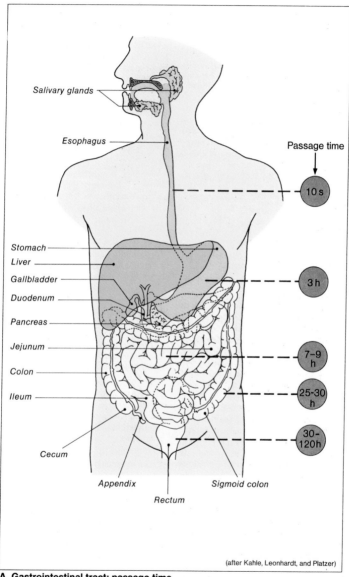

Salivary glands

Esophagus

Passage time

10 s

Stomach
Liver
Gallbladder
Duodenum
Pancreas
Jejunum
Colon
Ileum

3 h

7–9 h

25-30 h

30–120h

Cecum

Appendix
Rectum
Sigmoid colon

(after Kahle, Leonhardt, and Platzer)

A. Gastrointestinal tract; passage time

Saliva

Saliva facilitates chewing, swallowing, and speech, and is produced by the **parotid** and **submandibular** glands, which respond to reflex stimuli and produce 25% and 70% of the total salivary secretion. The **sublingual** gland and the mucus glands of the oral cavity secrete continuously. The salivary glands are representative of most exocrine glands. They secrete a mixture of enzymes, glycoproteins (mucins), inorganic ions, organic compounds, and water (c 1.5 l/ day). They are normally in a resting state in which baseline flow is quite small. Since they secrete according to need, they are individually innervated. Secretion is often well adapted to the function it must perform. Dogs, for example, secrete a viscous saliva to help in swallowing a piece of meat, or a watery saliva for washing dry food out of the mouth. When the secretory stimulus is over, the glands revert to the resting state.

Salivary flow is stimulated ($\rightarrow$ **A**) by touch, taste, and smell and by thought of food, i.e., by mechanical, physical, and psychic stimuli. The **reflexes** involved are partly *unconditioned* and do not need higher nerve centers. In addition, there are salivary reflexes which involve higher centers of the brain: *Conditioned reflexes* (Pavlov). The reflex stimulus is carried via cranial nerves V and IX for taste and touch and by higher centers for smell to an ill-defined "salivation center" ($\rightarrow$ **A, 1**). From the "center" the paths lead to sympathetic ($\rightarrow$ **A, 7**) and parasympathetic nerves ($\rightarrow$ **A, 2, 3**). Preganglionic fibers from VII run in the chorda tympani ($\rightarrow$ **A, 4**) to the lingual nerve (V). In the submandibular ganglion ($\rightarrow$ **A, 5**), they separate to supply the submandibular and sublingual glands. Fibers from IX synapse in the otic ganglion ($\rightarrow$ **A, 6**) and supply the parotid gland. Sympathetic fibers synapse in the superior cervical ganglion ($\rightarrow$ **A, 7**) and run along the arteries to the salivary glands. Efferent impulses activate cholinergic and both α- and β-adrenergic receptors ($\rightarrow$ p. 54). β-Activation increases intracellular cAMP ($\rightarrow$ p. 222) and results in secretion of enzymes, glycoproteins, and other macromolecules in a viscous saliva. (The parotid secretes less mucin than the other glands.) Cholinergic and α-adrenergic activation increase intracellular Ca^{2+} and result in rapid secretion of water and ions. (Ca^{2+} in the saliva complexes with the glycoproteins and increases viscosity.)

The fluid that is excreted in response to cholinergic stimulation is isosmotic and corresponds to the ECF. As the fluid passes through the salivary duct, ions are actively absorbed, and the final solution is hyposmotic. Na^+, H^+, and to a lesser extent Ca^{2+}, are resorbed at the expense of K^+ ($\rightarrow$ **B**). HCO_3^- is added by the action of carbonic anhydrase ($\rightarrow$ p. 130). The final composition depends on the rate of salivary flow and on the gland of origin. Osmolarity varies from 80 mOsm/l at low rates of flow to c. 300 mOsm/l at high rates. Salivary secretion depends in part on total body water balance. In water deprivation, flow is reduced (xerostomia), and the dryness of the mouth contributes to the sensation of *thirst* ($\rightarrow$ pp. 124, 172).

Parasympathetic activity via VII and IX activates the enzyme *kallikrein*, which cleaves bradykinin from a precursor in the capillaries. This action results in vasodilation and increased production of a copious watery saliva; it is blocked by atropine.

Saliva maintains the pH of the mouth at 7.0. When the mouth becomes acid, Ca^{2+} is lost from the teeth. Saliva also contains α-**amylase** (a starch-splitting enzyme), blood group substances ($\rightarrow$ p. 60), fluoride (which reduces incidence of caries), and thiocyanate (which has a disinfectant action). The salivary glands also take part in excreting inorganic and organic substances from the blood, such as iodide ion and some therapeutic drugs.

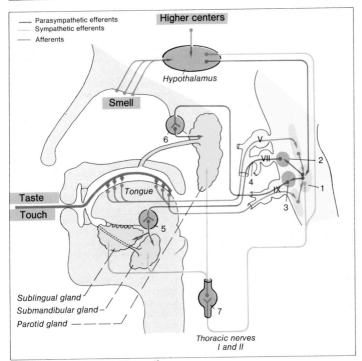

A. Salivary secretion; reflex pathways

Parasympathetic efferents
Sympathetic efferents
Afferents

Higher centers

Hypothalamus

Smell

Tongue

Taste

Touch

Sublingual gland
Submandibular gland
Parotid gland

Thoracic nerves
I and II

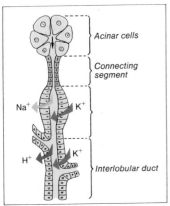

B. Formation of saliva

Acinar cells

Connecting
segment

Na⁺   K⁺

H⁺   K⁺

Interlobular duct

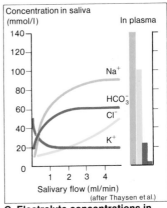

C. Electrolyte concentrations in saliva

Concentration in saliva
(mmol/l)

In plasma

Na^+

HCO_3^-

Cl^-

K^+

Salivary flow (ml/min)
(after Thaysen et al.)

Deglutition – Vomiting

Swallowing (**deglutition**) (→ **A**, **1–10**) is a **reflex** process. It is initiated by the voluntary act of collecting food on the tongue, raising the tongue, and propelling the food into the pharynx. The mouth is closed and the tongue presses against the gums (→ **A**, **1**); the soft palate is raised (→ **A**, **2**) to close the nasopharynx (→ **A**, **3**). Respiration is inhibited, and the glottis is closed to prevent passage of particles into the respiratory tract (→ **A**, **5**). Pressure from the tongue pushes the contents into the pharynx. Sphincter muscles of the lower pharynx relax (→ **A**, **6**). Resting tension of the 3-cm segment at the junction of pharynx and esophagus normally is high, but relaxation occurs reflexly to allow access to the esophagus. In the esophagus, an annular contraction forms behind the food bolus and propels it toward the stomach. As the bolus clears the pharynx, the glottis opens and respiration resumes (→ **A**, **7–10**). The esophagus is 25 to 30 cm long; its musculature at the upper end is striated, but the remainder is smooth muscle. At the junction of the esophagus with the stomach (cardia), the muscle has a high tonic activity; it acts as a sphincter and relaxes as part of the swallowing process.

Motility of the intestinal smooth muscle is closely related to function of the *myenteric plexus* (→ p. 196); defects of the plexus result in disordered peristalsis. In the esophagus, defects lead to failure of relaxation of the cardia; food accumulates and the esophagus dilates to accomodate the mass (*achalasia*).

Vomiting is a reflex response that is integrated in the medulla oblongata. Many visceral stimuli can induce vomiting: dilation of the stomach from overeating; gastric mucosal irritation (e.g., by alcohol); unpleasant sights, smells, and ideas contribute a psychic stimulus; touching the pharyngeal mucosa or stimulating the equilibration organs of the inner ear (*motion sickness*); pregnancy, in which normal "*morning sickness*" may progress to a threatening *hyperemesis gravidarum*; severe pain; poisons; toxins and medications (*apomorphine* is one of the most potent stimuli to vomiting and is used on occasion for the treatment of poisoning); X-ray irradiation; elevated intracranial pressure, as in a brain tumor. These stimuli activate the vomiting "center" in the medulla oblongata (→ **B**), which lies between the olive (→ **B**, **1**) and the solitary tract (→ **B**, **2**) in the area of the reticular formation (→ **B**, **3**). Occasionally, chemoreceptors in the area postrema (→ **B**, **4**) also play a role. **Prodromes of vomiting** include the sensation of nausea, hypersalivation, rapid heart rate, and altered respiration, paleness, sweating, and dilation of the pupils.

In vomiting proper, the diaphragm is fixed in midexcursion and the glottis is closed. The abdominal wall muscles contract and increase intra-abdominal pressure. The duodenum and pylorus constrict; the cardia simultaneously relaxes, allowing the gastric contents to be passed into the esophagus. The esophagopharyngeal sphincter relaxes, and the soft palate rises to permit the bolus to be expelled through the mouth.

Vomiting is a **protective reflex** to hinder damage to the stomach and to the total organism by poisoning. Prolonged vomiting is associated with losses of acidic gastric juice and leads to metabolic alkalosis (→ p. 102) and disturbances in fluid balance (→ pp. 124–129).

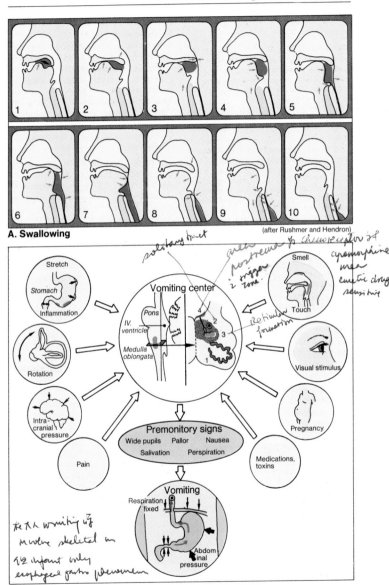

A. Swallowing

(after Rushmer and Hendron)

B. Vomiting

Stomach: Structure and Motility

The esophagus (→ **A**), opens at the **cardia** into the **fundus**, the upper one third of the stomach; the body or **corpus** connects with the **antrum**; the **pylorus** is the portion of the stomach at the transition to the duodenum. In general, the structure of the gastric wall resembles that of the small intestine (→ p. 197, **A**, **B**). The chief difference is in the mucosa, which includes three types of cells (→ **A**): throughout the stomach, there are **mucus-secreting cells** as in other parts of the gastrointestinal tract, but in the fundus there are in addition **chief** or peptic cells (CC) and **parietal** cells (PC). The mucosa is able to maintain the low pH of gastric juice (→ p. 194), because it is virtually impermeable to passive diffusion of ions. It is insensitive to extremes of temperature (ice cream, hot soup) and of osmolarity.

The chief digestive function of the stomach is to furnish a **reservoir** for liquefaction of solids in the diet. Although liquids pass through the stomach relatively quickly, solid matter may be retained in the stomach for hours.

The volume or capacity of the stomach (→ **B**) is controlled by tonic contraction of the longitudinal smooth muscle (LSM) layer that relaxes under influence of the vagus (→ p. 48) or of **gastrin** (→ **D**). When food enters the upper esophagus, a vagal reflex allows the stomach to increase its capacity (**receptive relaxation**). The LSM also influences intragastric pressure and establishes a pressure gradient across the pylorus to drive fluid contents into the duodenum.

The circular smooth muscle (CSM) layer of the corpus and antrum contracts concentrically and occludes the gastric lumen either partially or completely (→ **C**). These contractions move as peristaltic waves only in a caudad direction; there is no retropulsion. Concentric contraction does not take place in the fundus. During digestion, a band of partial concentric contraction moves cau-

dad, pressing gastric contents toward the pylorus. At the proximal antrum, the contraction wave halts, and the pyloric sphincter closes. The antral unit then undergoes systolic contraction, propelling the contents as a jet through the constriction back into the corpus. Repetition of the sequence at three cycles/min ultimately liquefies the gastric contents. Digestive contraction of the CSM and mixing activity are stimulated by the vagus and by gastrin.

The pylorus remains open between cycles to allow fluids to enter the duodenum. The duodenum contracts to propel the contents only when the pylorus is closed, thereby avoiding reflux into the stomach. Fluid is delivered to the duodenum in small spurts, the contents of which are monitored in the duodenal bulb. A fatty meal reduces gastric motility (→ **D**) and prolongs emptying time to allow a longer period for emulsification of fat in the lower duodenum and beyond. The fat, H^+ and other stimuli liberate the tissue hormones, **secretin, somatostatin (SIH)** and **gastric inhibitory peptide (GIP)** (→ **D**), which inhibit motility and acid secretion of the stomach to some extent. **Cholecystokinin-pancreozymin (CCK-PZ)** and **motilin** increase gastric motility.

Tissue hormones. Dispersed throughout the gastrointestinal tract (mainly duodenum) in the submucosa are hormone-producing cells. The hormones are: gastric and duodenal gastrin, secretin, CCK-PZ, GIP, motilin, enteroglucagon, SIH, vasointestinal peptide (VIP, increases blood flow of GI-tract), and hormones not well defined, like enterogastrone and villikinin. The hormones act both locally (paracrine) and distantly (endocrine). They are relatively small-chain peptides (up to c. 30 to 40 amino acids), which influence motility and secretion of the GI tract. Other peptides with endocrine and paracrine function have been identified in the brain and kidney. Many of these peptides have intramolecular regions with similar amino acid sequences.

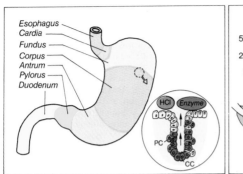

A. Anatomy of the stomach

Esophagus
Cardia
Fundus
Corpus
Antrum
Pylorus
Duodenum

HCl Enzyme

PC

CC

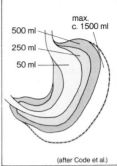

B. Filling conditions

max.
c. 1500 ml
500 ml
250 ml
50 ml

(after Code et al.)

C. Antral cycles (cineradiography)

Bulbus
duodeni
Duodenum
Pyloric
canal
Antrum

(after Carlson et al.)

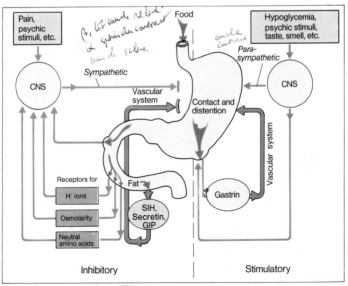

D. Influences on gastric motility

Pain,
psychic
stimuli, etc.

Food

Hypoglycemia,
psychic stimuli,
taste, smell, etc.

CNS

Sympathetic

Para-
sympathetic

CNS

Vascular
system

Contact and
distention

Vascular system

Receptors for

H⁺ ions

Fat

Gastrin

Osmolarity

SIH,
Secretin,
GIP

Neutral
amino acids

Inhibitory

Stimulatory

Gastric Secretion

The stomach secretes up to 3 l of gastric juice daily. The chief components are: **pepsins** (proteolytic enzymes or proteases of low specificity), **mucin, HCl,** and **intrinsic factor,** which is essential for absorption of the cobalt-containing vitamin B_{12}. Pepsins (approximately eight different molecules) are produced in the gastric chief cells as inactive precursors, **pepsinogens.** They are activated by the strong acid of the gastric juice, which provides an optimum pH for their action. They hydrolyze proteins at peptide bonds to form lower molecular weight peptides. Gastric acid is produced in the oxyntic, or **parietal cells,** which have three different stimulatory receptors (for gastrin, AcCh and histamine). The secretion of these cells is equivalent to HCl 0.17 N (pH < 1) but gastric pH rises to 2 to 4 when food dilutes and buffers the acid. Secretion of H^+ into the gastric lumen requires carbonic anhydrase ($\rightarrow$ p. 130) which splits water to H^+ and OH^- ($\rightarrow$ **B**). The H^+ is actively transported into canaliculi of the parietal cells to a transmembrane concentration ratio of 10^6 to 10^9. The pump is driven by ATP ($\rightarrow$ p. 14) and exchanges H^+ for K^+. K^+ comes into the gastric lumen by a K^+/Cl^- symport mechanism (K^+ recycling). Driving force is the K^+ gradient between cell and lumen which is established, as in any other cell, by an active ATP-driven Na^+ for K^+ exchange pump. The OH^- that is liberated forms HCO_3^- and moves out the blood side of the cell in exchange for Cl^- ($\rightarrow$ **B**). After eating, secretory activity rises to a maximum. Increased quantities of HCO_3^- are transferred to the blood and are excreted by the kidney producing the postprandial alkaline tide.

Stimulation of gastric secretion by a meal has three phases:

1. **Cephalic** phase (psychic-neural). Food in the mouth initiates reflex stimulation of gastric secretion. Emotions (anger, aggression), taste, sight, and smell contribute to the **afferent** limb of the reflex. The **efferent** limb in each case is the vagus (X). Its cholinergic (AcCh) action releases **gastrin.** It also acts directly on the parietal cells to release HCl and to sensitize them to gastrin. 2. **Gastric** (local) phase. Distension of the stomach stimulates gastric secretion (independent of gastrin), whereas chemical stimulation is mediated by gastrin (see below). Gastrin and AcCh cause intragastric release of **histamine** which also stimulates HCl secretion. 3. **Intestinal** phase. As the meal passes serially through the stomach and duodenum, distension of the walls and chemical stimuli combine to release **gastrin.** Inhibition of gastric motility and HCl secretion occurs when acid and dietary fats reach the duodenum. Release of *secretin, gastric inhibitory peptide (GIP), SIH* and *vasoactive intestinal peptide (VIP)* does not fully explain this inhibition.

Gastrins ($\rightarrow$ **A**; p. 193, D) are a family of tissue hormones ($\rightarrow$ p. 192) (c. 17 amino acids) with similar amino acid sequences. They are produced in the G cells of the gastric antrum and in the proximal duodenum. Very little gastrin is found in the acid-secreting portions of the stomach. Gastrin release ($\rightarrow$ **A**) is stimulated by amino acids and peptides bathing the pyloric gland mucosa (chemical), by insulin hypoglycemia (humoral), and by the vagus (neural). Gastrin stimulates parietal cells to secrete HCl ($\rightarrow$ **B**) and mucosal cells to grow. Gastrin release is modified by negative feedback ($\rightarrow$ **A**, 3); HCl production reduces gastrin release. In gastric hypochlorhydria, the feedback loop is interrupted and does not operate. The G cells enlarge and gastrin continues to be released, but no HCl is produced. The Zollinger-Ellison syndrome ($\rightarrow$ **C**) is fulminant peptic ulceration in association with an islet cell (non-β) tumor of the pancreas. Excess gastrin release from the tumor stimulates the parietal cells to hyperplasia and hypersecretion.

The gastric mucosal cells secrete **mucin,** which covers the stomach surface and protects it against self-digestion.

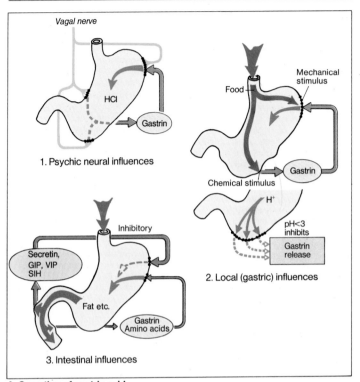

A. Secretion of gastric acid

Vagal nerve

HCl

Gastrin

1. Psychic neural influences

Mechanical stimulus

Food

Chemical stimulus

Gastrin

H⁺

pH<3 inhibits

Gastrin release

2. Local (gastric) influences

Inhibitory

Secretin, GIP, VIP SIH

Fat etc.

Gastrin
Amino acids

3. Intestinal influences

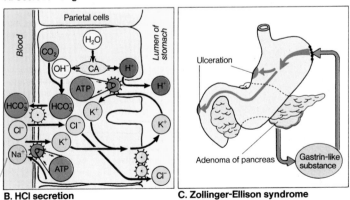

B. HCl secretion

Parietal cells

Blood

Lumen of stomach

CO_2

H_2O

OH^- CA H^+

ATP

H^+

HCO_3^- HCO_3^-

K^+

Cl^- Cl^-

K^+

Cl^- K^+

Na^+

ATP Cl^-

C. Zollinger-Ellison syndrome

Ulceration

Adenoma of pancreas

Gastrin-like substance

Small Intestine: Structure and Motility

The small intestine (c. 3 m long) comprises three segments: **duodenum**, **jejunum**, and **ileum**. The anatomic layers of the gastrointestinal tube are (→ **A**, **B**): serosa (→ **A**, **1**), longitudinal smooth muscle (LSM) (→ **A**, **2**), circular smooth muscle (CSM) (→ **A**, **3**), submucosa (→ **A**, **5**), and mucosa (→ **A**, **4**). Although the small intestine completes the digestive process by reducing the size of large molecules, its principal function is absorptive. Carbohydrates and proteins are absorbed into the hepatic portal system (→ **A**, **14**), and reach the liver before they enter the systemic circulation. Lipids (except: short chain fatty acids, → p. 204) are absorbed into the *lacteals*, small lymph vessels that drain each villus into the lymphatic system (→ **A**, **13**) and bypass the liver, entering the circulation via the thoracic duct.

The mucosal surface of the small intestine is specialized for both active and passive absorption (→ **A**, **B**). Annular folds in the mucosa (→ **A**) increase the surface area 1.3 times over that of a cylinder. Finger-like projections, the villi (→ **A**, **9**) (0.5 to 1.0 mm long; 20 to 40 villi/mm^2 of surface), increase the surface an additional 5 times, and the microvilli of the **brush border** another 30 times. The total surface area of the small intestine is c. 100 m^2. The villi are bordered by the crypts of Lieberkühn (→ **A**, **8**), which produce an isotonic secretion, and by the glands of Brunner, which produce a mucinous secretion. They are in a continuous growth cycle as the cells at the tips of the villi slough off and are replaced from the crypts in a 2-day cycle.

The duodenum **adjusts the osmolarity** of fluid from the stomach by adding or removing appropriate amounts of water or electrolytes. It thereby protects the cells of the jejunum and ileum, which can be damaged by extreme hyperosmolar or hypo-osmolar solutions. The ileum **adjusts the volume** of the intestinal contents by completing the absorptive process in the small intestine. Throughout the small intestine, large two-way fluxes of water and electrolytes may occur.

In the interdigestive period, the small intestine undergoes cycles of activity: a rest period (phase 1) of low activity is followed by an increasing frequency of firing of action potentials (AP) (phase 2), which are superimposed on the pacesetter potential.

The pacesetter is a slow (8/min) controlling potential whereas the faster spike potentials initiate the contractile response. In phase 2, there are segmenting annular contractions of the CSM (→ **C**, **2**); they are bidirectional for mixing of the intestinal contents. Forward propulsion is slow. In phase 3, AP activity peaks and rapid propulsive peristalses (→ **C**, **3**) take place followed by a decline of activity in phase 4. This cycle repeats every 12 min. The duodenum initiates 60 to 90 peristaltic waves per cycle. The stomach shows similar interdigestive motility with 10 to 15 peristaltic waves/cycle. The cycles are associated with a rise (phase 2) and fall (phase 4) in secretion of the tissue hormone **motilin**. Motilin release is inhibited by **somatostatin (SIH)** whereas **enteroglucagon** (stimulus: fat and glucose in duodenum) directly inhibits propulsive motility. Peristaltic contractions (phase 3) are not seen during the digestive period (except in the duodenum). Enteroglucagon prolongs the transit time, thus permitting complete digestion and absorption of the meal.

Innervation: The outer layer of nerve cells (Plexus myentericus) mainly controls motility (LSM, CSM); the inner layer (Plexus submucosus) is more involved in afferent and secretory activity (→ **B**). Autonomic nerves modulate these activities (→ pp. 49–57) and conduct visceral reflexes via the CNS.

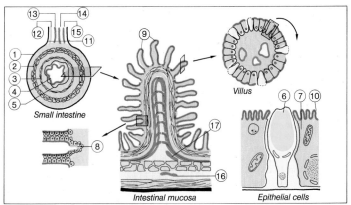

A. Structure of the intestine (schematic)

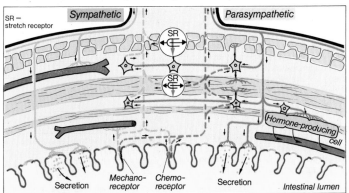

SR = stretch receptor

Sympathetic

Parasympathetic

Hormone-producing cell

Secretion   Mechano-receptor   Chemo-receptor   Secretion   Intestinal lumen

B. Innervation of the intestine (schematic)

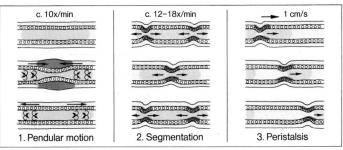

c. 10x/min   c. 12–18x/min   1 cm/s

1. Pendular motion   2. Segmentation   3. Peristalsis

C. Intestinal motility

Pancreatic Juice – Bile

Pancreas. Secretory activity of the **exocrine** pancreatic cells (82% of pancreas) may be basal or stimulated (→ **A**; p. 188). Secretion of c. 2 l of alkaline fluid daily is controlled partly by reflexes: visual, olfactory, and gustatory stimuli initiate the cephalic phase. Mechanical or chemical stimuli in the upper gut start reflexes in the vagus or sympathetic nerves. The chief stimuli, however, are the tissue hormones (→ p. 192) secretin (S) and cholecystokinin-pancreozymin (CCK-PZ). Other polypeptides (gastrin [→ pp. 192–195], VIP, glucagon, motilin [→ p. 196], somatostatin) may also play a role. The exocrine cells contain **zymogen** granules, which contain the pancreatic enzymes for splitting proteins, carbohydrates, and fats. Amino acids, fatty acids, and HCl in the gut stimulate release of CCK-PZ, which is carried by the blood to the pancreas where it activates exocytosis of the zymogen granules and release of enzymes. Acid in the duodenum stimulates release of S, which activates secretion of water and electrolytes into the pancreatic juice. HCO_3^- secretion also increases while Cl^- secretion decreases correspondingly (→ **B**). The alkaline fluid contributes to neutralizing gastric acid in the duodenum.

Protein digestion. Proteolytic enzymes are secreted as inactive precursors, **trypsinogen** and **chymotrypsinogen**. In the duodenum, **enterokinase** activates them to **trypsin** and **chymotrypsin** (→ **A**), which hydrolyze peptide bonds. If the activation occurs within the pancreas, autodigestion takes place and produces pancreatitis and cell necrosis. **Carboxypeptidase**, another pancreatic enzyme, splits peptide bonds at the −COOH end of proteins.

Carbohydrate digestion. Amylase splits starches and glycogens, and **maltase, lactase** and **saccharase** split

disaccharides to monosaccharides (→ pp. 182, 206).

Fat digestion. Pancreatic **lipase** splits triglycerides to monoglycerides and free fatty acids (→ pp. 182, 204).

The **bile** is the excretory fluid of the liver (c. 0.7 l/day). Hepatic bile is stored in the gall bladder (GB) where 80% to 90% of the water is removed when NaCl is actively resorbed across the GB wall (→ p. 201, D). GB bile is therefore more viscous and concentrated than hepatic bile. The GB is stimulated by CCK-PZ to contract and deliver its bile through the sphincter of Oddi into the duodenum (→ **A**). A **choleretic** substance stimulates bile **formation** in the liver (e.g., secretin, bile salts) whereas a **cholagogue** stimulates bile **delivery** by contraction of the GB (e.g., fats, egg yolk).

The major components of bile are: water, bile acids, lecithin, bilirubin, and cholesterol. The **bile acids** (= bile salts) are anionic steroids that are synthesized in liver cells from cholesterol (→ p. 201, A). Before being excreted into the bile, the **primary** bile acids, cholic and chenodeoxycholic acids, are conjugated with either glycine or taurine to form glycocholate, taurocholate, etc. Conjugation takes place in microsomal and lysosomal fractions of liver homogenates and is activated by coenzyme A. Bile salts function as detergents for absorption of lipids in the gut. Intestinal bacteria hydrolyze the conjugated bile acids and form **secondary** bile acids, deoxycholate and lithocholate. The deconjugated bile acids are returned to the liver by absorption in the terminal ileum in an **enterohepatic circulation**. In the liver, they inhibit synthesis of bile salts by a negative feedback mechanism. Daily, c. 24 g of bile salts are secreted and all but 0.5 g is resorbed from the gut. The liver normally synthesizes only 0.5 g/day of new bile salts, which exactly compensates for fecal losses.

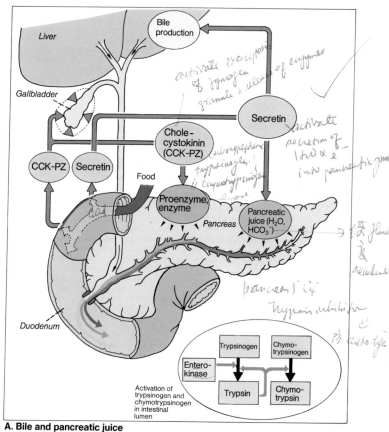

A. Bile and pancreatic juice

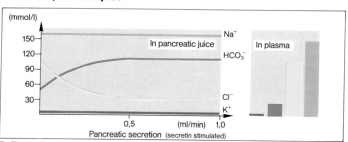

B. Electrolyte concentrations in pancreatic juice

Bile Flow — Hepatic Excretion

Bile flow originates at the bile canaliculus within the liver cell and also at the bile ductules. In the liver cell, there are two components to bile formation ($\to$ **B**): bile acid-dependent (BAD) and bile acid-independent (BAI) flow. For **BAD flow**, hepatocytes efficiently extract bile acids from blood in the sinusoids and actively transport them in conjugated form into the canaliculi. Concentrations of bile acids in the canaliculi may be more than a 1000 times greater than in the ECF. An osmotic gradient is thus established for diffusion of water into the canaliculi. Eventually, the bile acids reach the duodenum and are returned to the liver by **enterohepatic circulation** ($\to$ p. 198). BAD flow depends upon availability of bile acids and on the rate of their recycling. In man, bile acids are sidetracked and stored in the gall bladder, which superimposes an intermittency on bile flow every time it delivers bile acids to the duodenum.

BAI flow depends on liver cell mass. Membrane ATPase in the canaliculi pumps electrolytes into the lumen and water follows in osmotic equivalence. Bile flow is further modified by secretion and resorption beyond the canaliculus at the bile ductules and ducts. This portion of flow responds to secretin ($\to$ **B**), requires the action of carbonic anhydrase, and is rich in HCO_3^-.

Substances that undergo **hepatic excretion** may be divided into three classes: class A compounds have a bile/plasma concentration ratio of 1.0 (e.g., Na^+, K^+, Cl^-, glucose); class B compounds have a ratio greater than 1 (e.g., dyes, steroids, bile acids, bilirubin); class C compounds have a ratio less than 1 (e.g., macromolecules, inulin). Class B compounds comprise the largest group. With the exception of the bile acids, class B compounds share a common transport mechanism in liver and kidney.

Hepatic excretion of class B substances involves a series of steps. **Hepatic uptake** ($\to$ p. 203, A) from the sinusoids separates compounds from their binding sites on circulating proteins and takes them into the liver cell. The process can be saturated, and it shows competitive inhibition among the various substrates (except for bile acids). **Intracellular binding** is distinct from the initial uptake process. It takes place with high affinity to macromolecular receptors (not albumin). There are two receptor fractions: Y and Z. Y is actually a family of proteins, one of which, **ligandin**, is identical with glutathione-S-transferase B. Ligandin has a high affinity for a variety of dyes, drugs, and metabolic end products. It is the principal protein for intracellular binding. Z is most closely related to fatty acid transport. For **metabolic alteration**, compounds are transferred to the endoplasmic reticulum, which possesses mixed oxidases. These enzymes prepare a number of molecules for conjugation reactions by adding reactive groups, principally $-OH$ ($\to$ p. 116). **Conjugation** ($\to$ **A**) of most but not all molecules takes place chiefly with glucuronic acid but also with glutathione, glycine, and other amino acids and sulfate. Conjugation confers a structure that allows active transport into the canaliculi. Many of these compounds escape into the blood via the portal triads. They then recirculate both to the kidney and the liver and are actively excreted in both organs by a similar process. **Transport** of many class B substances can achieve bile to plasma concentration ratios well over 1,000. Such compounds establish an osmotic gradient and exert a transient choleretic action. In the gut, some compounds are deconjugated by intestinal bacteria and are returned to the liver by enterohepatic circulation. In the portal triads, bile and blood circulate in opposite directions. The conditions are appropriate for **countercurrent exchange** ($\to$ pp. 116, 120) and some substances are recirculated to the liver by this means as well.

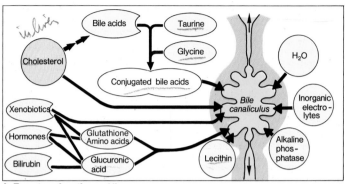

A. Excretory functions of liver

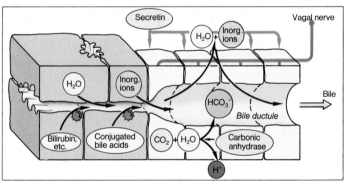

B. Transport in bile formation

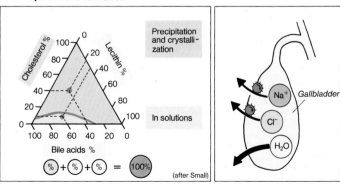

C. Micellar solution of cholesterol

D. Concentration of bile

Bilirubin – Jaundice

Bilirubin is derived (a) from the death of aging red blood cells when they are recycled in the reticuloendothelial system (80%, 4 mg/kg body weight/day), (b) from ineffective erythropoiesis (5%, 0.25 mg/kg/day), and (c) from other sources, chiefly the liver (15%, 0.8 mg/kg/day).

One gram hemoglobin generates 35 mg bilirubin (BR). In plasma, BR is poorly soluble and is transported tightly bound to albumin (→ **A**). The concentration of BR in plasma is a balance between rate of synthesis and rate of hepatic removal from plasma. BR is excreted in the liver as a class B (→ p. 200) substance by hepatic uptake, intracellular binding, conjugation, and active transport of the conjugated form into the bile. Some BR that enters the liver (approximately one third) or is synthesized in the liver returns to plasma unaltered. For the remainder, a monoglucuronide (BR-Glu) is synthesized by the action of glucuronyl transferase (UDPGT), but BR is excreted chiefly as a diglucuronide (BR-Glu$_2$) at a rate of 250 to 300 mg/day. Both glucuronides are water-soluble. In the gut, bacterial glucuronidases regenerate free, BR, some of which returns to the liver by enterohepatic circulation. Bacteria also reduce BR to colorless **urobilinogens** (→ **B**), a class of compounds including **D-urobilinogen**, mesobilirubinogen, and **stercobilinogen**. These form corresponding yellow-orange pigments (e.g., D-urobilin, stercobilin) on oxidation. Stercobilinogen is the predominant metabolite in the lower colon where it is resorbed into the systemic circulation. It bypasses the liver and is excreted chiefly in urine. Urobilinogens that are resorbed in the upper gut are excreted chiefly in bile, but when hepatic excretion becomes impaired, renal excretion increases.

Hyperbilirubinemia occurs when free or conjugated BR accumulates in the plasma. Normal plasma level for total BR = 3 to 20 µmol/l. At concentrations above 20 µmol/l, **jaundice (icterus)** develops, and the skin, scleras, and mucus membranes appear yellow. Jaundice may develop because of (1) a deficit in bile acid transport that reduces bile flow (= cholestasis) or (2) a deficit in BR transport. Jaundice may be prehepatic, intrahepatic, or posthepatic. These types of jaundice should be distinguished because treatment and prognosis depend upon the diagnosis; posthepatic jaundice more often requires surgical treatment, while the other two more often require medical treatment.

Prehepatic. Increased production of BR because of hemolysis may exceed the capacity of the liver to conjugate and excrete BR. The bottleneck is at UDPGT (→ **A**) so that unconjugated BR increases in the plasma.

Intrahepatic. Hepatic uptake and conjugation of BR are decreased when liver cells are impaired or when availability of UDPGT is reduced. Jaundice in the newborn develops because UDPGT is deficient. Many drugs, including steroids, can inhibit conjugation and transport of BR.

Posthepatic. Obstruction of bile ducts allows BR to reflux into the blood (countercurrent exchange), as occurs with tumors, gall stones, or inflammations of the bile ducts. BR-Glu and BR-Glu$_2$ are absent from the gut, resulting in a colorless (clay-colored) stool. They accumulate in blood, are excreted in the kidney, and color the urine dark orange.

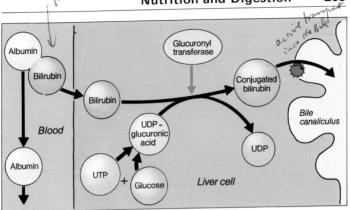

A. Conjugation of bilirubin

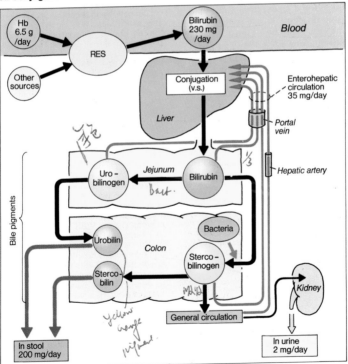

B. Metabolism and excretion of bilirubin

Digestion of Lipids

Lipids are poorly soluble in body water and must be pretreated before they become available to the cells. In man, absorption mechanisms are so efficient that 95% of dietary lipids are assimilated. Lipid digestion takes place in three phases:

1. **Intraluminal phase**. Dietary triglycerides (TG) are esters of glycerol with three fatty acids (FA) and are poorly soluble in water ($\rightarrow$ p. 183). Pancreatic **lipase** ($\rightarrow$ **B**) acts on TG emulsions at the oil-water interface. Its activity is increased by a co-lipase secreted by the pancreas and by bile acids that disperse the TG and render them vulnerable to enzyme action. Lipase specifically hydrolyzes TG at the 1- and 3- ester linkages to form FA and monoglycerides. The shorter FA (less than ten carbons) are absorbed into portal venous blood ($\rightarrow$ **B**) and are carried to the liver bound to albumin. Longer FA and monoglycerides must be incorporated into micelles for absorption.

Bile acids are **amphipaths**; they possess hydrophobic and hydrophilic regions. At low concentrations, they exist as monomers, but at high concentrations, they aggregate to form complexes called **micelles**. These are spherical or cylindric (3 to 10 nm diameter) with the hydrophobic portions oriented inward and the hydrophilic outward, facing the aqueous environment. The bile acid concentration at which micelles are formed is the **critical micellar concentration** (CMC); it varies with pH, Na^+ concentration, and temperature. The Krafft point is the temperature below which micelles do not form. The CMC for bile acids is 1 to 2 mmol/l. Stability of the micelle is increased by incorporating the phospholipid **lecithin** ($\rightarrow$ p. 201, A), which is synthesized chiefly in the liver. Aggregates of bile salts with lecithin, FA, cholesterol, or fat-soluble vitamins (A, D, E) form a larger **mixed micelle**. Water-insoluble lipids gain access to the circulation only in combination with micelles.

2. **Mucosal phase**. Micelles are neither bound to the microvilli nor are they absorbed intact. Instead, they contact the villus and facilitate diffusion of lipids across the unstirred water layer at the mucosal wall and into the cell. In the absence of micelles, this water layer poses resistance to diffusion and reduces absorption.

3. **Delivery phase**. Within the cell, long chain FA (over 14 carbons) form CoA derivatives and attach to an FA-binding protein. They are reesterified to TG by (a) a monoglyceride pathway (which is preferred) and (b) a glycerophosphate pathway. The TG, phospholipids, proteins (apolipoprotein B and others) and cholesterol form **chylomicrons** (MW 10^8–10^9) for transport of exogenous TG and very low density lipoproteins (VLDL, MW $5 \cdot 10^6$ – 10^8) for transport of endogenous TG. Chylomicrons are particles that increase in concentration and become visible in blood after a fatty meal. They are extruded from the cell, diffuse to the lacteals, and reach the blood via the lymphatics. They are metabolized by **lipoproteinlipase** (LPL) and other enzymes in muscle and fat cells (80%) and in the liver (20%) to lipoproteins with lower lipid concentration (= higher density). LPL is synthesized and activated in fat cells, bound to the endothelium and displaced linearly by *heparin*.

Cholesterol (CH) is a sterol (a steroidal alcohol). It is a component of cell membranes and can be synthesized by all cells but is synthesized chiefly by liver for bile acids and by endocrine glands for steroids. The liver controls esterification of CH with FA; free CH is found mostly in the plasma, CH esters in plasma. All CH in plasma is carried by lipoproteins (LP) for which it is a structural component. Beside the chylomicrons and the VLDL there are two further kinds of LP: low density lipoproteins (LDL) carry CH to the peripheral tissues and high density lipoproteins (HDL) carry CH to the liver. CH is partitioned 70% as hydrophobic esters inside the LP and 30% as free CH on the outer surface. Hepatic synthesis of CH is regulated by feedback from CH absorption in the gut. CH is excreted into the bile where it is held in solution by bile acids and lecithin. As long as the proportions of the three components (CH, BA, lecithin) fall within the micellar range ($\rightarrow$ p. 201, C), CH remains dissolved; if they fall outside this range, CH crystals form in the bile fluid and permit gall stones to develop.

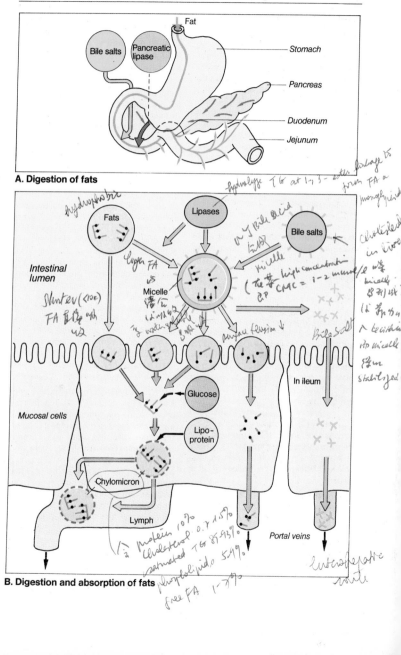

A. Digestion of fats

B. Digestion and absorption of fats

Digestion of Carbohydrates and Proteins

Carbohydrates. The diet contains polysaccharides, disaccharides (lactose, maltose, and sucrose), and monosaccharides. Starches, which are glucose polymers, are the only polysaccharides digested to any great extent in man; they comprise the largest portion of ingested carbohydrate. Digestion of starch by α-amylase (ptyalin) ($\rightarrow$ **A**) begins in the mouth ($\rightarrow$ p. 188) at an optimal pH close to 7; digestion is inhibited in the stomach. In the duodenum, the acid pH is neutralized, and digestion continues with the addition of pancreatic amylase to the intestinal contents. End products of starch breakdown include maltose and other small molecule saccharides (maltotriose and α-limit dextrin).

Assimilation of carbohydrate begins with absorption from the intestine. The intestinal active transport process for carbohydrates, however, is specific only for certain **mono**saccharides, especially glucose and galactose; fructose absorption appears to be passive. The end products of α-amylase digestion must therefore be broken down further. The enzymes for this hydrolysis (maltase and isomaltase) are located in the intestinal mucosa of the ileum. Glucose, the product of their activity, is absorbed into the portal blood by an active transport process that is coupled to Na$^+$ transport (Co-transport; $\rightarrow$ p. 112). There are corresponding enzymes, sucrase and lactase, for hydrolysis of other disaccharides.

In the absence of lactase, hydrolysis of lactose cannot occur. Some newborns have a congenital lactase deficiency and cannot assimilate lactose from milk; they develop diarrhea, bloating, and flatulence because of the osmotic effect of unabsorbed lactose and the production of toxic metabolites and gases by intestinal bacteria. Lactase activity is high at birth but declines during maturation in some genetic populations. Low activity is associated with intolerance to milk and is more common in black adults than in white adults.

Protein digestion begins in the stomach ($\rightarrow$ **B**), where gastric acid activates the three pepsinogens (produced by the chief cells) to eight different **pepsins** ($\rightarrow$ p. 194). Pepsins split proteins at bonds with tyrosine or phenylalanine; the pH optimum is 2 to 4. In the duodenum, where pancreatic and bile HCO_3^- neutralize gastric acid to pH c. 6.5, the pepsins are inactivated. Protein digestion continues by the action of pancreatic trypsinogen and chymotrypsinogen, which are activated by duodenal enterokinase ($\rightarrow$ p. 194). Trypsin and chymotrypsin split polypeptides and proteins to oligopeptides and dipeptides. Pancreatic carboxypeptidase and intestinal aminopeptidases split free amino acids from the ends of peptide chains. The amino acids are absorbed through the intestinal mucosa by several specific active transport mechanisms ($\rightarrow$ **B**). Absorption is more rapid in the upper intestinal tract. The number of different mechanisms has not yet been determined, but it appears that structurally similar amino acids participate in common transport systems.

Amino acids are co-transported with Na$^+$ ($\rightarrow$ p. 112). The dicarboxylic amino acids glutamate and aspartate, are metabolized in the mucosal cells rather than absorbed. Also intact dipeptides seem to be reabsorbed. There are a number of congenital defects in amino acid absorption; both renal and intestinal transport mechanisms may be defective; amino acids are lost into the urine (failure of resorption) and are not replaced by the intestine (failure of absorption).

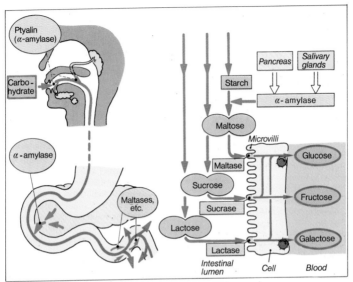

A. Digestion of carbohydrates

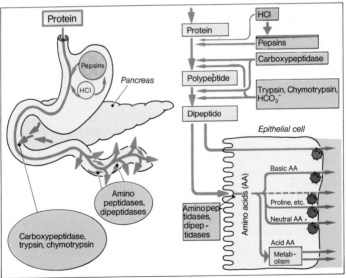

B. Digestion of proteins

Absorption of Water and Minerals

Daily intake of **water** is c. 1.5 l. Daily efflux of water into the GI tract as saliva, gastric juice, bile, and pancreatic and intestinal secretions is c. 6 l. Since only c. 0.1 l/day is excreted in the feces, the remaining 7.4 l must be absorbed. This water absorption takes place primarily in the jejunum and ileum and to a lesser extent in the colon. This **net absorption** is the resultant of water efflux and influx through the intestinal epithelium, where water moves freely in either direction. In the duodenum, efflux and influx of water are about equal and net absorption is small. In the ileum, net absorption is quite large ($\rightarrow$ **D**).

Na⁺ diffuses readily into the lumen when its contents are hypotonic and out of the lumen when they are hypertonic. Na⁺ **transport** out of the lumen increases when the contents are hypotonic. Na⁺ transport furnishes the driving force for the net absorption of water. It is quantitatively small in the duodenum and larger in the jejunum. To maintain electric neutrality, Na⁺ can be absorbed from the intestine only if accompanied by an anion (Cl⁻) or if a cation (K⁺) enters the intestine in exchange for Na⁺. Both events take place ($\rightarrow$ **A, C**); Cl⁻ absorption is prominent in the jejunum and K⁺ exchange is prominent in the ileum and large intestine. The driving force for Na⁺ absorption is the Na-K-ATPase pump in the cell membrane ($\rightarrow$ **B**). Its activity is influenced by aldosterone ($\rightarrow$ p. 137); the mechanism for absorption is similar to that in the kidney. Very little Na⁺ is excreted in the feces. Na⁺ transport is believed to be coupled with transport of glucose and amino acids. The process appears to be one of cotransport: absence of glucose or amino acids is associated with reduced Na⁺ transport, whereas absence of Na⁺ from the intestinal lumen hinders absorption of glucose or amino acids.

If the intestine contains a poorly absorbable ion or salt such as sulfate, an osmotically equivalent amount of water is obligated to remain in the intestine. This osmotic influence has a purgative action since it increases intestinal water excretion.

HCO₃⁻ fluxes are incompletely understood. HCO₃⁻ is absorbed where H⁺ efflux is greatest. In diarrhea, the loss of HCO₃⁻ and K⁺ can result in metabolic acidosis ($\rightarrow$ p. 102) and hypokalemia.

Calcium. 30% to 80% of Ca²⁺ in the diet is absorbed ($\rightarrow$ pp. 234–237) by active transport in the upper intestine. Absorption is facilitated by vitamin D and is reduced by formation of insoluble Ca²⁺ salts. Spinach and milk are an unfavorable dietary combination for children, as spinach contains oxalates that prevent absorption of Ca²⁺ from the milk. Mg²⁺ absorption is similar to that of Ca²⁺.

Dietary Fe (III) cannot be absorbed; it must first be reduced (by gastric HCl) to **Fe (II)**. Intestinal absorption appears to be adjusted to the body's requirement for Fe (*mucosal block*). The availability of free *transferrin*, a β-globulin which transports Fe in the plasma, seems to be regulatory for Fe absorption. Daily Fe intake is c. 10 to 15 mg; absorption depends on the dietary source (eggs 4%, meat 20%, iron salts 25%). Daily Fe losses are c 0.6 mg in feces and c. 1.3 mg in menstrual blood. *Iron deficiency* can occur (a) if dietary intake is insufficient, (b) if intestinal absorption is reduced (as in *achlorhydria*), or (c) if Fe requirements increase (pregnancy, chronic infection, hemorrhage). Contrary to Popeye and popular opinion, spinach is a poor source of Fe.

Total **body content of Fe** amounts to c. 4–5 g; of this about 65% are bound to hemoglobin, 4% to myoglobin, 1% to several enzymes, and 0.1% to transferrin. The remaining Fe is stored as *ferritin* and *hemosiderin* in liver and spleen, and, in the case of ferritin, in the gut mucosa.

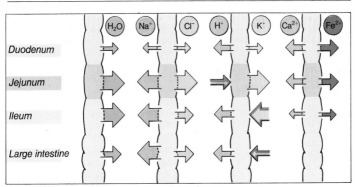

A. Intestinal absorption of water and electrolytes

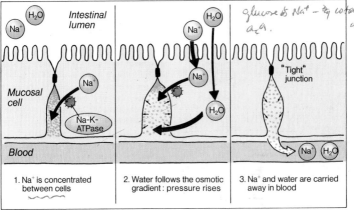

1. Na$^+$ is concentrated between cells

2. Water follows the osmotic gradient: pressure rises

3. Na$^+$ and water are carried away in blood

B. Absorption of salt and water

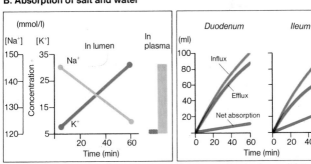

C. Na$^+$ - K$^+$ exchange in ileum
(after Code et al.)

D. Fluxes of water in small intestine
(after Code et al.)

Colon, Rectum, Defecation

The main **function of the colon** (1.3 m long) is to produce c. 100 to 200 g of feces from c. 500 to 1 500 ml of chyme. It continues the absorption of water and electrolytes that began in the upper intestinal tract. Vitamins are absorbed, some of which are synthesized by colonic bacteria. The colon has a larger diameter than the ileum. Its longitudinal muscles form three bands (teniae coli) that create pouches (haustra) along the length of the colon. The mucosa has no villi; there are, however, deep crypts in the mucosa lined predominantly with goblet cells that produce mucus. Some of the superficial cells have a brush border for absorption.

Motility. When food enters the stomach, the ileocecal valve relaxes allowing the small intestine to empty its contents into the large intestine (gastroileal reflex). Several mixing movements of the colon can be distinguished, chief among which is *haustration* or formation of pouches along the length of the colon. *Mass movements*, strong peristaltic waves, occur repeatedly throughout the day (4- to 6-h intervals). Mass movements require integrity of the myenteric plexus. Normally, three to four such movements are needed to propel the colonic contents to the rectum; however, they serve only for transfer and are not directly related to defecation.

A typical sequence is shown in Fig. **A**, which depicts the movement of a barium mass as seen during an X-ray examination. The meal is taken at 7 :00 A. M. At 12 :00 noon, the barium contrast medium has passed to the terminal ileum and cecum. The noon meal hastens emptying into the cecum. An annular constriction develops ($\rightarrow$ **A**, **3**). The transverse colon is filled ($\rightarrow$ **A**, **4**). Haustrations and annular contractions are seen throughout the length of the colon; their action mixes the colonic contents ($\rightarrow$ **A**, **5**). Some minutes later,

a gastrocolic reflex initiates a mass movement ($\rightarrow$ **A**, **6**), which moves the contents into the sigmoid ($\rightarrow$ **A**, **7**) and rectum ($\rightarrow$ **A**, **8**).

Final water resorption from the intestinal contents takes place in the **rectum**. Absorbable substances introduced into the rectum by clysis or by suppository enter the systemic venous drainage and bypass the liver. This method of drug administration is preferred for infants.

When the upper rectum (ampulla) ($\rightarrow$ **B**, **6**) becomes distended by its contents, pressure receptors ($\rightarrow$ **B**, **7**) stimulate a sensation of urgency for **defecation**. The act of defecation is initiated voluntarily (in most cases). The longitudinal muscles ($\rightarrow$ **B**, **8**) of the rectum contract. The two anal sphincters, the inner ($\rightarrow$ **B**, **3**) (involuntary) and the outer ($\rightarrow$ **B**, **4**) (voluntary), and the puborectal muscles ($\rightarrow$ **B**, **2**) relax. The rectum shortens, and the contents are expressed by annular contractions ($\rightarrow$ **B**, **9**) assisted by increased abdominal pressure ($\rightarrow$ **B**, **10**).

Dry weight of **feces** is 25% of its total weight ($\rightarrow$ **C**); one third of the dry weight is contributed by bacteria. Frequency of defecation varies widely, from three times daily to three times weekly, and depends upon the bulk content of the diet (chiefly cellulose). Cellulose is metabolized by intestinal bacteria to methane and other gases, which accounts for the flatulence that follows a meal of beans.

Diarrhea, if extreme, can lead to water and K^+ losses and acid-base disturbances (HCO_3^- loss; $\rightarrow$ p. 102). The severe diarrhea of **cholera** patients is caused by cholera enterotoxin. It elevates cyclic AMP ($\rightarrow$ p. 222) in the ileal mucosa, thus causing active NaCl secretion which is followed by water secretion into the gut lumen.

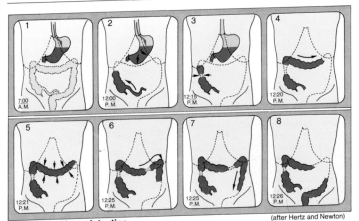

A. Motility of large intestine

(after Hertz and Newton)

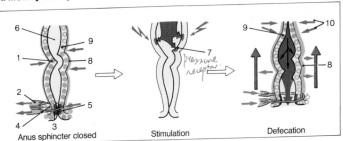

Anus sphincter closed   Stimulation   Defecation

B. Defecation

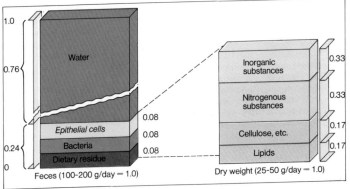

Water

0.76

Epithelial cells   0.08

Bacteria   0.08

Dietary residue   0.08

0.24

Feces (100–200 g/day = 1.0)

Inorganic substances   0.33

Nitrogenous substances   0.33

Cellulose, etc.   0.17

Lipids   0.17

Dry weight (25–50 g/day = 1.0)

C. Composition of feces

Physiologic Integration

In a single cell organism, such as a bacterium or a protozoan, the entire cell responds to stimuli that change its environment. In contrast, multicellular animals with their many specialized groups of cells, each with individual functions, cannot produce a uniform response from all cells. Some means of central integration and coordination is therefore essential. In mammals, three overlapping systems have been developed to process physiologic information. Each system has a characteristic receiver and sender, as well as a messenger. These systems are geared to organize responses to internal and external stimuli to meet the prime objective: *maintenance of uniformity and constancy of the cellular environment*, the ECF, by integration of specialized responses of individual organs. In this way, the composition of the ECF is held within limits compatible with viability of the cells. In addition, *growth, maturation* and *reproduction* are controlled by these systems.

Characteristics of Integrating Systems

1. The **nervous system** is specialized for acute rapid responses. It has two branches: (a) the **somatic** controls mainly voluntary muscle and (b) the **autonomic** ($\to$ pp. 48–57) controls involuntary functions. It receives input from the external environment over sensory afferent nerves (centripetal) and transmits messages over motor efferent nerves (centrifugal). Contacts between nerve units (synapses) are well defined and are insulated from neighboring cells. The messenger (neurotransmitter) ($\to$ pp. 52–55) is a charged water-soluble molecule of small molecular weight. It is released into the synapse by an electric stimulus and carries messages over microscopically small distances to a small number of receiver (target) cells.

2. The **endocrine system** is specialized for chronic responses. It has two components: (a) glands of internal secretion and (b) specialized, endocrine cells diffusely distributed within tissues with nonendocrine function. It controls nutrition, growth, reproduction, and internal homeostasis. It receives input from changes in the ECF. Its messenger (hormone) may be lipid- or water-soluble and may have a small or large molecular weight. It is released into the ECF by a chemical stimulus and is carried by blood to a distant site of action. It has the potential of contacting all the cells in the body.

3. The **neuroendocrine system** has features of both the nervous and the endocrine system. It is special in being able to convert neural (electric) into chemical (humoral) signals. The messenger (hormone) is synthesized in specialized nerve cells; it is released by an electric stimulus into the blood and acts at a distant site. Cells of the neuroendocrine system form two large populations: (a) neurons with large cell bodies in the supraoptic and paraventricular nuclei of the hypothalamus; their axons end in the posterior pituitary where they release oxytocin or ADH to act chiefly on uterus, breast or kidney; (b) neurons with small cell bodies, also in the hypothalamus; they synthesize and secrete releasing and inhibiting hormones, which regulate secretions of the anterior pituitary. Additional neuroendocrine overlap occurs at the adrenal medulla and the pineal body.

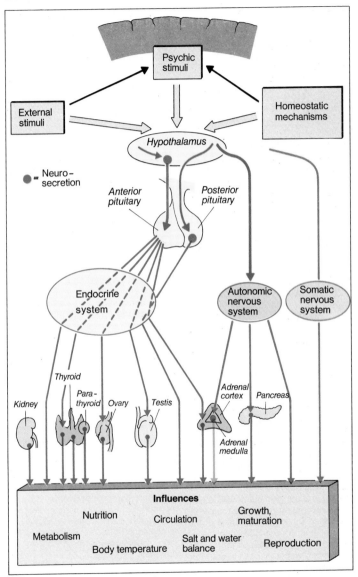

A. Neural and endocrine interrelations

Hormones

The **hormones** are endocrine messengers that evoke chemical responses in their target cells. They contrast with neurotransmitters, which are neural messengers evoking electric responses. There are two large classes of hormones: **steroids** (→ **A**, yellow), synthesized in glands of mesodermal origin, and **peptides** (including glycoproteins) (→ **A**, blue), synthesized in glands of ectodermal or endodermal origin. Peptides are produced as **prohormones** with low activity and are activated by hydrolytic splitting. Hormones of the thyroid and adrenal medulla (→ **A**, orange) are derived from tyrosine and form a third class. Watersoluble peptides are stored in intracellular granules, but fat-soluble steroids are not. Many hormones have short half-lives but long duration of action. Hormones are transported in the plasma and ECF bound to specific **hormone-binding proteins**, e.g., *transcortin* for cortisol. The binding proteins have a high affinity but low capacity for binding. The binding in the ECF does not initiate any biochemical response. Concentrations of hormones in the ECF are c. 10^{-8} to 10^{-12} mol/l.

Since hormones are carried in the blood, all cells are exposed to all hormones yet each hormone specifically influences only selected target cells. Each target cell must therefore possess a recognition site, the **receptor**, for its own hormone. Receptors have a high affinity and are stereospecific. Steroids are small molecules and bind with their receptors inside the cell (→ p. 224). Peptide hormones are large molecules and do not penetrate the cell membrane; receptors for peptide hormones are located on the outer surface of the cell membrane (→ p. 222). Binding of hormone with receptor activates the receptor's "executive site" and stimulates intracellular production of a second messenger. The second messenger initiates reactions that constitute the "hormone effect„. Different target tissues may have similar or different receptors for the same hormone, or a single cell may have receptors for more than one hormone. The hormone effect may be (1) to elicit genetic expression and induce synthesis of specific cell regulatory proteins, or (2) to produce allosteric changes in enzymes that alter their activity, or (3) to influence delivery of a cell substrate by changing membrane permeability.

Hormonal hierarchy (→ **A**). In many cases, the intial stimulus for a hormonal response begins in the central nervous system (CNS). Input to the higher brain centers is connected synaptically with the **hypothalamus** (→ p. 218) and initiates a neurosecretory output (→ p. 220). The hypothalamic message is amplified in the **anterior** or **posterior pituitary** (APit or PPit) (→ p. 218), which produce a second hormonal secretion. Hormones from the PPit influence their **target cells** directly. Most hormones from the APit stimulate a **target endocrine gland** to secrete a third hormone, the **end hormone**. The end hormone reacts with a receptor on its **target cell** directly. In each of these steps, a small stimulus produces a larger response and achieves an amplification through as many as three stages. The cascade TRH : TSH : TH amplifies from 10^3 to 10^6 times.

The **hypothalamus** produces small molecular weight peptides that act on the APit as releasing (RH) or inhibiting (IH) hormones. Some APit hormones (TSH) are controlled by RH, some (PRL) by IH, and some (STH) by both. The hypothalamic hormones, RH or IH, influence synthesis and release of APit hormones, but their actions are not completely specific since a single hypothalamic hormone, either RH or IH, may affect more than one APit hormone.

The **anterior pituitary** (APit) secretes **tropic** peptide hormones that act generally to stimulate their target glands.

Several hormones of the APit (TSH, FSH, LH) are highly complicated **glycoprotein** structures. The APit also secretes the simpler peptides, STH and PRL which are both end hormones.

Target glands, thyroid, adrenal and gonads, produce an end hormone in response to a tropic hormone from the APit.

The **end hormones** have, in general, anabolic or catabolic effects.

Catabolic	Anabolic
Cortisol	Androgens
Catecholamines	Estrogens
Thyroid	Gestagens
Parathyrin	Insulin
Glucagon	Somatotropin

The distinction is by no means rigid, however, because many hormones have **both** anabolic and catabolic effects. Secretion of the catabolic hormones tends to increase during emotional arousal ("fight or flight") or in response to physical stimuli (exercise, heat, cold, fasting). At the same time, secretion of anabolic hormones declines but later increases during recovery.

The endocrine system is thus characterized by multiple simultaneous responses to specific stimuli. The profile of responses is different and distinctive according to the stimulus since no single endocrine gland ever functions in isolation; interdependence and interaction characterize the endocrine system.

Tissue hormones are produced outside of the classic endocrine system. Some have specific, others have quite general effects. Most have a wider range of target tissues than the conventional hormones. These hormones include **prostaglandins, angiotensin, bradykinins, histamine, serotonin**, and intestinal hormones: **gastrin, secretin**, and **pancreozymin-cholecystokinin**. The tissue hormones produce effects both on distant cells (endocrine) and on local cells in their immediate vicinity (paracrine).

Nomenclature of hypothalamic and pituitary hormones and abbreviations

Recommended name*	Other names	Abbreviations
Hypothalamus		

The ending-**liberin** is used for releasing hormones (RH) or factors (RF)
The ending-**statin** is used for release-inhibiting hormones (IH) or factors (IF)

Corticoliberin	Corticotropin RF	**CRF**, CRH
Folliberin**	Follicle-stimulating hormone RH	FSH-RH
Luliberin**	Luteinizing hormone RH	**LRH**, LHRH
Gonadoliberin**	Gonadotropic hormone RH	**FSH/LH-RH**, GNRH
Melanostatin	Melanotropin release IF	MIH, **MIF**
Melanoliberin	Melanotropin RH	MRH, **MRF**
Prolactostatin	Prolactin release IF	PHI, **PIF**
Somatostatin	Somatotropin or growth hormone release IH	STS, SIH
Somatoliberin	Somatotropin or growth hormone RF	SRH, **SRF**, GH-RH
Thyroliberin	Thyrotropin RH	**TRH**, TRF

Pituitary   The ending-**tropin** is used for all anterior pituitary hormones

Corticotropin	Adrenocorticotropic hormone	ACTH
Follitropin	Follicle-stimulating hormone	FSH
Lutropin	Luteinizing hormone or interstitial cell-stimulating hormone	LH ICSH
Melanotropin	Melanocyte-stimulating hormone	MSH
Somatotropin	Growth hormone	**STH**, GH
Thyrotropin	Thyroid-stimulating hormone	TSH
Prolactin	Luteotropic or mammatropic hormone	**PRL**, LTH
Oxytocin	Ocytocin	
Vasopressin	Antidiuretic hormone, adiuretin	ADH

* Recommended by IUPAC-IUB Comm. on Biochemical Nomenclature (1974).
**probably identical

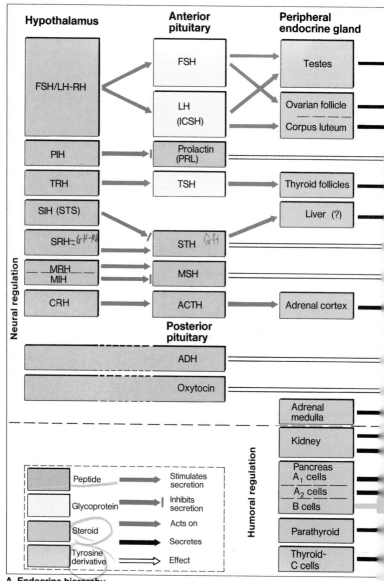

A. Endocrine hierarchy

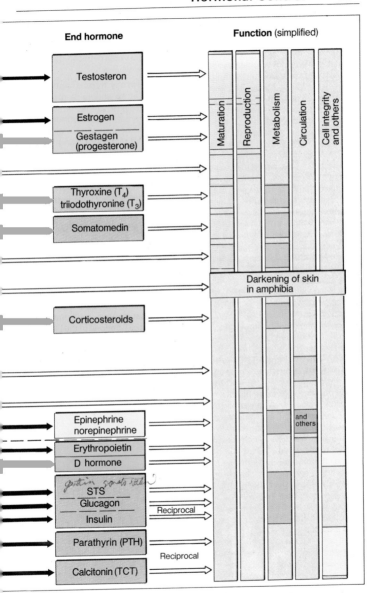

Feedback Control

Feedback is the circumstance in which the response of a cell influences the stimulus that produced that response. In **positive feedback**, the stimulus is amplified by the response. This results in a larger subsequent response, which feeds back to produce an even greater stimulus. Positive feedback is regenerative and, unless interrupted, the system is self-destructive. In **negative feedback**, the stimulus is decreased by the response. This results in a smaller subsequent response, which allows the stimulus to return to its original magnitude. Negative feedback has many biologic applications that maintain homeostasis. Hormones of the hypothalamus (HT) and anterior pituitary (APit) interact to achieve a relatively constant level of end hormone secretion. When the APit is switched on by the HT, it releases one of its tropic hormones (e.g., ACTH) (→ **A, 1**). The target gland responds to its tropic hormone with increased cell growth and secretion (→ **A, 2, 3**) in proportion to the tropic stimulation (positive limb of the feedback loop) (→ **A**, green arrows). The end hormone returns to the HT and/or APit via the blood stream and reduces secretion of the tropic hormone (→ **A, 4**) (inverse proportionality; negative limb of the feedback loop) (→ **A**, brown arrows). As secretion of the end hormone declines, its restraining influence on the HT and APit is reduced and more tropic hormone is secreted (→ **A, 6, 7**). By this means, fluctuations in hormonal concentrations are kept at a minimum. Feedback loops exist between HT, APit, and end hormone in several combinations (→ **A, 8**). Changes in serum osmolarity control ADH secretion by negative feedback (→ p. 126). The two hormone pairs, insulin-glucagon and parathormone-calcitonin, are reciprocating systems that regulate blood levels of glucose and calcium, respectively (→ pp. 216–217; 226–229; 234–237). A falling level of glucose feeds back to in-crease glucagon and decrease insulin secretion simultaneously; a rising level causes the converse response. A similar relationship exists between Ca^{2+} and its hormones.

Compensatory hypertrophy. When a portion of an endocrine gland is removed, as in thyroidectomy, the reduced secretion of end hormone allows more tropic hormone to be released. As a result, the remaining gland increases both in size and function until its original rate of secretion is restored. If the gland is unable to compensate fully, secretion of the tropic hormone is maintained at an elevated level (e. g., gonadotropin secretion after menopause).

Compensatory atrophy (iatrogenic). When a hormone is given as medication, it reduces secretion of tropic hormone in the same way as does the natural hormone. This reduced secretion has the effect of decreasing the size of the gland and reducing its natural secretion by an amount equivalent to the medication. As a result, the concentration of the hormone in the blood does not change. Only after the amount of medication is equal to the natural secretion, additional hormone given as medication will raise the plasma concentration. Chronic administration of a hormone thus results in atrophy of the gland that produces it.

Rebound. When hormonal medication is stopped, there is a time lag before the feedback system fully recovers. During this period when feedback is inoperative, one limb of the loop may become hyperactive because it is no longer under control, and it will produce larger than normal amounts of hormone.

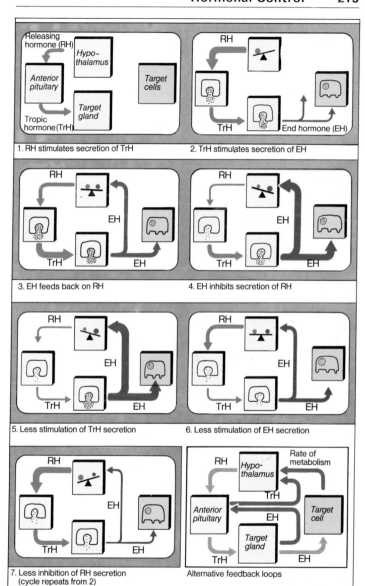

1. RH stimulates secretion of TrH

2. TrH stimulates secretion of EH

3. EH feeds back on RH

4. EH inhibits secretion of RH

5. Less stimulation of TrH secretion

6. Less stimulation of EH secretion

7. Less inhibition of RH secretion
   (cycle repeats from 2)

Alternative feedback loops

A. Negative feedback control of secretion

Neuroendocrinology

Neuroendocrinology concerns that part of the endocrine system that is under neural control. Secretion is an important aspect of the function of all nerves; neurosecretory nerves (NSN) are neurons that are specialized to synthesize and secrete humoral messengers. They have all the characteristics of regular nerves but do not form synapses; their messengers are released directly into the blood. Neurosecretion also refers to the direct neural control of hormone secretion as occurs in the adrenal medulla.

In the hypothalamus, NSN occur in more or less diffuse centers or nuclei. The NSN synthesize hormones on the endoplasmic reticulum of the soma and pass them on to the Golgi apparatus where the hormones are incorporated into granules (100 to 300 nm). Some synthesis also occurs in the axon. The granules are transported by axoplasmic flow to the nerve terminals. The posterior pituitary (PPit) is largely a collection of swollen axon terminals containing neurosecretory granules. The hormones are released by exocytosis ($\rightarrow$ p. 12) after an action potential (AP) reaches the nerve terminal. The AP of NSN is two to ten times longer in duration than in other nerves to allow sustained release of the hormone. Hormones of the PPit are released directly into the general circulation. Other neurosecretions that control the anterior pituitary (APit) are released into the veins of the pituitary portal circulation ($\rightarrow$ **A**) and are carried to the cells of the APit. Hormones produced by the hypothalamic NSN have three classes of action:

1. **Releasing hormones** (RH) stimulate release of tropic hormones in the APit. They act within minutes but have a short half-life. RH react with membrane receptors in the APit and stimulate synthesis of cAMP ($\rightarrow$ p. 222). Calcium is required. The cell membrane depolar-izes, and tropic hormone is released. Simultaneously, synthesis of new hormone is stimulated. Release of RH from NSN is under feedback control either by end hormone or by tropic hormone.

2. **Release-inhibiting hormones** (IH) suppress release of hormones of the APit. Only when secretion of the IH is blocked, is the tropic hormone released, e.g., PIF retards liberation of prolactin (PRL) ($\rightarrow$ pp. 216 – 217). Estrogen inhibits release of PIF and allows secretion of PRL. Somatostatin (SIH) is another IH; it suppresses release of several hormones: GH, TSH, but also glucagon, insulin, gastrin and motilin ($\rightarrow$ p. 192–196). IH have an action in the cell independent of cAMP activation.

3. **PPit hormones**: oxytocin ($\rightarrow$ p. 244) and ADH ($\rightarrow$ p. 126) are released into the general circulation and affect their target cells directly.

A number of peptides act outside of the endocrine system on higher nerve centers. SIH and TRH seem to be neurotransmitters as well as hormones and may influence behavioral and mood expressions. These actions may explain the intimate interactions between emotions, psychosomatic disturbances, and homeostatic mechanisms.

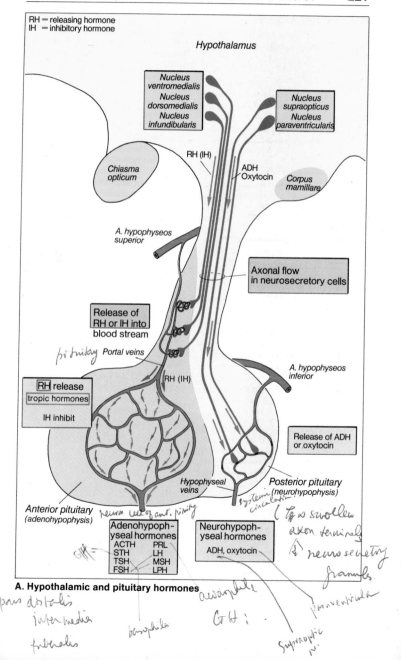

RH = releasing hormone
IH = inhibitory hormone

Hypothalamus

Nucleus ventromedialis
Nucleus dorsomedialis
Nucleus infundibularis

Nucleus supraopticus
Nucleus paraventricularis

Chiasma opticum

RH (IH)

ADH
Oxytocin

Corpus mamillare

A. hypophyseos superior

Axonal flow in neurosecretory cells

Release of RH or IH into blood stream

pituitary Portal veins

RH (IH)

A. hypophyseos inferior

RH release
tropic hormones

IH inhibit

Release of ADH or oxytocin

Hypophyseal veins

Posterior pituitary (neurohypophysis)

systemic circulation

Anterior pituitary (adenohypophysis)

neuron cell of ant. pituitary

(To a swollen axon terminals & neurosecretory granules

Adenohypoph-yseal hormones	
ACTH	PRL
STH	LH
TSH	MSH
FSH	LPH

GHH =

aerophile

Neurohypoph-yseal hormones
ADH, oxytocin

paraventricular

A. Hypothalamic and pituitary hormones

pars distalis
intermedia
tuberalis

basophile

GH:

Supraoptic n.

Cyclic Adenosine Monophosphate (cAMP) as a Second Messenger of Hormonal Control

The hormones, as first messenger molecules, are secreted into the blood stream and are carried to their target cells. At the target ($\to$ **A**), each hormone combines with a specific receptor and stimulates production of a second messenger that is exclusively intracellular. The second messenger initiates a sequence of reactions that result in expressing the characteristic cell response. **Peptide, glycoprotein, and catecholamine hormones** react with a receptor that is located on the outer surface of the membrane and commonly stimulate synthesis of **cAMP** as a second messenger. Synthesis of cAMP requires **adenylate cyclase (AC)**, the active site of which is located on the inner surface of the cell membrane. When the hormone combines with its receptor on the outer membrane surface, it activates AC. Although many hormones act via AC and cAMP, their action is specific since they only react with their own receptors on their target cells. Nevertheless, a cell may have receptors for more than one hormone. Fat cells, for example, respond to ACTH, insulin, glucagon, epinephrine, and secretin.

cAMP is found in all animal cells and in bacteria. The only other cyclic nucleotide universally distributed in mammalian cells is the guanyl analog, **cGMP**, the synthesis of which is catalyzed by guanylate cyclase. Synthesis of cGMP is activated by catecholamines (α) ($\to$ pp. 54–55), histamine H_1 receptors, acetylcholine (muscarinic) ($\to$ p. 52), and glutamate; cAMP is activated by catecholamines (β), dopamine, histamine H_2-receptors, serotonin, peptide, or protein hormones and prostaglandin E. In many cells, cAMP and cGMP give rise to reciprocal or antagonistic effects in a yin-yang relationship.

The action of cAMP is terminated by **phosphodiesterase (PDE)**, which breaks down cAMP ($\to$ **B**). PDE in turn is inactivated by a variety of compounds such as methylated xanthines (theophylline, caffeine), which prolong the life of cAMP in the cell and extend the duration of its effects.

The primary intracellular receptors for the second messenger (cAMP or cGMP or intracellular bound steroid [$\to$ p. 224]) are the **protein kinases**. These are normally inactive, but the presence of a second messenger liberates an active subunit that catalyzes phosphorylation of a specific protein. *Phosphorylated proteins* appear to be the *universal links* between any hormonal stimulation and the cellular response. In primitive cells, cAMP-regulated protein phosphorylation is associated with accumulation and utilization of energy-yielding substrates. In more complex tissues, the processes that are influenced are as diverse as glucose and fat metabolism, protein synthesis, cell mitosis and differentiation, secretion, membrane permeability, membrane potential, and muscle contraction. Biochemical specificity occurs because each cell has different receptors for the first messenger and different protein kinases and effector proteins. The action of phosphorylated proteins is terminated by *protein phosphatases*.

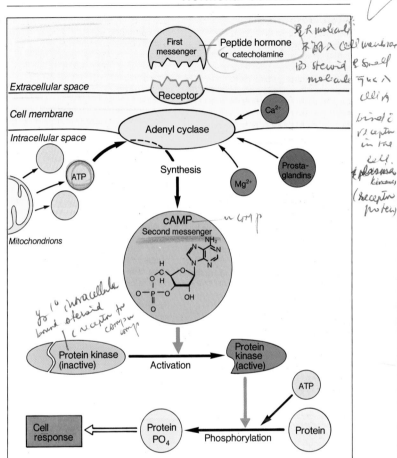

A. Second messenger - cAMP

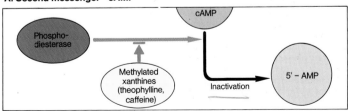

B. Inactivation of cAMP

Steroid Hormones

Steroid hormones and derivatives of vitamin D are similar to the peptide hormones in that they also initiate intracellular reactions, which bring about a characteristic cellular response, but the biochemical sequences in the two cases differ in several important respects (→ **B**). A steroid reacts first in the cytoplasm rather than in the membrane. The target cells for a given steroid contain binding proteins with a high affinity for the hormone; its combination with these specific **receptor proteins** (→ **A**) precedes appearance of the hormone effect. It is not known whether receptors for the same hormone are the same in each of the target tissues. Within the same cell, however, more than one receptor has been identified for a single hormone (estradiol), and some cells even have receptors for more than one steroid (estradiol and progesterone).

Receptors reside in the cytoplasm pending availability of their steroid. The number of receptors does not remain fixed; estrogen, for example, induces the generation of increased amounts of progesterone receptors in progesterone target cells. Receptivity to a given hormone may depend upon the amount of corresponding receptor in the cell. Deficiency of receptor may account for lack of responsiveness in some individuals.

Once the steroid-receptor complex is formed, it migrates and attaches to chromatin sites in the nucleus. Neither the steroid nor the receptor alone is capable of activating the nucleus. In an as yet undefined way, formation of the complex produces a conformational change in the receptor, which allows its translocation to DNA-acceptor sites on the chromatin of target cell nuclei. There are, for example, c. 500 sites for estradiol per nucleus, but they are not necessarily specific. Specificity resides, instead, in the presence or absence of receptors. Transcription processes are initiated in the nucleus within minutes, with an accumulation of mRNA as a key event (→ pp. 15–16). There follows an activation of rRNA and tRNA synthesis, which results in an increased capacity for **translation** when nuclear RNA is transported into the cytoplasm. New protein is synthesized on the ribosomes. This induced protein then initiates the characteristic steroid-mediated response of that target cell. Induced hormones are therefore specific for hormone and for the particular target cell. Glucocorticoids induce tryptophane pyrrolase, tyrosine aminotransferase, and glutamine synthetase; a vitamin D metabolite induces synthesis of a calcium-transport protein; aldosterone induces a protein that increases membrane transport of Na^+. The action of aldosterone can be inhibited by spirolactone. Spirolactone can bind with the aldosterone-binding receptor and thus prevents activation by aldosterone. Since the spirolactone-protein complex cannot bind to nuclear receptor sites, induction of effector protein does not occur.

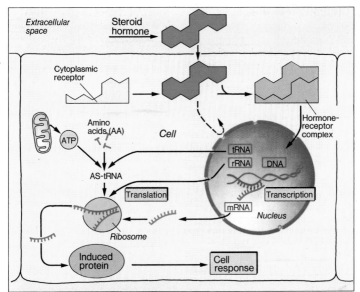

A. Mechanism of action of steroid hormones

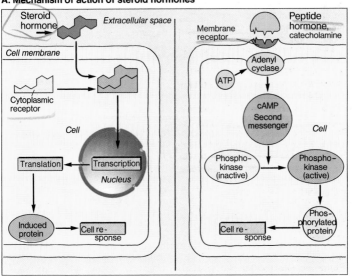

B. Comparison of action mechanisms for steroid and peptide hormones

226 Hormonal Control

Carbohydrate Metabolism

Glucose is the major source of metabolic energy (38 mol ATP/mol glucose). The concentration of glucose in the blood is kept constant by hormonal interactions that influence the balance between **glucose utilization** and **glucose production** ($\rightarrow$ **A**).

Definitions ($\rightarrow$ **C**)

Glycolysis: the anaerobic breakdown of glucose to lactate. In the process, α-glycerophosphate is formed and reacts with free fatty acids (FFA) to form triglycerides (TG) ($\rightarrow$ p. 182).

Glycogenesis: formation of glycogen from glucose; occurs chiefly in liver but also in muscle and kidney. It is a mechanism for storing excess glucose and it keeps glucose levels from rising.

Glycogenolysis: breakdown of glycogen to form glucose. It is the reverse of glycogenesis.

Gluconeogenesis: formation of new glucose, especially from amino acids. Long-chain fatty acids also stimulate synthesis of glucose from pyruvate by reverse glycolysis.

Lipolysis: breakdown of tissue fat and TG to form FFA.

Lipogenesis: extraction of FFA from blood to convert them to TG for storage of fat in the cells.

Ketogenesis: hepatic production of the "ketone bodies", acetoacetate, β-OH-butyrate, and acetone from fatty acids (and some amino acids). If synthesis exceeds breakdown, **ketosis** (and acidosis) develops.

cAMP ($\rightarrow$ p. 222) has many roles in carbohydrate metabolism. An increase in cAMP is associated with decreased glycogenesis, increased gluconeogenesis, proteolysis, hyperglycemia, and lipolysis.

Pancreas: has exocrine ($\rightarrow$ p. 198), paracrine, and endocrine functions. The islands of Langerhans (c. 3% of pancreas) contain several types of hormone-producing cells: A_1 cells produce gastrin; A_2 cells produce **glucagon**; B cells produce **insulin**; D cells produce somatostatin (STS). A cells comprise c. 25% and B cells c. 75% of the islands. Secretion from one type influences secretion from another type (paracrine function). **Functions**: (a) to resist changes in blood glucose levels; (b) to facilitate storage of energy sources after a meal by stimulating glycogenesis and lipogenesis; (c) to mobilize energy sources during fasting or exercise by reversing these processes.

Insulin: formed in B cells from a precursor, proinsulin (84 amino acids). Proinsulin has three components: an acidic A chain (21 amino acids), a basic B chain (30 amino acids), and a connecting C peptide. Insulin is formed by splitting the C peptide from proinsulin. Insulin possesses a complex three-dimensional configuration that is sustained by three sulfhydryl bonds; distortion of this structure reduces insulin activity. Insulin occurs in several polymeric forms, but the monomer is the active form. Insulin is stored in granules that are released by exocytosis; it has a half-life of 30 min and is inactivated by insulinases, chiefly in the liver. The pancreas stores 6 to 10 mg insulin and releases c. 2 mg/day. A dose of 4 µg/kg reduces blood sugar by 50%.

Regulation of insulin secretion. The chief stimulus for release is an elevation of blood glucose levels ($\rightarrow$ **B**). Stimulation releases 2% to 5% of the B cell content. Sensitivity of the B cells is greatest at a blood sugar of 100 to 300 mg% (0.6 to 1.7 mmol/l) and is enhanced by glucagon, gastrin, or secretin. Amino acids (lysine, arginine, leucine) and hormones (STH, ACTH, TSH, adrenal, and gonadal steroids) also are stimulatory. Insulin secretion is inhibited by epinephrine (Ep) via α-receptors and by STS. When insulin reduces the blood glucose level, the

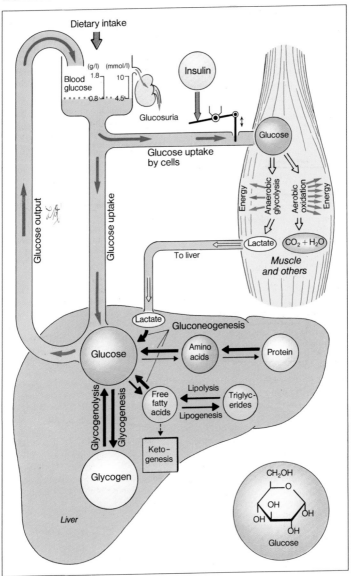

A. Glucose metabolism

change is detected in the CNS, and Ep is released to block insulin secretion. Some chemicals (alloxan) destroy B cells and produce insulin deficiency diabetes.

Actions (→ **A**). Insulin lowers blood sugar by increasing cellular uptake and decreasing hepatic production of glucose. Its effects are chiefly anabolic. It acts on muscle, fat, and liver; in other tissues, especially brain, it has little or no effect. Insulin reacts with its target cell membrane receptors to produce responses equivalent to those of reduced cAMP. The major responses are (1) increased flux of glucose and amino acids from blood into cells and (2) altered enzyme activity in muscle, liver (glycogenesis), and fat (lipogenesis). Insulin deficit results in a form of diabetes with hyperglycemia, increased lipolysis, and elevated plasma levels of FFA. Hepatic oxidation of FFA is increased, ketone bodies are generated in excess, and metabolic acidosis may develop.

Glucagon increases blood sugar by increasing hepatic production of glucose.

Chemistry. It is a single-chain peptide (29 amino acids), synthesized in the A_2 cells, and stored in granules. In some species, enteroglucagon is produced in the intestinal tract.

Control. Hypoglycemia and amino acids (alanine) are the chief stimuli for release. Starvation, sympathetic stimuli, and lowered plasma levels of FFA are also effective. Secretion is inhibited by insulin, STS and by hyperglycemia.

Actions. Liver and fat are the major target cells; glucagon is ketogenic in liver and lipolytic in fat. In the heart, it has a positive inotropic effect by stimulating β-receptors. It stimulates insulin release, but the significance of this paracrine effect is not known.

Blood sugar. Glucagon stimulates hepatic glycogenolysis to mobilize glucose (hyperglycemia) but is antagonized by insulin, which activates cellular utilization of glucose (hypoglycemia). The end effect is maintenance of a constant blood sugar concentration. During an emergency (fight-or-flight), glucagon actions are supported by epinephrine (Ep), which stimulates glycolysis both in liver and muscle.

Net hepatic glycogenesis is determined by the balance among insulin, glucagon, and Ep. The action of insulin dominates over glucagon. Net gluconeogenesis is influenced by the balance among insulin and adrenal corticosteroids. STH and STS are also involved in blood sugar homeostasis.

Somatostatin (STS), or growth hormone release-inhibiting hormone (SIH), is a recently discovered peptide (14 amino acids). It is found in the CNS, the hypothalamus, D or A_1 cells of the pancreas, stomach, and thyroid. It is also found in the soma of many adrenergic neurons.

Control. Glucose, amino acids, cholecystokinin-pancreozymin, secretin as well as glucagon stimulate pancreatic STS release.

Actions. It suppresses cAMP levels and calcium influx at the cell. It inhibits release of STH, TSH, gastrin, insulin, glucagon, and enteroglucagon. It reduces GI activity probably by inhibiting the release of acetylcholine by parasympathetic nerves, and it retards the entry of ingested nutrients into the circulation. These actions may permit the pancreas to coordinate the uptake of nutrients from the gut (STS) with their efflux from ECF to ICF (insulin).

Adrenal steroids and catecholamines also influence blood sugar (→ pp. 240; 54–57).

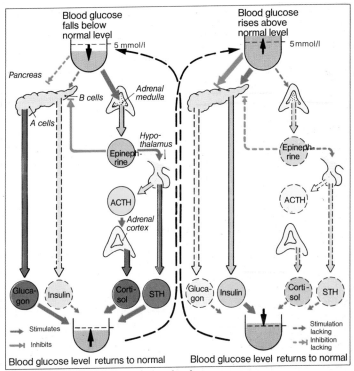

B. Hormonal control of blood glucose level

Hormone	Insulin	Glucagon	Epinephrine	Cortisol
Function	Regulatory		Emergency	Restorative
Glucose				
Peripheral utilization	+	−	+	−
Glycolysis	+	−	+	−
Gluconeogenesis	−	+	+	+
Glycogen				
Synthesis ⇌ lysis	←	→	→	←
Fat				
Synthesis ⇌ lysis	←	→	→	→

C. Hormonal influences on carbohydrate and fat metabolism

Thyroid Hormones

The thyroid gland contains spherical follicles (50 to 500 μm diameter), the cells of which synthesize the thyroid hormones (TH): **thyroxine** (T_4) and **triiodothyronine** (T_3). The thyroid also contains clear cells (parafollicular or C cells), which synthesize calcitonin ($\to$ p. 236). The TH influence growth, differentiation, and metabolism. They are stored in **colloid**, a viscous fluid that fills the space inside the follicles.

Chemistry

Synthesis of thyroglobulin (TGB) ($\to$ **A**). TGB is a nonhormonal glycoprotein specialized for storing TH. It is synthesized on ribosomes from the amino acid pool; carbohydrates are added at the Golgi apparatus. The completed TGB is transferred by exocytosis to the follicular space where it is incorporated into colloid. Plasma level of TGB = 5 μg/l.

Synthesis of TH ($\to$ **B, D**): There are four steps in synthesis. They occur principally on the membrane surface. Step 1: iodide (I^-) is **actively** accumulated at the basal cell membrane. This active transport is stimulated by TSH (via cAMP) ($\to$ p. 222). Anions such as perchlorate and thiocyanate compete with I^- for transport. In follicle cells, I^- can be concentrated to 50 times the plasma concentration. Active accumulation of I^- occurs also in other tissues: gastric mucosa, choroid plexus, salivary glands, etc. Step 2: **oxidation** ($I^- \to I°$) takes place at the apical membrane and requires a peroxidase. Step 3: **organic binding**. The $I°$ thus formed binds to tyrosyl residues on TGB and forms monoiodotyrosine or diiodotyrosine (MIT or DIT). These steps are stimulated by TSH and are inhibited by perchlorate, thiouracils, and large levels of I^-. Step 4: **coupling** of iodinated tyrosines generates the TH. DIT

+ DIT $\to T_4$; MIT + DIT $\to T_3$. Both T_3 and T_4 are stored in colloid on TGB.

Release of TH ($\to$ **C**): TSH from the pituitary simultaneously stimulates resorption of TGB from the follicular space (by endocytosis) and migration of lysosomes from the basal region of the cell toward the colloid droplets. The lysosomes fuse with the TGB and liberate T_3 and T_4 (by hydrolysis) for release into the blood. TGB is thus both substrate for hormone synthesis and storage form of the hormones. MIT and DIT are also generated during hydrolysis. These compounds are deiodinated and the products, tyrosine and I are returned to the cell pool. The I pool of the thyroid ($\to$ **E**) is unusually large. Even under maximal stimulation by TSH, no more than 1% to 2% of the total I content is depleted per hour.

Iodine metabolism ($\to$ **E**): Circulating I exists in three forms: (1) inorganic at a plasma level of 2 to 10 μg/l, (2) organic nonhormonal in the form of TGB, MIT, and DIT in minute amounts, and (3) organic hormonal in the form of protein-bound iodine (PBI) at a plasma level of 35 to 80 μg/l. Some 90% of the PBI is T_4 and can be extracted as butanol-extractable iodine (BEI). Daily, 50 to 100 μg of hormonal I is utilized and must be replaced by synthesis of an equivalent amount. Iodine that is lost by excretion is replaced from dietary sources such as sea food, iodized salt, and bread.

Thyroid hormone metabolism. T_4 has been considered to be a prohormone that generates T_3 as the active hormone at the target cell. T_3 is two to four times as potent as T_4. T_3 acts promptly, producing its effects within hours, whereas T_4 requires days for maximal response. Of the circulating T_3, c. 20% is derived from secretion by the thyroid gland and 80% from deiodination of T_4 at the target cells. Of circulating T_4, 1/4 to 1/3 is converted to T_3, 1/4 is excreted in bile and urine as a glucuronide, and c. 1/2 exerts hormonal activity. In plasma, the ratio T_3/T_4 = c. 0.01. Both hormones are

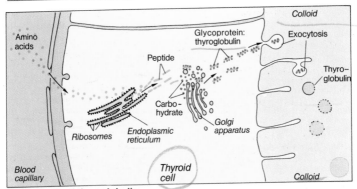

A. Synthesis of thyroglobulin

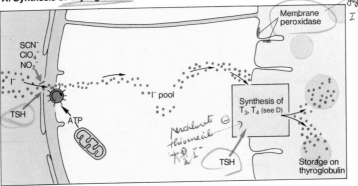

B. Synthesis of thyroid hormones

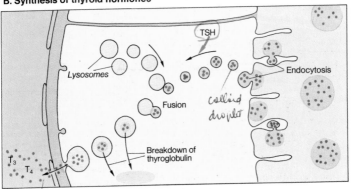

C. Secretion of thyroid hormones

circulated in the plasma in protein-bound form. Several proteins contribute to the binding: (a) thyroid-binding globulin (TGB) is specific, has high affinity, is present in low concentration; it has a much higher affinity for T_4 than for T_3 but it binds both hormones, and its concentration is increased by estrogen; (b) thyroid-binding prealbumin (TBPA) binds only T_4; (c) serum albumin (HSA) binds the remainder of T_3 and T_4; it is present in high concentration but has a low affinity for the TH.

Control of TH secretion. To meet changing requirements, most hormones undergo large fluctuations in their plasma concentrations. The TH are unusual in that their plasma concentrations remain remarkably constant; when increased utilization of TH begins to lower plasma concentration, restorative mechanisms in hypothalamus and anterior pituitary (APit) are immediately initiated. The system for TH regulation has several components:

1. The **hypothalamus** contains **TRH-synthetase**, which generates TRH (pGlu-His-Pro-amide). The enzyme is stimulated by acute cold. Neurosecretory release of TRH is influenced by monoamines (dopamine and perhaps norepinephrine). TRH action in the APit is not specific; in addition to effects on TSH, it also increases circulating PRL and STH.

2. The anterior pituitary contains cells specialized for synthesis of **TSH**, the *thyrotropes*. TRH binds to their membranes, activates cAMP production, and thereby increases both release and synthesis of TSH by apparently independent mechanisms. Thyrotrope function is inhibited by a negative feedback loop involving TH. T_3 binds to thyrotrope nuclei (affinity for T_3 is ten times greater than for T_4) and influences production of a protein that blocks the response of the thyrotrope to TRH.

3. The **thyroid** responds to TSH via increased synthesis of cAMP ($\rightarrow$ p. 222). Within 2 min, I^- uptake and synthesis of TH increase; colloid resorption and lysosome migration occur within 5 min; TH are released into blood within 10 min.

Goiter is enlargement of the thyroid gland. It may be diffuse or nodular. Diffuse goiter occurs following interference with the negative feedback loop of TH control. If this causes a chronic increase in release of TSH from the APit, hypertrophy and hyperplasia of the follicular cells in the thyroid gland occur. In goiter, thyroid function may be increased (Graves' disease), normal (I deficiency), or reduced (goitrogen ingestion).

Actions of TH. The target tissue for TH is poorly defined; many cells are affected. Unlike peptide hormones, TH bind intracellularly, at the nuclei, but unlike steroids, a cytoplasmic receptor is not required for translocation. The simplest hormone structures are the TH and the catecholamines (Ep and NE); they are hydroxylated aromatic amino acids and have several functional similarities. The TH, like Ep, stimulate O_2 **utilization** (metabolic rate) and **heat production**. The TH have a **permissive** action for other hormones by changing affinity of membrane receptors. In hypothyroidism, calorigenic effects of insulin, glucagon, STH, Ep, and F are minimal. In hyperthyroidism, sensitivity to Ep is increased; in fact, many of the symptoms resemble a state of epinephrine excess. The TH also affect **growth** and **maturation**, especially of brain and bone. In the newborn, TH deficiency, if not corrected, results in **cretinism** with mental retardation (irreversible after 6 months), delayed maturation of bones and teeth, stunted growth, and retarded sexual development. The best current concept to explain the multiple and varied actions of TH is that they alter membrane characteristics by changing permeability or receptor affinity.

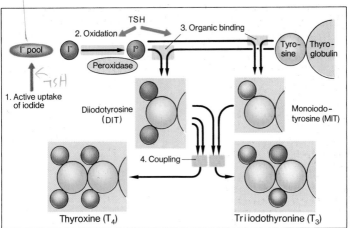

D. Synthesis of T₃ and T₄

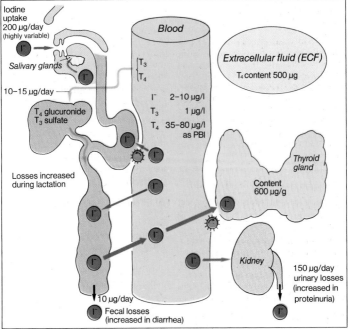

E. Iodine balance

Calcium Homeostasis and Bone Metabolism

Calcium (Ca) is critical for cell growth and development. It influences membrane permeability, blood clotting, neuromuscular transmission ($\rightarrow$ p. 30), excitation-contraction ($\rightarrow$ p. 36), and excitation-excretion coupling, sol-gel transformations, enzyme reactions, and cAMP-dependent functions ($\rightarrow$ p. 222).

Ca metabolism ($\rightarrow$ **A**). Ca is the fifth most abundant element in the body comprising 2 % of body weight; 99 % is in bone, 1 % in body fluids. Its concentration in plasma = 2.3 to 2.7 mmol/l (9.2 to 10.8 mg%). In plasma, Ca exists in three forms: (1) ionized (c. 50 %), (2) complexed with anions (c. 10 %) and (3) bound to plasma proteins (c. 40 %). Protein binding varies with protein concentration and blood pH (0.2 mmol Ca^{2+}/l for each pH unit). Binding increases in alkalosis and decreases in acidosis. **Phosphorus** (P) metabolism is intimately related to Ca. Plasma level of inorganic P = 0.8 to 1.4 mmol/l (2.5 to 4.3 mg%); levels are higher during growth. Total body P = c. 225 mmol, 80% to 85% being in the bones and teeth. Salts of Ca and P are poorly soluble in water. In a nearly saturated solution as in blood, the product of Ca and P concentrations is constant. Thus, if P increases, Ca *must* fall. Clinically, an infusion of P will drive Ca into the bone and will cause hypocalcemia. Conversely, if plasma levels of P decrease, bone dissolves and Ca is liberated to raise the Ca level.

Bone is composed of an organic matrix (30%), inorganic minerals (45%), including Ca, P, Mg, and Na, and water (25%). The Na content of bone contributes significantly to Na^+ balance in the body. Matrix is largely collagen that contains a large fraction of hydroxyproline (OH-Pro), which is found only in collagen. When bone is resorbed, urinary excretion of OH-Pro increases.

Mature bone is in a dynamic equilibrium of continuous resorption and deposition. The bone surface contains undifferentiated cells. These are transformed to preosteo**clasts**, which are **activated** to osteoclasts to resorb bone surface. When osteoclastic activity subsides, the cells are **modulated** to preosteo**blasts** and then to osteoblasts that lay down new bone surface. Osteoblasts either **mature** to osteo**cytes** or **revert** to undifferentiated cells. Activation to osteoclasts is stimulated by PTH with vitamin D, STH, and TH and is inhibited by TCT, E_2, and F. Modulation is stimulated by physical stress, TCT, P, STH, and E_2 and is prevented by PTH with vitamin D. **Alkaline phosphatase** in the osteoblasts creates a high local concentration of P, which exceeds the solubility product for Ca and P and precipitates Ca on the bone surface.

Ca balance ($\rightarrow$ **B**) is the result of intake and loss of Ca. It involves shifts of Ca across five compartments: ECF, ICF, bone, GI tract, and kidney under the influence of three hormones: PTH, TCT, $1,25\text{-}(OH)_2\text{-}D_3$. Ca enters plasma by GI absorption and bone resorption and is removed from plasma by bone deposition and renal excretion. Daily Ca requirement 0.25 to 0.5 mmol/ day/kg (c. 1 g/day for an adult). Only 15% to 40% of dietary Ca is absorbed. Fecal Ca is 10 to 20 mmol/day and is mostly unabsorbed dietary Ca ($\rightarrow$ **A**). Urinary excretion is c. 2 to 3 mmol/day ($\rightarrow$ **A**). In pregnancy, the fetus absorbs 500 to 800 mmol Ca from the mother; the newborn takes as much as 2000 mmol Ca in milk from the mother. Negative Ca balance is common in lactation.

Parathyroid Hormone (PTH) ($\rightarrow$ D)

Chemistry. PTH is synthesized on ribosomes of the parathyroid (PT) gland, first as a prohormone (109 amino acids), which is hydrolyzed to PTH (84 amino acids) and is stored in granules. Secretion rate = 0.1 to 1.0 pmol/kg

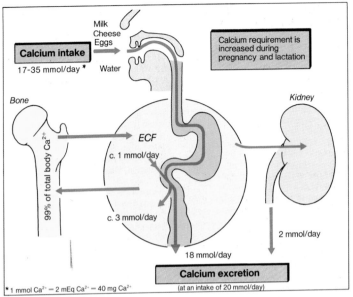

A. Calcium balance

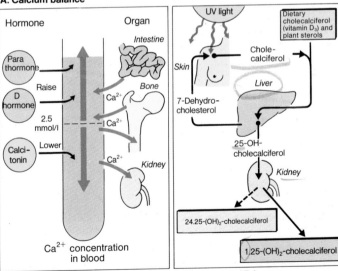

B. Influences on Ca²⁺ blood levels

C. Synthesis of D hormone

/min; gland content = 10 pmol/kg.

Regulation. Synthesis and secretion are inversely related to Ca^{2+} levels in plasma. Ca^{2+} regulates uptake of amino acids into pro-PTH. In the PT gland, a high Ca^{2+} level inhibits cAMP. In most other cells, Ca stimulates cAMP ($\rightarrow$ p. 222) synthesis.

Actions and effects. PTH is released in hypocalcemia ($\rightarrow$ **D**, left side) to restore Ca^{2+} levels. PTH acts via cAMP ($\rightarrow$ p. 222). **Bone**: PTH stimulates osteoclasts to resorb bone Ca. Bone matrix is destroyed and OH-Pro is excreted. **GI tract**: PTH increases Ca^{2+} absorption in the presence of vitamin D. **Kidney**: PTH (a) reduces tubular resorption of P, causing phosphaturia and (b) increases tubular resorption ($\rightarrow$ p. 137) of Ca^{2+}. In hypocalcemia, as in treatment of hypoparathyroidism, PTH first causes a small reduction of renal Ca^{2+} excretion, but as the plasma level of Ca^{2+} rises, filtered Ca^{2+} increases and eventually overwhelms tubular resorption so that in the end Ca^{2+} excretion increases. PTH also affects renal synthesis of 1,25-$(OH)_2$-D_3 ($\rightarrow$ **C**).

Calcitonin (TCT) ($\rightarrow$ D)

Chemistry. TCT is produced in the parafollicular C cells of the thyroid. C cells comprise 20% of the gland. TCT (32 amino acids) is secreted at 10 pmol/kg/h; gland content = 2 nmol/kg.

Regulation. Secretion and synthesis are increased 10 to 100 times by elevated Ca^{2+} ($\rightarrow$ **D**, right side) via cAMP. Ca^{2+} stimulates C cells to produce cAMP.

Actions. In hypercalcemia, TCT is released to lower plasma Ca^{2+}. **Bone**: TCT inhibits bone resorption and excretion of OH-Pro (hypocalcemic action). **GI tract**: no important action. **Kidney**: increased excretion of Ca^{2+} and P (hypophosphatemic action). TCT inhibits conversion of 25-OH-D_3 to 1,25-$(OH)_2$-D_3 ($\rightarrow$ **C**).

Vitamin D

Chemistry ($\rightarrow$ **C**). Vitamin D is taken in the diet as irradiated plant sterols (calciferol, vitamin D_2) or is synthesized from the provitamin 7-dehydrocholesterol by the action of UV light on the skin to form vitamin D_3. In liver microsomes, D_3 is hydroxylated to 25-OH-D_3, which reaches the kidney where it is again hydroxylated to form either 1,25-$(OH)_2$-D_3 (**D-hormone**) or 24,25-$(OH)_2$-D_3. Synthesis of the two forms is reciprocal; when more 1,25-$(OH)_2$-D_3 is formed, less 24,25-$(OH)_2$-D_3 is synthesized. D_3 is a vitamin because its synthesis is controlled by UV light instead of an enzyme. 1,25-$(OH)_2$-D_3 is a hormone because it acts at a site distant from where it is synthesized. D hormone is transported in plasma bound to specific proteins.

Regulation. Synthesis of D hormone is stimulated by low intracellular P or by Ca^{2+}. When PTH enhances renal excretion of P, it enhances D hormone synthesis. Hypocalcemia stimulates synthesis indirectly by calling forth PTH secretion.

Actions ($\rightarrow$ **D**). D hormone acts intracellularly (like steroids) by binding to a receptor and acting on the nucleus. **Bone**: D hormone is a cofactor for PTH to stimulate bone resorption. Both the 1,25- and 24,25-$(OH)_2$ forms are necessary for bone mineralization. A deficit results in rickets in children. The 1,25-$(OH)_2$ form maintains blood levels of Ca and P; the 24,25-$(OH)_2$ form contributes to mineralization of bone matrix. **GI tract**: increases absorption of Ca and P independent of PTH. Induces synthesis of a Ca-binding protein. **Kidney**: stimulates tubular resorption of Ca and P. It can antagonize the phosphaturic action of PTH.

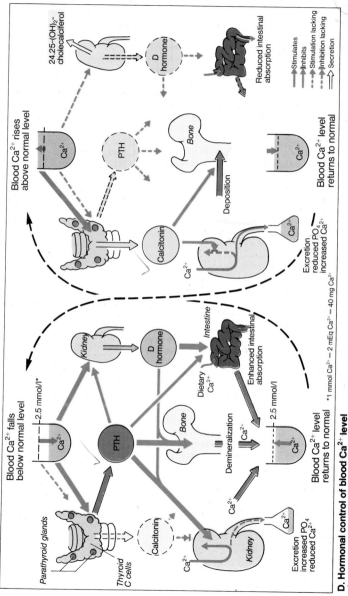

D. Hormonal control of blood Ca²⁺ level

*1 mmol Ca²⁺ = 2 mEq Ca²⁺ = 40 mg Ca²⁺

Blood Ca²⁺ rises above normal level

24.25-(OH)₂-cholecalciferol

D hormone

Reduced intestinal absorption

Ca²⁺

PTH

Bone

Calcitonin

Deposition

Ca²⁺

Blood Ca²⁺ level returns to normal

Excretion reduced PO₄, increased Ca²⁺

Stimulates
Inhibits
Stimulation lacking
Inhibition lacking
Secretion

Blood Ca²⁺ falls below normal level

2.5 mmol/l*

Kidney

D hormone

Intestine

Dietary Ca²⁺

Enhanced intestinal absorption

Ca²⁺

PTH

Bone

Demineralization

2.5 mmol/l

Blood Ca²⁺ level returns to normal

Ca²⁺

Parathyroid glands

Thyroid C cells

Calcitonin

Kidney

Ca²⁺

Excretion increased PO₄, reduced Ca²⁺

Biosynthesis of Steroid Hormones

Cholesterol is the source of all steroids (→ p. 224). It is derived ultimately from acetate and is synthesized chiefly in the liver but also in steroid-producing glands. The placenta (→ p. 248), which also produces steroids, is unable to synthesize cholesterol from acetate and must acquire it directly from the circulation. All steroid-producing glands contain stores of cholesterol and of ascorbic acid. When the glands are stimulated to steroid biosynthesis, cholesterol and ascorbic acid content decline in proportion to the stimulus.

The keystone of steroid biosynthesis is **progesterone (P)**, a C-21 steroid; almost all steroid hormones and intermediates can be derived from P. Control of steroid synthesis by ACTH, FSH and LH is achieved in large part by influencing transformation of cholesterol to P (→ **A, a, b**). In principle, all biosynthetic pathways are common to all steroid-producing glands, i.e., each gland is potentially capable of producing all of the steroid hormones. In practice, however, each gland is specialized to produce principally its own characteristic hormones. **Adrenocortical steroids** contain 21 carbon atoms, **male sex steroids** 19 carbons, and **female sex hormones** 18 carbons. Both testis and ovary are specialized to split the C-20, 21 side chain (→ **A, m, l**). The ovary carries the synthesis one stage farther than the testis to produce estrogens. The adrenal cortex performs a number of hydroxylations involving carbons at C-11, 17, and 21 to generate three categories of corticosteroids. In general, hydroxylation at C-21 (→ **A, c, h**) must precede that at C-11 (→ **A, d, j**). If C-17 is hydroxylated (→ **A, f**), synthesis proceeds toward **cortisol** (→ **A, j**), the major adrenal cortical hormone. Hydroxylation at C-17 occurs in the zona fasciculata and reticularis of the cortex but not in the glomerulosa (→ p.

240). In the glomerulosa, **desoxycorticosterone** and **aldosterone** (→ **A, c, d, e**) are synthesized rather than cortisol (→ p. 240).

From P, synthesis can go via androstenedione (→ **A, l**) to **testosterone** (→ **A, g**) (male sex hormone) or to **estradiol** (→ **A, o, p**) (female sex hormone). An alternate sequence via dehydroepiandrosterone (DHEA) (→ **A, m**) also leads to sex steroids. The chief precursors for sex steroids are 17-OH progesterone (→ **A, f**) or 17-OH pregnenolone (→ **A, g**). From these are generated the **17-ketosteroids** (→ **A, l**), DHEA, etiocholanolone, androstenedione, and androsterone. The 17-ketosteroids are synthesized in adrenal cortex as well as in gonads and are excreted into the urine. Note that male sex hormones are intermediates in the synthesis of female sex steroids (→ **A, r**).

Steroids are rapidly removed from the blood by the liver. After oral intake of steroids, as much as 90% to 95% of the dose can be extracted from portal blood in one passage through the liver (first pass effect). In the liver, hydroxy groups are conjugated with sulfate or with glucuronic acid (→ p. 116), and the products are rapidly excreted into bile or urine. **Estriol** is the principal excretory product of estrogens and **pregnandiol** (→ p. 248) of gestagens.

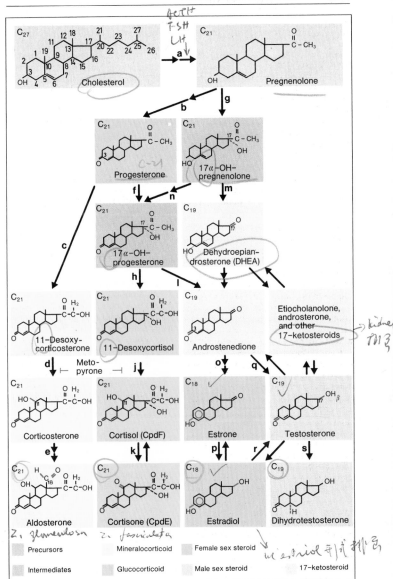

A. Biosynthesis of steroid hormones

Adrenal Cortex: Glucocorticoids

Chemistry. The adrenal cortex contains three zones: zona glomerulosa (ZG), fasciculata (ZF), and reticularis (ZR) ($\rightarrow$ **A**). ZG produces salt-retaining or mineralocorticoids; ZF produces glucocorticoids, the sugar or S hormones; ZR produces anabolic or nitrogen-retaining N hormones ($\rightarrow$ p. 250). In man, 17-hydroxylase, which is necessary for synthesis of cortisol (F), does not occur in the ZG; for this reason, ZG cannot synthesize F and other glucocorticoids. An essential characteristic of glucocorticoids is presence of an OH or keto-group at positions 11 and 17 ($\rightarrow$ p. 239). Glucocorticoids are transported principally in protein-bound form. Transcortin, a globulin, binds specifically with high affinity but low capacity; albumin binds non-specifically with low affinity but high capacity.

Regulation ($\rightarrow$ **A**). Corticotropin-releasing hormone (CRH) from the hypothalamus stimulates synthesis and release of ACTH in the anterior pituitary. ACTH activates all adrenal cortical zones but mostly ZF and ZR. The chief hormone produced is **cortisol** (hydrocortisone or **F**). Release of ACTH is increased by physiologic "stress" and by epinephrine (Ep); it is under negative feedback control by F ($\rightarrow$ p. 218). Spontaneous ACTH release is minimal at night and maximal in the early morning ($\rightarrow$ **B**). Feedback inhibition by F is weakest at the time of the morning ACTH peak. If F synthesis is blocked (because of inborn enzyme deficiency or during metopyrone medication, $\rightarrow$ p. 239, A, d, j), ACTH production increases and causes adrenal hyperactivity and subsequent hypertrophy (compensatory hypertrophy) ($\rightarrow$ p. 218).

Action. In response to physiologic "stress" (fight-or-flight), catecholamines are released from the adrenal medulla ($\rightarrow$ pp. 56–57). Within minutes, they stimulate mobilization of defenses and release of F via ACTH. Effects of F become maximal in 2 to 4 h; they support defense mechanisms and simultaneously replenish energy stores depleted by Ep.

(a) **Carbohydrates**: F blocks peripheral utilization of glucose and causes hyperglycemia ($\rightarrow$ p. 229, C). It restores depleted hepatic glycogen by stimulating gluconeogenesis and glycogenesis. Since gluconeogenesis requires mobilization of amino acids from tissue proteins, F has a **catabolic** effect and leads to nitrogen loss. These actions may be described as **diabetogenic** or anti-insulin effects. (b) **Antiphlogistic**: F suppresses inflammatory responses by inhibiting prostaglandin synthesis and by stabilizing lysosomal membranes. Proliferation of fibroblasts is reduced and less scar tissue is formed, but wound healing is retarded. F can hinder localization of infections by the body's defenses. Lymphocytes, lymph tissue, and eosinophils contribute to immunologic defenses by cellular disruption. This destruction of selected cell types causes increased excretion of purines, chiefly uric acid, in urine. (c) **Minerals**: all corticosteroids have some salt-retaining effect, but, with the exception of aldosterone and desoxycorticosterone, the effect is insignificant at physiologic levels. At higher concentrations, salt retention becomes clinically apparent and is often associated with the development of hypertension. (d) **Circulation**: glucocorticoids have profound effects on blood vessels; they antagonize the vasodilating effect of intrinsic histamine (also effective in antiphlogistic action) and have a permissive effect on the vasoconstrictive response of small vessels to catecholamines. This permissive effect supports the circulation and retards the development of shock ($\rightarrow$ p. 172).

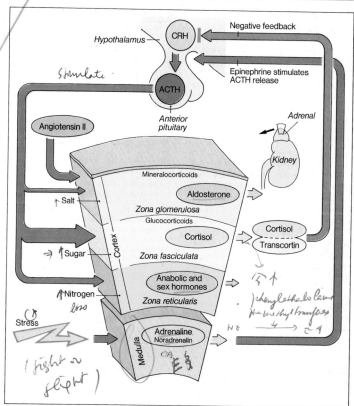

A. Adrenal

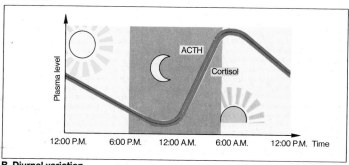

B. Diurnal variation

Menstrual Cycle

Human female sexuality cannot be understood without a thorough consideration of the menstrual cycle. Only primates menstruate; in other animals, vaginal bleeding does not have a sexual significance. Several hormones are involved in the menstrual cycle: in the hypothalamus – FSH/LH-RH (GNRH) and PIF; in the anterior pituitary – **FSH, LH**, and **PRL**; in the ovary – **estradiol** (E_2) and **progesterone (P)** ($\rightarrow$ pp. 215–217). Cyclic production of these hormones is regulated by both positive and negative feedback loops. The cycle is conventionally described as having 28 days ($\rightarrow$ **A**), but it may vary from 21 to over 35 days. The second, or **secretory**, phase of the cycle is relatively uniform in duration (c. 14 days), but the first, or **proliferative**, phase is variable and determines whether the cycle will be long or short.

Day 1: onset of menstrual bleeding. Duration c. 5 days.

Days 5 to 14: the **proliferative phase** begins at the end of menstruation and continues till ovulation. Uterine endometrium is built up in preparation for reception of the fertilized ovum.

Ovary: increasing secretion of FSH stimulates development and maturation of a single ovarian **follicle**. At this stage, FSH secretion exceeds LH. Ovarian blood flow increases, and the maturing follicle produces increasing amounts of E_2, which also contributes to maturation.

Uterus: endometrial proliferation, increased mitosis, and lengthening of endometrial glands and arterioles progress. Cell water content increases.

Cervix: the cervical os is small and tight. Mucus shows a fern pattern on drying because of its high Na^+ content; it can be drawn out in long threads (spinnbarkeit).

Vagina: cells are acidophilic with pyknotic nuclei.

Day 14: ovulation: production of E_2 peaks on day 13 and, by a positive feedback effect (Hohlweg effect), stimulates a surge in GNRH and LH release, without which ovulation cannot occur. Although FSH secretion also increases, LH now exceeds FSH. Basal body temperature rises about 0.5 °C. Cervical mucus is thin and watery and the os is enlarged to receive sperm.

Days 14 to 28: luteal or **secretory phase** is characterized by development of the corpus luteum and the secretory endometrium. LH activates luteinization.

Ovary: **corpus luteum** is formed and secretes E_2 and P.

Uterus: action of E_2 during the proliferative phase has made the endometrium responsive to P ($\rightarrow$ p. 224). Endometrial glands become tortuous, and the arterioles become spiral. Mitotic figures are fewer. Cell content of glycogen and phosphatase is increased. Uterine motility is decreased. Toward the end of this phase, decidual cells form in the endometrium. Maximal uterine response to P occurs at day 22, when implantation would occur. If it does not occur, E_2 and P secretion continues and inhibits GNRH secretion, which leads to involution of the corpus luteum. There follows a decline in hormonal support for the endometrium.

Cervix: mucus becomes thick and viscid, its volume is reduced, and it no longer shows its fern pattern. Sperm cannot be received anymore.

Vagina: cells become basophilic, are folded, and form clumps.

Days 27 to 28: a sudden decline in E_2 and P secretion is followed by constriction of spiral arterioles and development of endometrial ischemia. These events are the precursors of desquamation (*menstruation*).

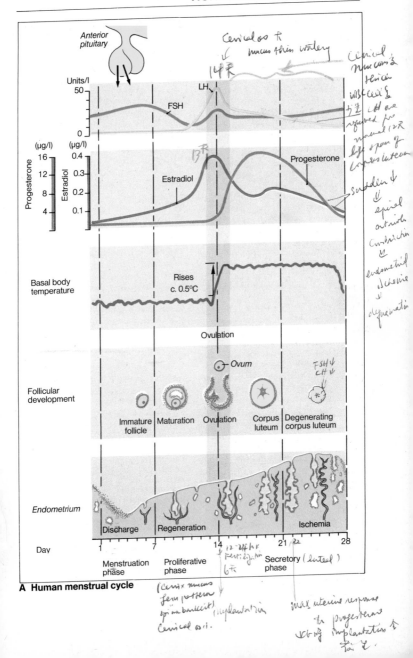

A Human menstrual cycle

Hormonal Interactions During the Menstrual Cycle

Secretion of gonadotropic hormones by the anterior pituitary (APit) becomes cyclic in the female at puberty; in the male, secretion continues at constant levels. The female cyclic pattern of secretion is probably basic, but treatment of immature females with male hormone during the perinatal period can prevent development of cyclic secretion at puberty.

LHRH, a hypothalamic decapeptide, stimulates secretion of both **FSH** and **LH** by the APit, but existence of a separate FSH-RH has not yet been excluded. A common name, therefore, has been introduced: *Gonadoliberin* or FSH/LH-RH or GNRH. Secretion of FSH and LH relative to each other changes continuously during the menstrual cycle ($\rightarrow$ p. 243) so that influences other than FSH/LH-RH must also be operative. Of the ovarian hormones, *estradiol* (E_2) has important modulating effects, but the actions of E_2 are themselves influenced by the relative concentration of *progesterone* (**P**). Cells of the APit have more than one class of binding site for FSH/LH-RH, which determine whether FSH or LH will be released. Relative affinities of these sites are affected by E_2 alone or in combination with P.

During the proliferative phase of the menstrual cycle, FSH secretion rises while LH remains low ($\rightarrow$ p. 243, A). At days 12 to 13 ($\rightarrow$ **A**), E_2 concentration rises rapidly. After a lag of 24 h, positive feedback stimulates release of FSH/LH-RH (Hohlweg effect) and produces a surge of FSH and LH on day 14 ($\rightarrow$ **A**). If the surge fails to occur or if it is of insufficient magnitude, ovulation does not take place. In the period just after ovulation, increasing production of P assists in enhancing release of FSH and LH. Simultaneously, following rupture of the follicle, secretion of E_2 declines. This fall removes a negative feedback influence that results in an increase in FSH release and accounts for the second peak in E_2 secretion during the secretory phase of the cycle ($\rightarrow$ p. 243). At this stage, increasing levels of E_2 and P exert a negative feedback influence on FSH/LH-RH ($\rightarrow$ **A**, day 20). It is not clear what initiates the events just prior to menstruation but a characteristic sudden fall in circulating levels of E_2 and P seems to act as a trigger.

Role of Prolactin

Prolactin (PRL) or luteotropin has a diversity of action in vertebrates that is unmatched by any other hormone. A *lactogenic* action has been convincingly demonstrated in human females. During pregnancy, hypersecretion of PRL is progressive and contributes to breast growth and milk production. (*Milk ejection* is regulated by *oxytocin*, $\rightarrow$ p. 248). Under normal conditions, PRL secretion is held in check by a release-inhibiting factor (**PIF**) ($\rightarrow$ **A**). TRH increases PRL releases. PIF is under feedback control by E_2 ($\rightarrow$ **A**, days 13, 14) and under neural control from higher centers. Suppression of PIF allows release of PRL. A number of factors including emotion, stress (hypoglycemia), drugs (morphine, reserpine, phenothiazine, tranquilizers, etc.), sleep, intercourse and nursing, suppress PIF secretion. The physiological afferent stimulus for PIF suppression which leads to an elevated PRL release is sucking at the maternal mamilla (sucking reflex).

Pathologic *hyperprolactinemia* may occur under certain drugs (see above), in patients with a PRL producing tumor of the anterior pituitary, or in severe hypothyreosis where elevated TRH production stimulates PRL release. In females hyperprolactinemia leads not only to galactorrhea but also to amenorrhea and anovulatoric infertility. Also in males, which have about the same PRL plasma levels as nonpregnant females, hyperprolactinemia leads to hypogonadism and impairs sexual activity and libido. These symptoms show that PRL has other effects beside that on lactogenesis.

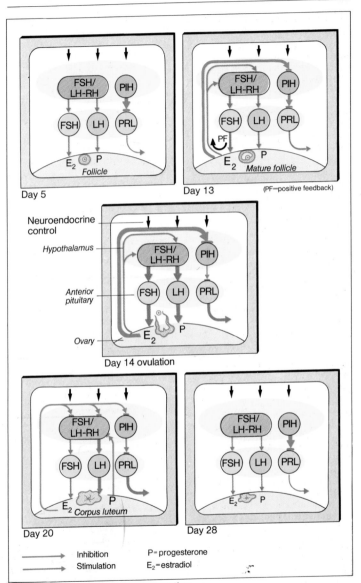

Day 5

Day 13

(PF=positive feedback)

Neuroendocrine control

Hypothalamus

Anterior pituitary

Ovary

Day 14 ovulation

Day 20

Day 28

→ Inhibition   P= progesterone
→ Stimulation   E₂=estradiol

A. Hormonal interactions in menstruation

Estrogens

Estrogenic hormone is responsible for development of female sexual characteristics and for preparation of the genital tract for response to progesterone. It functions as a female "ripening" hormone.

Chemistry. Estrogens (C-18 steroids) arise from androgens (C-19 steroids) ($\rightarrow$ p. 239), principally from the 17-ketosteroid (17-KS) androstenedione but also by aromatization of testosterone. Estrogens are synthesized in ovary (granulosa and theca cells), adrenal cortex, testis (Leydig cells), and placenta. Relative potencies of the main estrogens are: estradiol (E_2) 1.0, estrone (E_1) 0.5, and estriol (E_3) 0.1. E_2 is transported in the blood on a specific binding globulin. E_3 is the main metabolic product. E_3 may be antiestrogenic since it competes with E_2 for receptors but has a lesser potency. By this means, it may inhibit ovulation. Estrogens, like other natural sex steroids, are poorly effective when taken orally because a large fraction of the dose is extracted during the first passage through the liver.

Secretion rates (E_2) (mg/day)

Males	0.1
Females, follicular phase	0.15
Ovulation	0.6–0.8
Luteal phase	0.3
Menstrual phase	0.08
Pregnancy 3rd trimester	8–15

Actions. **Genital tract**. *Ovary*: E_2 increases blood flow and contributes to ripening of the ovum and maturation of the follicle. Size and volume of all genital tissues are increased. *Uterus*: mitosis is increased. The endometrium proliferates. Myometrial contractions are enhanced; frequency is faster (2 to 4/min), duration of contraction is shorter, and amplitude is smaller. *Vagina*: mucosa thickens and surface cells become cornified. Glycogen content of the mucosal cells is increased, which permits more lactic acid production by Döderlein bacilli and a more acid pH. These actions of E_2 increase resistance to vaginal infection. *Cervix*: mucus becomes copious and shows a fern pat-tern. *Breast*: initial growth is stimulated, mainly in the ductal tissue.

These actions have the function of preparing the female for reception of sperm to fertilize the ovum and for the pregnancy. Capacitation of sperm and passage to the ovum are enhanced. Estrogen directs fat deposition in sites characteristic for the female (breasts, hips, thighs, etc.).

Endocrines. E_2 is the principal female hormone in feedback modulation of FSH and LH. It modulates tissue responses to P by controlling the number of receptor binding sites for P inside the cell. Concentrations of specific binding globulins in the plasma are increased by E_2 so that plasma concentrations of T_4 and T_3 are influenced.

Other tissues. E_2 stimulates *salt and water retention* both systemically and locally. Local application causes an increase in cellular salt and water content with localized edema. This action has been the basis of some cosmetic creams intended to eradicate wrinkles. In epileptic patients, water and salt retention in the presence of E_2 lowers the threshhold for convulsions.

Bone. Rate of linear growth is retarded and epiphyseal closure is hastened. Since the effect is most prominent on the long bones, the height of young women can be influenced by the time of onset of puberty. E_2 has a specific anabolic action on genital tissues but also influences anabolism systemically. It is important for laying down of bone matrix; it stimulates osteoblasts to deposit Ca^{2+} ($\rightarrow$ p. 234). After menopause, women are prone to develop osteomalacia. Protein synthesis is enhanced and angiotensinogen and carrier globulins for vitamin A, iron, and copper are increased. Proteins of the blood clotting system are synthesized, and clotting mechanisms may become more effective; clotting time may be decreased.

Lipids. Atherosclerosis and myocardial infarction are less frequent in females before menopause. E_2 reduces plasma cholesterol and β-lipoproteins and increases high density lipoproteins and triglycerides.

Skin. Sebaceous glands regress. The epidermis becomes thinner and softer and more subcutaneous fat is deposited.

CNS. Important behavioral (sexual and social) effects occur. Emotional depression may be antagonized.

Progesterone (P)

Progesterone (P) is produced principally during the second or luteal phase of the menstrual cycle. Its chief function is to prepare the genital tract for reception and maturation of the fertilized ovum. It influences three classes of tissue: the genital tract, endocrine glands, and several nonendocrine organs.

Chemistry. P is a 21-C steroid produced in the ovarian follicle, corpus luteum, placenta, and adrenal cortex of men and women. The synthetic pathway ($\rightarrow$ p. 239) is either from preformed cholesterol or from acetate to cholesterol and then to pregnenolone, the immediate precursor of P. P is almost completely metabolized in one passage through the liver and is therefore not effective by oral dosage. In plasma, it is transported on a glycoprotein, P-binding globulin. Excretion is by urine and bile. The chief metabolite is pregnandiol. Levels of urinary pregnandiol or of plasma P indicate the extent of corpus luteum function.

Actions. Most actions of P require previous priming with E_2. E_2 increases cell concentration of P-receptor binding sites ($\rightarrow$ p. 224) during the follicular phase of the cycle. As P begins to dominate the luteal phase, the concentration of free P-binding sites decreases.

Genital tract. The *uterus* is the chief target for P. After E_2 priming, P stimulates myometrial growth. In the endometrium, which has grown under E_2 influence, P causes stromal changes in the glands, vascular supply, and glycogen content. These changes peak on day 21 to 22, the same time as the plasma peak for P. By these changes, P converts proliferative to secretory endometrium. Prolonged treatment with P makes the uterus unsuitable for implantation because the endometrium regresses. P reduces muscular activity of the uterus during pregnancy. Myometrial contractions under P influence last 1 to 2 min at a frequency of 1 every 2 to 3 min. In the E_2-primed *vagina*, P causes desquamation of cell clumps and inhibits cornification produced by E_2. In the *breast*, P develops the lobuloalveolar system.

Endocrines. During the luteal phase, P levels increase ($\rightarrow$ p. 243). P binds to nuclei of selected hypothalamic cells and reduces secretion of FSH and LH (negative feedback) ($\rightarrow$ p. 245). Continuous administration of P throughout the cycle lowers the midcycle surge in LH release and inhibits often ovulation. This action together with the inhibitory effect of P on passage through cervix mucus ($\rightarrow$ p. 343) and capacitation of sperm is the basis of contraception by the "mini-pill." P also suppresses release of ACTH and GH.

Systemic effects. CNS: large doses of P produce anesthesia. The active agent is the P metabolite, pregnanolone. P raises the threshold for epileptic seizures; their frequency in patients decreases during the luteal phase. Behavioral disturbances and depression premenstrually and in the last trimester of pregnancy may be attributed to P. Its thermogenic effect raises the basal body temperature at midcycle ($\rightarrow$ p. 243). **Metabolic:** P causes negative N balance by increasing hepatic metabolism of amino acids. **Kidney:** P inhibits aldosterone action, probably by competing for aldosterone receptors. It causes increased Na^+ excretion, which is followed by an increase in activity of the angiotensin-renin system.

Source	Secretion rate (mg/day)	Concentration (μg/l plasma)
Males	0.75	0.3
Females, follicular phase	4	0.3
Luteal phase	30	10–20
Early pregnancy	90	40
Late pregnancy	320	130
First day postpartum	–	20

Endocrinology of Pregnancy

The **placenta** has several functions: to maintain pregnancy, to nourish the fetus and carry away its waste products, and to meet the greatly augmented hormonal needs of pregnancy, which cannot be supplied alone by the maternal endocrine system.

The human placenta produces several hormones: E_2, **P, chorionic gonadotropin (HCG)** (with actions of LH, PRL, and to a lesser extent FSH) and **chorionic growth hormone (HGH)** (with actions of GH and PRL). The ovary is essential for maintenance of the *fetoplacental unit* in the early weeks of pregnancy. By week 12 to 16, the placenta has developed sufficiently to function independently of the ovary.

The placenta has several unique characteristics: it is independent of the usual feedback control; it synthesizes both steroid *and* peptide hormones in the same tissue ($\rightarrow$ **A**, **B**); the peptide phase predominates during the 1st trimester, the steroid phase during the 2nd and 3rd trimesters. The placenta secretes its hormones into two distinct compartments, the fetal and the maternal ($\rightarrow$ **A**). Unlike the adrenal and gonads, the placenta cannot synthesize steroids de novo from acetate but receives precursors, cholesterol or DHEA, from the two compartments; for E_2, they are derived chiefly from the maternal circulation during the early stages of pregnancy and from the fetal circulation during the later stages; for P, they are derived only from the maternal circulation. **HCG** secretion ($\rightarrow$ **B**) starts immediately on implantation of the fertilized ovum and increases rapidly during the peptide phase of placental activity. It declines rapidly during the 3rd month and maintains a low rate throughout the remainder of pregnancy. HCG supports the maternal corpus luteum and stimulates the ovary to produce E_2 and P. It also stimulates the fetal adrenal to secrete DHEA and other steroid precursors

for the steroid phase of placental activity. **Early pregnancy tests** are based on the presence of HCG in the urine. The more sensitive the test, the earlier the diagnosis of pregnancy can be made, but in any case, the test is not fully reliable before days 7 to 14 because of the low levels of HCG secretion.

Adrenal cortex.

Fetal: grows early in pregnancy and rapidly; 80% of the cortex is in a special innermost zone, the fetal zone (FZ). At month 4, the adrenal is larger than the kidneys but regresses rapidly thereafter. The FZ is functional only during the intrauterine period. After birth, it involutes. It responds to ACTH and HCG and produces mainly 17-ketosteroids (DHEA) for steroid synthesis in the placenta. *Maternal:* Zona glomerulosa and Zona fasciculata ($\rightarrow$ p. 240) increase in size and produce more aldosterone and cortisol.

Anterior pituitary and thyroid.

Maternal: hypertrophy. FSH and LH release decreases during pregnancy. T_3 and T_4 secretion is elevated.

Ovary.

Maternal: GH and HCG stimulate and maintain the corpus luteum until week 9. Maturation of new follicles is suppressed, and secretion of E_2 and P is enhanced. Between the 3rd and 5th month, function and size decrease and the corpus luteum is no longer essential for maintenance of the pregnancy.

Parturition. The mechanism is obscure. **Oxytocin** from the posterior pituitary stimulates the uterus to strong contractions, but it is not essential for delivery. The uterus becomes sensitive to oxytocin only in the last days of pregnancy.

Oxytocin also regulates *milk ejection*. The lactogenic hormone is prolactin ($\rightarrow$ p. 244).

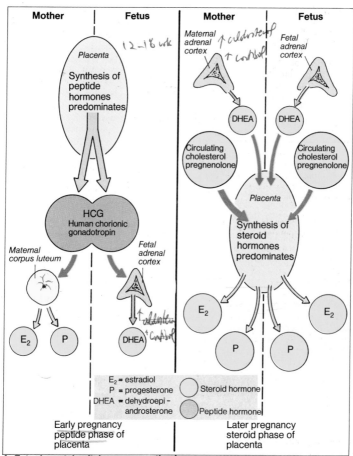

A. Fetoplacental unit; hormone synthesis

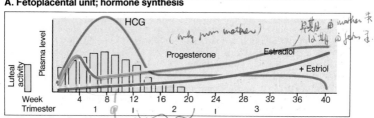

B. Plasma levels of hormone during pregnancy

Androgens

The chief function of the testes is to produce the male germ cell, the sperm (spermatogenesis). Functional cells in the testis are of two types: Sertoli cells in the seminiferous tubules produce sperm; Leydig (interstitial) cells produce **testosterone (T)** and estradiol (E_2).

Spermatogenesis ($\rightarrow$ **A**). FSH stimulates Sertoli cells to produce androgen-binding protein, the intracellular receptor for T. T is essential for all stages of spermatogenesis and is produced in the Leydig cells under the influence of LH (also called ICSH = interstitial cell-stimulating hormone). Vitamin A (retinol) also is essential; its metabolism is inhibited by alcohol, which accounts for the frequency of sterility in alcoholics.

Chemistry. Androgenic steroids (T) are C-19 steroids. They are synthesized in the testis, adrenal cortex, and ovary ($\rightarrow$ pp. 238–239). An O at C-17 is essential for activity. 17-ketosteroids, principally dehydroepiandrosterone (DHEA), are also androgenic, but they differ in their relative androgenic/anabolic ratios. Secretion rates for T: males = 7 mg/day; females = 1 to 2 mg/day. Plasma concentration males = 7 µg/l; females = 0.5 µg/l. Production of T in adult males decreases with age. Like other steroids, T circulates in bound form with a globulin. Orally administered T is extracted from portal vein blood by the liver and is metabolized by reduction at C-4.5 ($\rightarrow$ p. 239); the 5-β form is inactive, but the **5-α-dihydro-T** is active in the cell. In the hypothalamus, T is converted to 5-α-dihydro-T or to estradiol, which react with specific intracellular receptors.

Control. The male anterior pituitary synthesizes the same FSH, LH, and PRL as the female. Secretion in the male is relatively constant rather than cyclic. FSH stimulates Sertoli cells, but the negative limb of the feedback loop is not known. It has been postulated that **inhibin** is synthesized in Sertoli cells for feedback regulation of FSH ($\rightarrow$ **A**). LH (ICSH) stimulates T secretion; T inhibits LH secretion by negative feedback. Some T is converted to an estrogen by aromatization of its A ring ($\rightarrow$ p. 239). Both forms are effective in the feedback system.

Effects. T is responsible for spermatogenesis and for development of sexual maturation in the male.

Sex. The sexual phenotype develops in the 12th to 15th intrauterine week. The differences between the sexes, genetically, are the Y chromosome, which is male specific, and a variety of autosomal genes. In the absence of any hormones, the fetus develops as a female. T must be present for male characteristics to develop. T also conditions the hypothalamus to produce gonadotropins at a constant rate at puberty instead of cyclically; in the absence of T during the perinatal period, production of gonadotropins at puberty is cyclic ($\rightarrow$ p. 244). Secondary sex characteristics are also influenced by T: genital growth, size of larynx, facial hair, pubic and axillary hair, growth of prostate and seminal vesicles, skin thickness, and sebaceous gland activity. T promotes behavioral changes at puberty, such as sexual activity, libido, and aggressiveness.

Anabolic effects include increase in muscle mass and positive nitrogen balance. Before puberty, growth averages 5 cm/year; at puberty, T stimulates a growth spurt to 8 cm/year. Synthesis of erythropoietin ($\rightarrow$ p. 58) in the kidney is stimulated. This action of T accounts for the larger red blood cell mass in males. Low density lipoproteins are increased by T and are associated with a greater risk of vascular disease in males.

Sex determination ($\rightarrow$ **B, C**) is complex. It depends on several factors none of which need be consistent with any other: (1) sex chromosomes (in man, XX = female, XY = male), (2) gonads (ovary or testis), (3) external genitalia, (4) internal genitalia, (5) secondary sex characteristics, (6) habitus, (7) sex in which the child was reared, (8) sex hormone production, and (9) sex role in society.

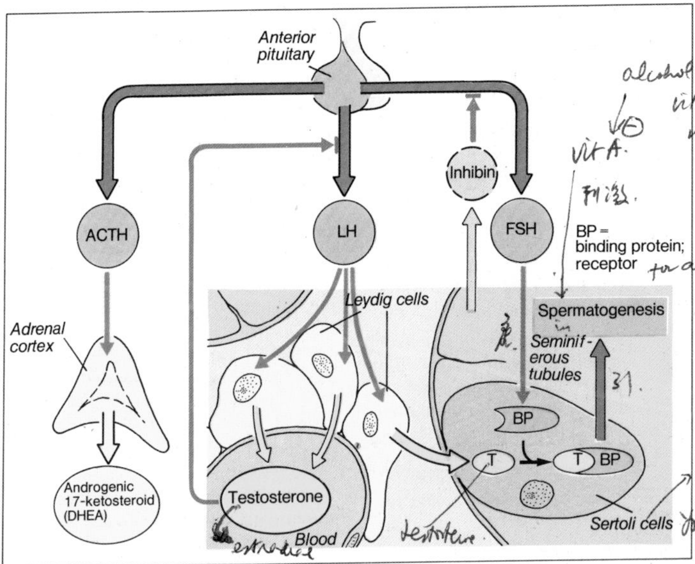
Anterior pituitary
ACTH
LH
Inhibin
FSH
BP = binding protein; receptor
Adrenal cortex
Androgenic 17-ketosteroid (DHEA)
Leydig cells
Spermatogenesis
Seminif-erous tubules
BP
T
T BP
Testosterone
Blood
Sertoli cells

A. Androgenic hormones

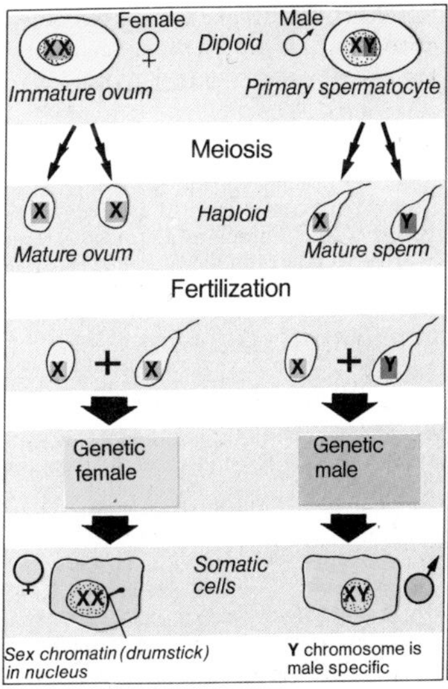
Female
Male
Diploid
XX
XY
Immature ovum
Primary spermatocyte
Meiosis
Haploid
X
X
X
Y
Mature ovum
Mature sperm
Fertilization
X + X
X + Y
Genetic female
Genetic male
Somatic cells
XX
XY
Sex chromatin (drumstick) in nucleus
Y chromosome is male specific

B. Genetic determination of sex

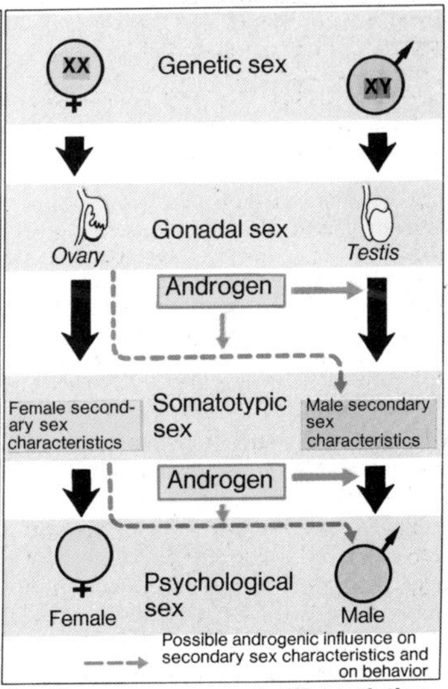
XX
Genetic sex
XY
Ovary
Gonadal sex
Testis
Androgen
Female second-ary sex characteristics
Somatotypic sex
Male secondary sex characteristics
Androgen
Female
Psychological sex
Male
Possible androgenic influence on secondary sex characteristics and on behavior

C. Influences on sexual differentiation

Central Nervous System (CNS)

The functional anatomy of the CNS involves (1) **information input** by sensory afferent nerves from the sense organs; (2) **integration of information** by interconnection of neurons; (3) **response to information** by motor efferent nerves to muscles and exocrine glands. Endocrine glands are controlled both by nerves and by humors via the neuroendocrine system.

The CNS in man is comprised of the **brain** and **spinal cord** ($\to$ **A**). The latter retains to some extent the segmental organization of the CNS of more primitive animals; spinal nerves enter and leave the spinal cord at the level of the corresponding vertebra. **Afferent** fibers enter the spinal cord at the posterior or dorsal root; **efferent** fibers exit at the anterior or ventral root ($\to$ **B**). Fibers from each ventral root also supply the autonomic nervous system ganglia ($\to$ p. 48). A cross section of the spinal cord ($\to$ **B**) reveals two fields: a central **gray matter** composed chiefly of nerve cell bodies and a peripheral **white matter** composed chiefly of ascending and descending nerve fibers. The ventral horns of the gray matter contain predominantly efferent motor neurons; the dorsal horns contain a large number of **interneurons** that permit sharing of information within the network. The cell bodies (soma) of afferent neurons lie outside the spinal cord ($\to$ **B**).

The brain is a specialized extension of the spinal cord. In ascending sequence, it is composed of the **medulla oblongata** ($\to$ **E, 1**), **pons** ($\to$ **E, 2**), **mesencephalon** (midbrain) ($\to$ **E, 3**), **diencephalon, telencephalon** (cerebral hemispheres) ($\to$ **D, E**), and the **cerebellum** ($\to$ **C, E, F**). The first three components make up the **brain stem**, which contains the nuclei of the cerebral nerves, the respiratory and cardiovascular control areas ($\to$ pp. 92, 165) and other centers.

The diencephalon includes the **thalamus** ($\to$ **D, 4**), an important switchboard for all afferent sensory inputs, and the **hypothalamus** ($\to$ **D, 5**), a control center for autonomic functions ($\to$ pp. 48–57) and for integration of the nervous system with the endocrine system ($\to$ p. 220).

The **telencephalon** contains centers for motor control (caudatum [$\to$ **D, 7**] putamen [**D, 8**], pallidum [$\to$**D, 9**]) and components of the **limbic system** (amygdala [$\to$ **D, 10**]). The external surface of the telencephalon is the **cortex**, which is divided by *sulci* (grooves; $\to$ **C, D, E, 12, 13**) into four major *lobes* ($\to$ **C, D, E**): frontal, parietal, occipital, and temporal.

The two hemispheres are connected by the **corpus callosum** ($\to$ **D, 14; E**), in which fibers from each hemisphere cross to the other. The **reticular activating system** (RAS) is a diffuse network of cells that connects the telencephalon to lower centers.

Cerebrospinal fluid

The brain and spinal cord are bathed in c. 150 ml of **cerebrospinal fluid (CSF)**, which is both nutritive and protective. The CSF is secreted by the *choroid plexus*, a collection of blood vessels and tissue, into the cavities of the brain: the two lateral *ventricles* ($\to$ **D, 15; F**), which join at the third ventricle and connect with the fourth ventricle ($\to$ **F**); c. 650 ml of CSF are produced daily. The CSF exits from the fourth ventricle to fill the surrounding subarachnoid space ($\to$ **F**). It is absorbed through arachnoid villi ($\to$ **F**) and is carried away by the vascular system. Secretion of CSF continues even if outflow of fluid is obstructed; the CSF then accumulates and increases pressure within the cranial vault (*hydrocephalus*). Except for CO_2, O_2, and H_2O, exchange of substances between CSF and blood is limited (*blood-brain barrier*).

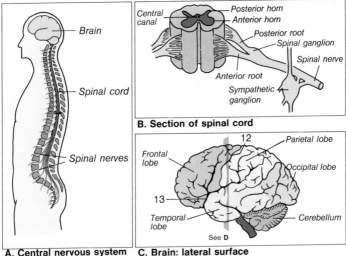

A. Central nervous system

B. Section of spinal cord

- Central canal
- Posterior horn
- Anterior horn
- Posterior root
- Spinal ganglion
- Spinal nerve
- Anterior root
- Sympathetic ganglion

Brain

Spinal cord

Spinal nerves

C. Brain: lateral surface

- 12
- Parietal lobe
- Frontal lobe
- Occipital lobe
- 13
- Temporal lobe
- Cerebellum
- See D

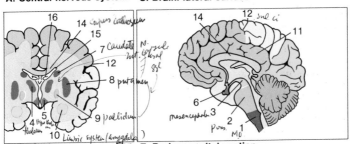

D. Brain: transverse section

- 16
- 14 *corpus callosum*
- 15
- 7 *Caudate N. corpus mot. dorsal gg*
- 12
- 8 *putamen*
- 9 *pallidum*
- 5
- 4 *Hypothalamus*
- 10 *Limbic system (amygdala)*

E. Brain: saggital section

- 14
- 12 *sul ci*
- 11
- 6
- *mesencephalon* 3
- 2
- *Pons* 1 *MO*

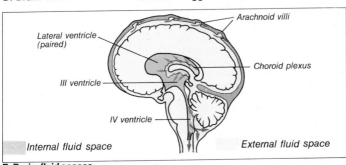

F. Brain: fluid spaces

- Arachnoid villi
- Lateral ventricle (paired)
- Choroid plexus
- III ventricle
- IV ventricle
- Internal fluid space
- External fluid space

Processing of Informational Input

The sensory receptors initiate most of the impulse traffic within the nervous system. They are organized to detect and to respond to stimuli arising in the environment; they signal the presence of a stimulus and its magnitude. The main types of receptors include mechanoreceptors, photoreceptors, thermoreceptors, and chemoreceptors. Each is specialized to select from the environment that stimulus which is adequate for its function; the ear responds to sound, the nose to smell, etc.

A characteristic of receptor stimulation is that different kinds of stimuli applied to the same receptor elicit the same sensation. The eye produces sensation of light whether the stimulus is light, an electric potential, or a mechanical blow ("seeing stars"). On the other hand, identical stimuli applied to different receptors elicit different sensations. The final interpretation of sensation depends as well on the neural path that the impulse takes as on its destination in the CNS. The chain of events in stimulus reception involves: (1) absorption of stimulus information by the receptor, (2) transduction of stimulus information to electric information, (3) production of a slow **generator potential** (GP) that (4) stimulates a spike discharge (action potential or AP) ($\rightarrow$ pp. 24–27) in the afferent axon. When a receptor is stimulated, it produces a graded depolarization potential, the GP ($\rightarrow$ **B**, **1**). The GP remains a *local potential* and is not propagated because it is quickly dissipated by decremental conduction. As magnitude of stimulus is increased, the GP also increases ($\rightarrow$ **C**, **1**) until the axon threshold ($\rightarrow$ pp. 24–25) is reached and a propagated AP is initiated ($\rightarrow$ **B**, **1**). At a still larger stimulus, the GP increases still more and the AP fires repetitively. In this way, stimulus magnitude is converted to a frequency that varies as a power function rather than as a linear function

($\rightarrow$ **C**, **1**). A linear function is not workable; it would exceed the capacity of a neuron to respond since the range of sensory stimulation is up to 10^{12} and the neuron is limited by its refractory period to a maximal frequency of 1,000/s. Encoding of information as a frequency instead of as a magnitude has an advantage: since information must travel over axons as long as 1 m, a magnitude would lose more informational content than a frequency. When a stimulus is maintained at a constant level, most receptors (except proportional receptors) show **adaptation**, and the frequency of AP spikes decreases ($\rightarrow$ p. 277, C). At the synapse, the neurotransmitter stimulates a potential on the postsynaptic membrane ($\rightarrow$ **B**, **2**). When this potential reaches threshold, the impulse is propagated further. Modulation of information takes place at the synapse by its ability to elicit either excitatory (EPSP) ($\rightarrow$ p. 28) or inhibitory (IPSP) impulses, which influence the **contrast** of information ($\rightarrow$ **D**).

The unit of information content is the binary digit, or **bit** ($\rightarrow$ **A**). Flow of information is a time function (bits/s). One printed letter has 4.5 bits, a book page c. 1,000 bits. If a page is read in 20 s, intake is 50 bits/s. Sensory intake is up to 10^9 bits/s but only a small fraction, 10 to 100 bits/s, is consciously recognized. Information output is c. 10^7 bits/s.

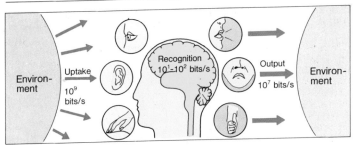

A. Information flow

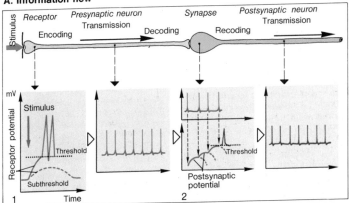

B. Processing of information; encoding

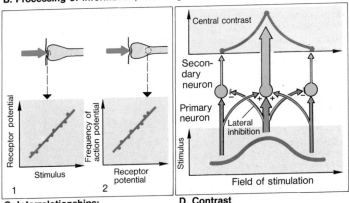

C. Interrelationships:
stimulus, potentials

D. Contrast

Superficial Senses (Skin) — Pain

Receptors of the skin constantly monitor the environment; they are basic to temperature control and protect the body from injury. Motor control of fine as well as gross movements and awareness of the environment all depend on skin receptors. All of these activities are based on reception of touch, temperature, vibration, and pain. These superficial sensations, together with the proprioceptive (deep) sensations (receptors in muscles, joints, and ligaments) ($\to$ p. 258), constitute the somatovisceral senses.

There are three principal types of stimulus detection:

1. **Proportional** receptors ($\to$ **B**, **1**) detect **differences in intensity** of the stimulus. They transmit signals as long as they are deformed

2. **Differential** receptors ($\to$ **B**, **2**) respond to a **change in intensity** of stimulus (velocity). They have a low threshold and discharge for a few hundredths of a millisecond, then adapt even though a stimulus continues.

3. **Proportional-differential** receptors respond with a burst of activity and then adapt to a lower but sustained level of discharge ($\to$ **D**).

Specialized **mechanoreceptors in the skin** include Meissner's corpuscles ($\to$ **A**, **1**), hair follicle receptors ($\to$ **A**, **4**), touch plates ($\to$ **A**, **5**), and Merkel's discs ($\to$ **A**, **2**). Mechanoreceptors respond to **pressure, touch,** and **vibration**. When Merkel's discs are distorted by a weight (pressure stimulus), the spike frequency that they produce is proportional to the pressure (intensity) ($\to$ **B**, **1**). Disturbance of Meissner's corpuscles, as in bending the hairs of the skin, is detected as a *rate of deformation* rather than as an intensity. The spike frequency is proportional to this rate of change of the stimulus; a constant stimulus is not registered ($\to$

B, **2**). Pacinian corpuscles ($\to$ **A**, **3**) are specialized to respond to *vibration*. A single cycle stimulus produces a *single* afferent impulse. The vibration *frequency* of the stimulus is matched by a corresponding spike frequency ($\to$ **B**, **3**). In the range 20 to 400 Hz, mechanical frequency and spike frequency are similar.

Velocity detectors occur not only in skin but also in muscle, tendons, and joints, where they function as proprioceptive detectors ($\to$ **D**).

Temperature (T) below normal body T is sensed by cold receptors and above body T by heat receptors. In the range of 28° to 36°C, the colder the T, the higher the spike frequency discharged by the cold receptors. Similarly, in the range of 36° to 43°C, heat receptors increase the spike frequency as T increases ($\to$ **C**). Between 20° to 40°C, T receptors show slow adaptation; bathing in cool water seems colder at first than it does after a short period of adaptation. For T above 45°C, there are independent heat receptors, but pain receptors also begin to play a part.

Pain is a message that damage to the organism has occurred ($\to$ p. 262). Identification of the cause is less important than recognition of the effect. Receptors report not only skin pain but also visceral pain (gallbladder or ureteral colic) and "deep" pain (headache). Skin pain is sensed in two stages: initially, there is a sensation of sharp pain, which stimulates a flight-or-escape response ($\to$ p. 260), and then a subsequent continuous ache, which provokes protection of the damaged part. The pain receptors are the free nerve endings. They do not adapt to pain stimuli, as is exemplified by an unremitting toothache ($\to$ p. 262).

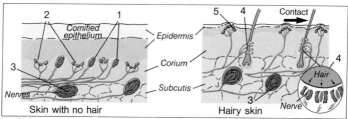

A. Skin receptors

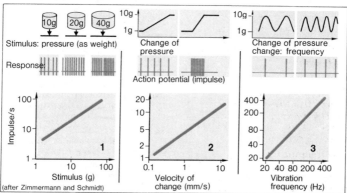

B. Responses of skin receptors to (1) pressure, (2) touch, (3) vibration

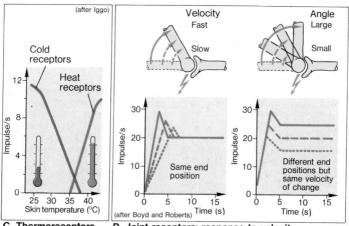

C. Thermoreceptors

D. Joint receptors: response to velocity and to joint position

Proprioception – Muscle Reflexes

Conscious muscular adaptation to the environment requires automatic input of proprioceptive information on joint position and muscle tension. Proprioceptive receptors are distributed in joint capsules, tendons, and muscles. Impulses pass up the dorsal spinal columns to the cerebellum and via thalamic radiations to the cortex. The supraspinal efferents (upper motor neurons (→ p. 268) can interrupt postural reflexes to initiate voluntary muscle activity. The proprioceptive organ in muscle is the **muscle spindle** (→ A), a fusiform capsule of connective tissue containing c. ten striated muscle fibers, the **intrafusal fibers** (IF). The ends of the IF attach to tendons or to regular contractile muscle fibers, the **extrafusal fibers** (EF). One muscle contains c. 10^2 (M. lumbricalis) to 10^3 (M. temporalis) motor units (→ p. 30). In the mean 0.15 – 0.65 muscle spindles are found per unit. Other receptors include the Golgi tendon organs, paciniform bodies, and free nerve endings. The spindles function as a **length** meter and the tendon organs as a **tension** meter.

Muscles take part in two classes of reflex. (1) **Phasic** reflexes are dynamic and monosynaptic. A brief stretch of the motor spindles, as in the knee jerk reflex (→ B), produces a synchronous neuronal discharge involving Ia neurons. (2) **Tonic** reflexes are static and polysynaptic. A prolonged asynchronous neuronal discharge causes sustained muscle contraction and accounts for muscle tone. The reflex travels via supraspinal centers; it involves group II afferents.

Afferent nerves from the muscles (→ C). (1) Group Ia fibers are the primary spindle nerve endings; they are thick (12 to 20 μm) and myelinated. The endings wrap around the central portion of the IF (→ A, C). When this annulospiral complex is stretched, it initiates a monosynaptic reflex via α-motor cells of the anterior spinal horn; the muscle contracts to counteract the

stretch. Simultaneously, via interneurons, a polysynaptic reflex inhibits the antagonist muscle (reciprocal inhibition). When the EF contract, tension on the IF is reduced and the rate of impulse firing from the IF decreases. In contrast, when the IF contract, they stretch the annulospiral complex and increase the firing rate. The spindles are velocity receptors (→ p. 256) and respond to rate of change of muscle length. (2) The Ib fibers from the tendon organs transmit excitatory impulses at most muscle tensions but at high tension they inhibit muscle contraction to protect against overload (autogenic inhibition). (3) Group II fibers (6 to 12 μm thick) are the secondary spindle endings. They activate flexor muscles and inhibit extensors.

Efferent nerves to the muscles (→ C). (1) γ-Fibers comprise 30% of the cells in the anterior horn. They end on motor end-plates of the IF and are called **fusimotor** fibers. They contract the IF, which activate a monosynaptic reflex via the Ia fibers (dynamic), or they adjust tension in the IF and their sensitivity to muscle stretch (tonic). Discharge of γ-fibers is controlled by supraspinal centers and is influenced by emotions (e.g., anxiety). (2) The α-motor neurons are the **final common path** to the EF. Important α-γ linkages influence muscle reflexes. (3) The **Renshaw cells** are recurrent collaterals from the α-motor axons. They synapse with inhibitory interneurons and feedback to the motor neuron (recurrent inhibition).

Walking is a practical example of these reflexes: the powerful extensor muscles support body posture. The supraspinal centers signal voluntary motion. A step is started by the flexors that lift the leg. Flexor contraction immediately inhibits extensors reciprocally. When the extensor spindles become stretched, they contract the extensor, and the leg swings forward. Overswing is prevented by tension on the tendon organs, which inhibit extensors and stimulate flexors reciprocally.

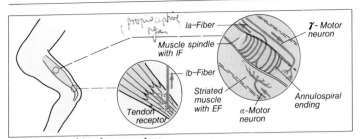

A. Spindle and tendon receptors

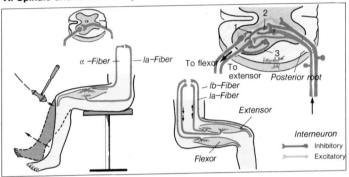

B. Monosynaptic reflex

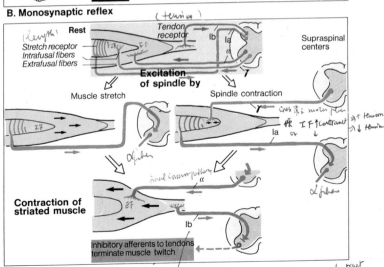

C. Stretch reflex

Polysynaptic Reflexes

In contrast to the simple reflex that originates and terminates in the same
muscle, the polysynaptic reflex originates in one organ and terminates in
another. Since several synapses are involved, there is more synaptic delay,
and the reflex takes longer than a monosynaptic reflex. In general, a monosyn-
aptic reflex activates α- and inhibits γ-
motor neurons; a polysynaptic reflex
activates both α- and γ-neurons.

A typical polysynaptic reflex is the
withdrawal reflex ($\rightarrow$ **A**). A painful
stimulus in the foot, as in stepping on a
tack, activates flexion and inhibits extension in all muscle groups in that leg.
The afferent impulses follow several
pathways: (1) via excitatory neurons to
the ipsilateral flexor muscles, which
contract ($\rightarrow$ **A, 1**); (2) via inhibitory interneurons ($\rightarrow$ **A, 2**) to the ipsilateral
extensor muscles, which relax ($\rightarrow$ **A, 3**),
(3) via excitatory interneurons ($\rightarrow$ **A, 4**)
to the contralateral extensors ($\rightarrow$ **A, 5**);
this **crossed extensor reflex** mag-
nifies withdrawal from the damaging
stimulus; (4) via inhibitory interneurons
to the contralateral flexors, which relax
($\rightarrow$ **A, 6**); (5) *via ascending and descending fibers to other spinal seg-
ments* ($\rightarrow$ **A, 7, 8**) whereby activity
spreads to all four extremities.
Simultaneously, pain sensations ($\rightarrow$ pp.
256, 262) are conducted to the cortex
where they are consciously recognized.
Withdrawal reflexes thus prepare the individual for flight. As the intensity of the
damaging stimulus is increased, the reaction time is shortened because of
summation of the EPSPs ($\rightarrow$ p. 28).

Polysynaptic reflexes include additional
protective responses, such as the corneal reflex, tear flow, coughing, and
sneezing. Similar reflexes are associated
with nutrition (sucking and swallowing). In the clinic, reflexes can be used
to test the integrity of neurologic pathways and to help to localize lesions in
the nervous system. Such diagnostic reflexes include Babinski, cremasteric,
and abdominal reflexes.

Synaptic Inhibition.

Transmission across the synapse can be
inhibited either at the presynaptic or the
postsynaptic membrane. The mechanisms differ.

Presynaptic inhibition ($\rightarrow$ **B**). An
impulse from an excited neuron ($\rightarrow$ **B,
c**) depolarizes the terminal portion of
the presynaptic axon ($\rightarrow$ **B, a**). This action reduces the amplitude of the action
potential, which is propagated in neuron **a** so that less neurotransmitter is liberated into the synaptic cleft ($\rightarrow$ **B, d**).
As a result, the postsynaptic membrane
produces a smaller EPSP ($\rightarrow$ pp. 28,
29). Inhibition by this process can reduce the EPSP enough to prevent generation of a postsynaptic action potential in neuron **b** ($\rightarrow$ **B**).

Postsynaptic inhibition ($\rightarrow$ **C**). An
impulse from an inhibitory interneuron
causes transient hyperpolarization of
the postsynaptic membrane (IPSP) and
increases the amount of neurotransmitter that must be released to generate an
action potential. Interneurons of this
class can be activated by collaterals ($\rightarrow$
C, 1) in recurrent or feedback inhibition
as by the Renshaw cells ($\rightarrow$ **C, 2**).. A
"feed-forward" inhibition that limits
duration of excitation or contrasts the
information ($\rightarrow$ p. 255, **D**) also occurs
($\rightarrow$ **C**).

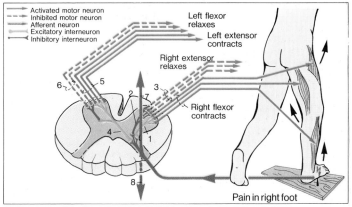

A. Polysynaptic reflex

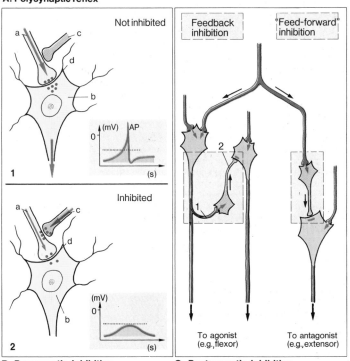

B. Presynaptic inhibition   **C. Postsynaptic inhibition**

Central Processing of Sensory Input

Information from the external environment reaches the central nervous system via receptors of the major senses, the proprioceptors and the skin. The major fraction of sensory impulses from the last two ultimately reach the sensory cortex in the **postcentral gyrus** (→ **B**) where each portion of the body is represented in proportion to the richness of its innervation. The impulses ascend to the cortex in discrete tracts in the posterior spinal cord.

1. Touch, pressure, and proprioception ascend in the dorsal columns (→ **C, 1**). The fibers reach the medulla before they synapse. Second order neurons pass to the cerebellum or cross the midline to reach the thalamus.

2. Some touch and pressure fibers and fibers for pain and temperature synapse in the dorsal horn (→ **C, 2**). The second order neurons cross and ascend in the ventral spinothalamic tract through the brain stem to the thalamus.

3. Deep sensation and other sensory inputs ascend to the cerebellum in the posterior spinocerebellar tract (→ **C, 3**), and in the anterior spinocerebellar tract (→ **C, 4**).

4. Sensory fibers from the head (trigeminal V) also end in the thalamus.

From the thalamus, third order neurons course to the various cortical sensory regions, chiefly to the postcentral gyrus. These are specific thalamocortical projections (→ **D**) and reach c. 20% of the cortex (cf. optical and acustic projections). In addition, there are nonspecific projections (→ **E**) via the reticular formation to all areas of the cortex. These tracts play a fundamental role in alertness and awareness. The reticular formation thus participates importantly in both sensory and motor control.

Lesions of the postcentral gyrus diminish but do not abolish sensation. Proprioception and fine touch are first affected, temperature less, and pain the least. Recovery from a lesion is in the reverse order with pain first to recover and proprioception last. Transections of half of the spinal cord (Brown-Sequard syndrome) causes loss of deep sensibility ipsilaterally and of pain and temperature contralaterally in the region caudal of that spinal segment, in addition to ipsilateral motor paralysis.

Superficial pain (→ p. 256). Receptors are the isolated nerve endings. There are two rates of transmission, fast pain in myelinated fibers and slow pain in unmyelinated fibers. These give the sensation of an immediate sharp localized pain, which is followed by a dull diffuse pain. The contribution of the cortex in pain sensation is to define the nature and source of pain, but pain sensations can be perceived at the thalamic level even in the absence of the cortex. Painful stimuli evoke potent protective responses, such as withdrawal or avoidance. Among the senses, pain is unusual in being linked to a strong emotional component with a connotation of unpleasantness. Damage to the thalamus can produce a syndrome in which minor sensations lead to prolonged and severe pain. In contrast, damage to the frontal cortex, as in lobotomy, dissociates the perception of pain from its unpleasant aspect. The patient claims that it hurts but it doesn't bother him.

Referred pain. Irritation of an internal organ may produce sensations of pain at a distant site. Thus, gallbladder pain may be referred to the back or to the shoulder. Cardiac pain may be referred to the inner aspect of the left upper arm with ulnar distribution to the fingers. A knowledge of these sites of referral has diagnostic importance to the physician.

Control of pain: Afferent pain signals can be inhibited at the level of the dorsal horn of the spinal cord by efferent neurons originating from areas of the brainstem (Nucleus raphe magnus, midbrain periaqueductal grey, and others). This **analgesic mechanism** is activated by endogenous *endorphin*, exogenous opiates and, via the medullary reticular formation, by noxious and thermal somatosensory input (e.g. *acupuncture*).

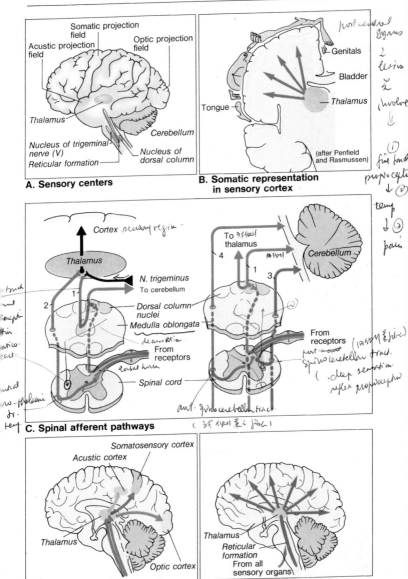

A. Sensory centers

Somatic projection field
Acustic projection field
Optic projection field
Thalamus
Cerebellum
Nucleus of trigeminal nerve (V)
Reticular formation
Nucleus of dorsal column

B. Somatic representation in sensory cortex

Genitals
Bladder
Thalamus
Tongue
(after Penfield and Rasmussen)

C. Spinal afferent pathways

Cortex
Thalamus
N. trigeminus
To cerebellum
Dorsal column nuclei
Medulla oblongata
From receptors
Spinal cord
To thalamus
Cerebellum
From receptors

D. Specific paths to cortex

Somatosensory cortex
Acustic cortex
Thalamus
Optic cortex

E. Nonspecific paths to cortex

Thalamus
Reticular formation
From all sensory organs

Motor Activity – Motor Hold System

With few exceptions, all activities of the CNS, receiving, processing, and integrating information, ultimately find expression in contraction of a muscle. Muscle activity may occur in terms of purposeful motion (motor move) or of postural control (**motor hold**). Skeletal muscles receive their motor impulses from the anterior horn cells of the spinal cord, the lower motor neurons. These, in turn, receive their inputs from the peripheral spinal segments (spinal reflexes, → pp. 258–261) and from supraspinal centers. The principal **supraspinal centers** are: (1) the *cortex* which initiates skilled motor activity (→ p. 268); (2) the *motor centers of the brain stem* (often called "extrapyramidal system"), which adjusts posture and balance; (3) the *cerebellum*, with its afferent and efferent connections, which coordinates muscle stimuli to produce smooth integrated muscle activity (→ p. 266) and in which "programs" of ballistic movements are stored (→ p. 268).

Thrombosis or hemorrhage involving blood vessels of the brain may damage motor tracts. The *internal capsule* is especially vulnerable since motor tract fibers are concentrated in this region. Damage to the spinal cord, as in *transection*, produces first a loss of peripheral reflexes (spinal shock) followed by recovery in spite of persistence of the spinal lesion.

The **motor hold system** supports involuntary muscular activity, such as maintaining *posture*, muscle *tonus*, and muscular support for the skeleton. Control of the motor hold system is primarily effected by the **motor centers of the brain stem** (→ **A**): *Nucleus ruber* (→ **A 1**), *vestibular nuclei* (→ **A4**) (particularly the lateral nucleus of Deiter, → p. 266) and parts of the *reticular formation* (→ **A2**, **A3**).
The main **afferents** to these centers come from the organs of balance (→ **A** and p. 278) from the proprioceptors (→

p. 258) of the neck (→ **A**) and from the cerebellum (→ p. 266). The **tracts descending** from the nucleus ruber and from the medullary parts of the reticular formation to the spinal cord (*tractus rubrospinalis* and *tractus reticulospinalis lateralis*, → **A**) exert a predominantly inhibitory influence on the α- and γ-motoneurons (→ p. 258) of the extensor muscles and a stimulating effect on the flexor muscles. Conversely, the pathways originating in Deiter's nucleus and in the pontine reticular formation (*tractus vestibulospinalis* and *tractus reticulospinalis medialis*) inhibit the flexor muscles and stimulate the extensors (→ **A**). Transections of the brain system below the nucleus ruber leads to a state of *decerebrate rigidity*, since the tonic extensor influence of Deiter's nucleus now predominates.

The motor centers of the brain stem coordinate the postural and positioning reflexes whose (involuntary) function is to maintain body posture and balance: Static and statokinetic reflexes. **Static reflexes**: (1) **Postural reflexes** serve to maintain muscular tone (→ p. 38) and coordinate eye movements. The sensory input comes from the organ of balance (tonic labyrinth reflexes) and from the neck proprioceptors (tonic neck reflexes). (2) The same afferents are involved in the **righting reflexes** which serve to return the body to its normal position. Firstly, the head (via the labyrinth afferents, → p. 278), then the rump (via the neck afferents) are returned to their normal position. In addition, afferents from the eyes, ears, olfactory organs and skin receptors modify these righting reflexes.
Statokinetic reflexes play an important role in posture and positioning of the moving body, e.g. in preparation for jumping, in rotatory nystagmus, or in extensor and flexor reflexes during vertical acceleration (elevator).

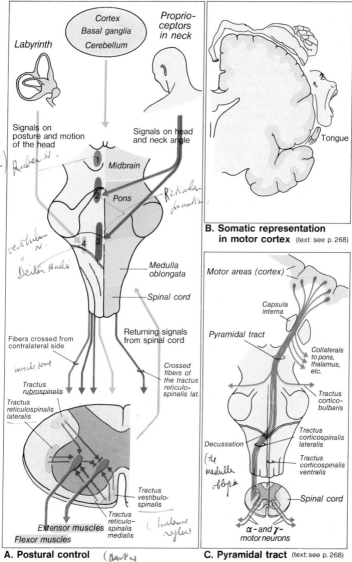

Cortex
Basal ganglia
Cerebellum

Proprio-
ceptors
in neck

Labyrinth

Signals on
posture and motion
of the head

Signals on head
and neck angle

1 Midbrain

2 Pons

4 3

Medulla
oblongata

Spinal cord

Returning signals
from spinal cord

Fibers crossed from
contralateral side

Crossed
fibers of
the tractus
reticulo-
spinalis lat.

Tractus
rubrospinalis

Tractus
reticulospinalis
lateralis

Tractus
vestibulo-
spinalis

Tractus
reticulo-
spinalis
medialis

Extensor muscles
Flexor muscles

A. Postural control

Tongue

**B. Somatic representation
in motor cortex** (text: see p. 268)

Motor areas (cortex)

Capsula
interna

Pyramidal tract

Collaterals
to pons,
thalamus,
etc.

Tractus
cortico-
bulbaris

Tractus
corticospinalis
lateralis

Decussation

Tractus
corticospinalis
ventralis

Spinal cord

α- and γ-
motor neurons

C. Pyramidal tract (text: see p. 268)

Function of the Cerebellum

The cerebellum fulfills a number of important functions in motor control. It optimizes the motor hold reflexes (**A1** and p. 264), it coordinates interplay between motor hold and motor move systems (→ **A2, B**), it plays an important role in correcting the course of slow motor movements (→ **A2** and p. 268), and it provides the programs for rapid motor movements (→ **A3** and p. 268).

Phylogenetically the cerebellum can be divided into three parts, the *archicerebellum*, *palaeocerebellum* and *neocerebellum* (→ **A**). Somewhat simplified, these can be considered as corresponding to three groups of **afferents**: a) the **archicerebellum** receives information concerning equilibrium and acceleration from the labyrinth (→ also p. 278) (partially relayed in the vestibular nuclei); b) the **palaeocerebellum** receives "copies" of the motor efferent impulses of the pyramidal tract (→ p. 268) and information from the organs of movement (proprioceptors, → p. 258) and from the body surface; c) The **neocerebellum** receives "plans" for movements from the associative areas of the cortex via the pons. Cerebellar **efferents** lead to the corresponding cerebellar nuclei (→ **A**, top) and thence to the *motosensory cortex* (via thalamus) and to the motor centers of the *brain stem* (→ p. 264). These tracts finally enter the motoneurons of the spinal cord (and the cranial motor nerves) via the pyramidal and "extrapyramidal" tracts. How these afferent and efferent pathways are involved in the motor hold and motor move functions of the cerebellum is shown in Fig. **A1–A3**.

The efferent pathways of the cerebellar cortex consist of neurites of the *Purkinje cells*. They exert an inhibitory effect on the cerebellar nuclei. Afferents from the spinal cord that have been relayed in the olive terminate in the cerebellum as *climbing fibres*. Via their stimulatory synapses they augment the inhibitory effect of the Purkinje cells. All other afferents to the cerebellum end as *moss fibres*. Via stimulation of *"granule"* cells and *parallel fibres* they can either augment the inhibitory action of the Purkinje cells or deinhibit them via intervening inhibitory *Golgi cells*. The presence of such multiple inhibitory circuits accounts for the fact that afferent impulses to the cerebellum are "extinguished" after about 0.1 s.

Disorders of the cerebellum lead to disturbances in muscular coordination and muscular tonus. Uncertain and exaggerated movements (atactic gait) result. The performance of antagonistic movements in rapid succession is no longer possible (adiadochokinesis). The disturbance in correction of motor move results in intention tremor. Further symptoms are abnormal eye movements (cerebellar nystagmus) and difficulties in speech.

The role of the cerebellum in integration and coordination with other motor centers (→ pp. 258, 260, 264, 268) and sense organs can be illustrated using a tennis player as an example (→ **B**). When one partner serves, the body of the other player moves to meet the ball (motor move), whereby adequate support (right leg) and balance (left arm) have to be maintained (motor hold). Eye movements keep the ball within view, the optical cortex analyses the path and speed of the ball. The "associative" cerebral cortex plans the return movements, taking into consideration the ball, net, opponent's position and, among other things, that the backlash resulting from impact of racket and ball must be compensated by motor hold movements. Movement programs from the cerebellar cortex and basal ganglia assist the motosensory cortex initiating the motion of hitting. The ball is not only hit and returned to the opponent's court, but is as a rule also rotated by a tangential stroke (cutting).

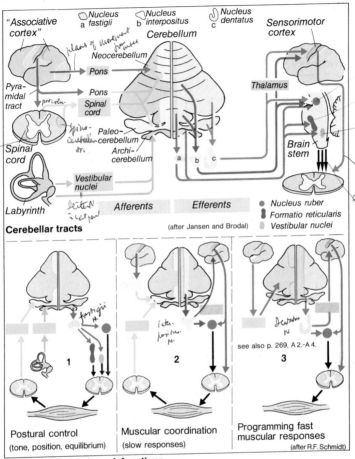

Cerebellum

"Associative cortex"

Nucleus
a fastigii

Nucleus
b interpositus

Nucleus
c dentatus

Sensorimotor cortex

Neocerebellum

Pons

Pyra-
midal
tract

Pons

Spinal
cord

Thalamus

Spinal
cord

Paleo-
cerebellum

Archi-
cerebellum

Brain
stem

Vestibular
nuclei

Labyrinth

Afferents   **Efferents**

● Nucleus ruber
❚ Formatio reticularis
Vestibular nuclei

Cerebellar tracts   (after Jansen and Brodal)

1

2

see also p. 269, A 2.-A 4.

3

Postural control
(tone, position, equilibrium)

Muscular coordination
(slow responses)

**Programming fast
muscular responses**
(after R.F.Schmidt)

A. Cerebellar tracts and functions

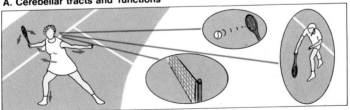

B. Sensorimotor integration (see text)

Motor Move System – Purposeful Motor Control

Motor control involves the participation of spinal cord, brain stem (→ p. 264), cerebral cortex, cerebellum (→ p. 266), basal ganglia (→ **B**), thalamus and other subcortical centers. Their joint action makes possible the performance of purposeful motor movement including the translation of thought and will into motor action. Motor move control (particularly at the level of the brain stem) is closely linked with motor hold movements, since every purposeful movement must be accompanied by readjustment by the motor hold system (→ p. 267, B).

In the primary motosensory cerebral cortex (*Gyrus praecentralis*) each region of the body is represented by its own area (somatotopic representation, → p. 265, B). The same arrangement is found in the secondary motosensory cortex adjoining the primary motosensory area. Those parts of the body capable of delicate movements (fingers, face) are relatively better represented than others.

The **pyramidal cells** of the motosensory cortex are gathered together in columns. The nerve cells within any one column have the function of supplying the skeletal muscles responsible for the movement of, for example, one particular joint, i.e. the arrangement within the motosensory cortex is related to movements rather than to individual muscles.

Efferents lead from the motosensory areas and their associated sensomotor areas to the motoneurons of the spinal cord and to the nuclei of the cranial nerves (*Tractus corticospinalis lateralis et ventralis* [**Pyramidal tract**] and *tractus corticobulbaris*). Further efferents of the motosensory cortex extend to the brain stem, whose descending tracts (via nucleus ruber and reticular formation) also end on the motoneurons of the spinal cord (→ p. 265,

A). These connections are also known as the "extrapyramidal" tracts, as distinct from the pyramidal tract. Collateral (branching) fibers of the pyramidal tract go to the pons (and thence to the cerebellum, → p. 266), the thalamus, the dorsal column nuclei (→ p. 262) and others.

The **basal ganglia** (striatum, pallidum, substantia nigra, nucleus subthalamicus, → **B**) are important relay stations for passing on stimuli from the non specific associative areas of the motosensory areas of the cortex. The basal ganglia appear to fulfill the important role of providing the programs for slow, steady (ramp) movements.

How a stimulus develops from the moment of impetus up to the performance of a movement is not clearly understood. A simplified scheme is shown in Fig. **A**: The *impetus* to voluntary movement arises, in some unknown way, primarily in subcortical areas, and the message is passed on to the "associative" regions of the cortex. A corresponding potential can be registered over the entire brain. The *plan* for the movement is passed on from the "associative" cortical areas to the cerebellum and basal ganglia. The latter provides *programs* for the slow movements and the cerebellar hemispheres those for the rapid purposeful movements. The programs are then relayed via the thalamus to the motosensory areas of the cortex which, as the ultimate supraspinal relay station, finally sets off the *performance* of the movement.

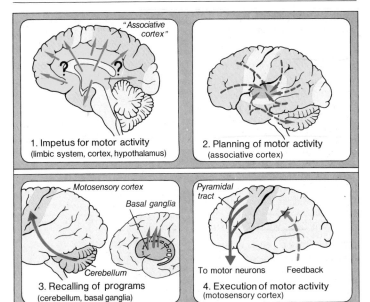

1. Impetus for motor activity
(limbic system, cortex, hypothalamus)

2. Planning of motor activity
(associative cortex)

3. Recalling of programs
(cerebellum, basal ganglia)

4. Execution of motor activity
(motosensory cortex)

A. From stimulus to execution

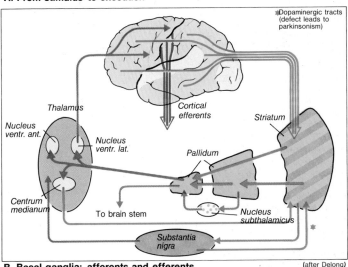

B. Basal ganglia: afferents and efferents   (after Delong)

Hypothalamus – Limbic System – Frontal Cortex

The **hypothalamus** (HT), a part of the diencephalon, is the principal structure in the CNS for *homeostatic control of the internal environment*. It integrates sensory, motor, autonomic, and endocrine functions and programs an overall response. It regulates thirst and appetite, blood pressure, reproduction, body temperature, endocrine and autonomic activity. Its anatomic boundaries are indistinct; these include the mamillary bodies, optic chiasm, walls of the third ventricle, median eminence, infundibulum, and posterior pituitary (PPit). *Afferent* fibers arrive from the globus pallidum, amygdaloid nucleus, and olfactory bulb. Cortical afferents come directly as well as indirectly via the thalamus. *Efferent* fibers connect with the mesencephalon, reticular formation, and the frontal lobes. Important efferents connect also with the vasomotor and respiratory control areas in the medulla (→ pp. 93, 165).

The HT contains two classes of neurosecretory nerves (→ p. 220): (a) from the supraoptic and paraventricular nuclei, fibers pass to the PPit where the nerve endings store ocytocin and vasopressin (ADH) and (b) fibers that end on capillaries of the portal system and secrete hormones for control of the anterior pituitary. The HT senses temperature via thermoreceptors (→ p. 180) and osmolality via osmoreceptors (→ p. 126). Its connections with the medulla influence blood pressure and heart rate.

Integrative functions may involve mobilization for the **fight-or-flight reaction** (increased blood flow in muscles, rise of blood pressure and respiratory rate, vasoconstriction in GI tract etc.), for **nutrition** (vasodilation in GI tract), or for **sexual response** and **reproduction** (control of courtship behavior, sexual erection, intercourse, pregnancy and parturition).

The **limbic system** consists of the amygdalus (one of the basal ganglia), hippocampus, cingulate gyrus, and septum. Its functions are not well understood, but it contributes to program selection by the HT. Stimulation of various parts of the limbic system evokes emotion, pleasure, grooming, sexual activity, and courtship behavior. Bursts of anger or rage follow stimulation of the amygdalus. The **rhinencephalon** overlaps some functions of the limbic system. Fibers of the olfactory bulb connect with a variety of structures concerned with nutrition and reproduction, including the limbic system. The rhinencephalon is not only an organ of smell, it also signals danger, sexual attractiveness, food, and other "desirables." Thus, the rhinencephalon functions to transmit signals from the external environment to the sensory cortex and to stimulate internal responses of affect, emotion, and attitude, in part via the limbic system.

The **frontal cortex** is one of the non-specific areas of the cerebral cortex. It is most highly developed in man. Its functions include abstract thinking, planning for future events, and control of instinctive behavior. Lesions in the frontal cortex may result in perseveration and in shortening of the attention span.

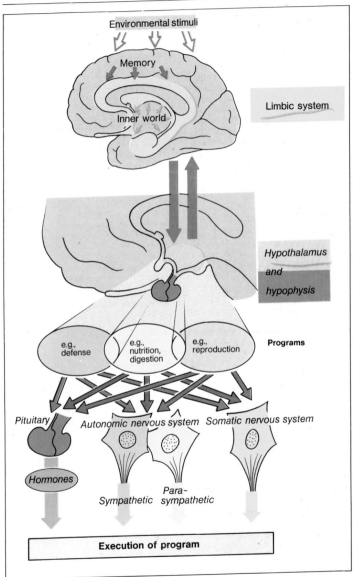

A. Limbic system and hypothalamus

EEG – Sleep

The **electroencephalogram**. The brain, like the heart (→ p. 154), generates cyclic organized electric potentials. The electronencephalogram, or EEG, is the record of these potentials (→ A). Normally, the EEG has three bands, α-, β-, and δ-waves, which vary in frequency and amplitude (→ B) : α-waves – 8 to 13 Hz at c. 50 mV; β-waves – 14 to 30 Hz (but up to 60 Hz) at 5 to 10 mV; δ-waves – 0.5 to 3 Hz at 20 to 200 mV.

α-Waves occur in the unalerted brain, as in light sleep or at rest; they are abolished by sensory stimuli (open eyes) (α-block) and are replaced by β-waves. α-Waves define a **synchronized** EEG, β-waves, a **desynchronized** EEG. Desynchronization correlates with arousal and alertness and is brought about by sensory stimulation. During light sleep, ϑ-waves with a low frequency appear. δ-Waves normally appear in deep sleep (stages E); when they appear in adults under conditions other than deep sleep, they suggest a pathologic state (tumor, intoxication, mental deficiency, or epilepsy).

The EEG is changed by hypoglycemia, hypoxia, and hypercapnia. It assists in diagnosis of brain death, brain tumor, and epilepsy; in the last, spike and dome waves occur characteristically at the onset of an epileptic attack (→ B). However, the EEG may be misleading since it can be normal in c. 30% of epileptics.

Sleep is not only a state of unconsciousness; it is a state of altered functional organization of the CNS. Thresholds for sensory stimulation are elevated and motor discharges are reduced. Muscle tone is lower and reflexes are altered. The metabolic rate and central body temperature fall, blood pressure and heart rate are lowered, and respiration changes. Waking involves arousal of the reticular formation.

Normal sleep is a cyclic sequence (→ C) progressing from light to deeper stages of sleep and returning to light sleep. The cycle repeats four to five times a night. In the waking state, the EEG is desynchronized (random waves, low voltage, higher frequency). Closing the eyes removes visual input and allows α-waves to appear (synchronized EEG). Drowsiness results in reduced prominence of α-waves and appearance of slower waves. In the lightest stage of sleep, α-waves disappear, and slow hippocampal β-waves and sleep spindles appear. As sleep becomes deeper, the spindles are replaced by slow waves with high amplitudes. The lightest stage of sleep (stage B) shows rapid EEG rhythms with low amplitude; rapid eye movements characterize this stage of **REM** or paradoxical sleep. It is associated with secretion of norepinephrine (from the locus ceruleus). Dreams take place during this stage, heart rate and respiratory rate increase, and, in males, penile erection occurs. Periods of REM sleep last c. 20 min and recur every 90 min; they become longer toward morning. Infants have more REM sleep than adults. All other stages of sleep show synchronized EEG rhythms and are non-REM in character. They are associated with localized secretion of serotonin.

The wake-sleep cycle normally is synchronized with the external day-night cycle at a 24 h rhythm. The **internal** biologic clock for diurnal rhythm functions approximately on an individual (mostly c. 25 h) wake-sleep cycle. This cycle becomes explicit when volunteers are isolated from external environmental signals (→ D). Synchronization to the normal 24-h day-night cycle takes place involuntarily but becomes temporarily disturbed by travel from one longitude to another ("jet lag").

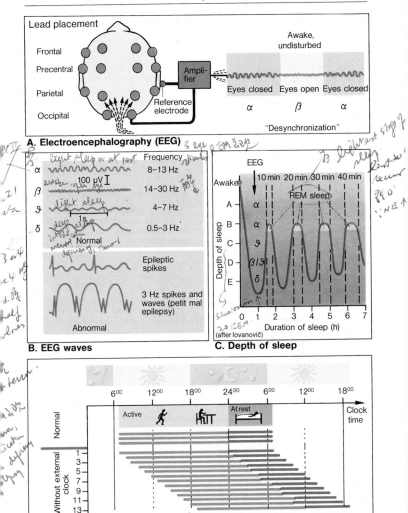

A. Electroencephalography (EEG)

B. EEG waves

		Frequency
α		8–13 Hz
β		14–30 Hz
ϑ		4–7 Hz
δ		0.5–3 Hz

100 µV

1 s.

Normal

Epileptic spikes

3 Hz spikes and waves (petit mal epilepsy)

Abnormal

C. Depth of sleep

EEG

Awake

10 min 20 min 30 min 40 min

REM sleep

A — α
B — α
C — ϑ
D — β/ϑ
E — δ

Depth of sleep

Duration of sleep (h)

(after Iovanovič)

D. Circadian periods

6⁰⁰ 12⁰⁰ 18⁰⁰ 24⁰⁰ 6⁰⁰ 12⁰⁰ 18⁰⁰ — Clock time

Normal

Active — At rest

Without external clock

Days

Circadian period of 25 h shifted 1 h each day for 12 days

(after Aschoff)

Speech and Memory

The human **conscious state** has several aspects, including directed and organized awareness, capacity to deal with abstractions, and ability to verbalize experience, to plan, and to extract new relationships from observations. Consciousness in these aspects occurs only in a highly developed nervous system; the cerebral cortex and its associations with the reticular formation are essential anatomic substrates. The survival value of consciousness is the ability to adapt to circumstances for which reflex responses would be insufficient.

The use of words in speech, reading, and writing is an elaborate means of communication seen only in man. **Speech** serves to create concepts, to fashion ideas, and then to express them; without conceptualization, economical storage in a memory would not be possible. Speech is based on complex neural networks that are asymmetrically distributed in the cerebral cortex. As a rule, one hemisphere, usually the left, becomes dominant in the acquisition of language and in manipulative skills. In some patients with severe epilepsy, it becomes necessary to sever the connections between the two hemispheres by cutting the corpus callosum (*split brain*). In such patients, pictures or words can be described or named only when explored by the *right* hand or when exposed in the right half of the visual field. The left hemisphere has a virtual monopoly in control of motor systems for linguistic expression (speech and writing). The right hemisphere has control of processing complex visual information where verbal coding is unnecessary. Right-sided cerebral lesions are associated with disorders of music performance and appreciation; such patients often cease to dream.

Only a fraction of the sensory input is stored permanently in **memory**. The storage process itself seems capable of modification after the occurrence of the learning event. For instance, a physical blow to the head, electroshock treatment, and administration of anesthetics prior to the establishment of long-term memory can prevent the storage of information received just before the trauma. It appears that memory cannot assume its permanent form immediately after an event. Instead, memory storage takes place over a time interval, during which there is consolidation from a transient to a permanent form. However, once in permanent storage, information can no longer be interfered with by physical or chemical treatment.

There are several phases in memory formation that occur in sequence (→ **A**). In the earliest phases, memory is retained transiently by electric activity of neurons. In this phase small pieces of information like a telephone number can be stored in the *primary memory* for several seconds. Storing information in this memory can be prolonged by reentry of the information (e. g. rehearsal of the telephone number if the line is busy). In this stage, memory is vulnerable to many influences.

If rehearsal is repeated often enough the information is stored in a long-term memory (*secondary memory*, → **A**). Storage here lasts minutes to years and speed of recall is slow. Some individually or generally important information (own name and birthday, writing and reading) are stored in another type of long-term memory (*tertiary memory*, → **A**). In this case the trace is stored for life and the speed of recall is very rapid. Long-term memory is based on protein synthesis rather than on electrical activity.

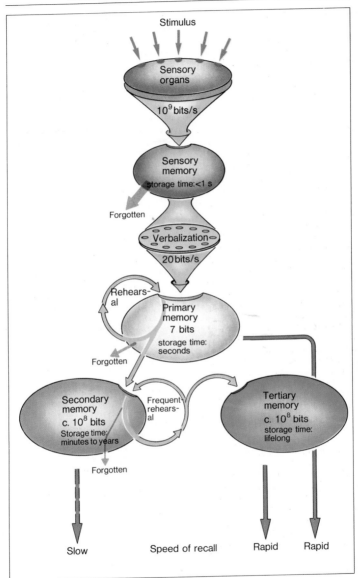

A. Information storage (memory)

Smell

Smell and taste are chemical senses. The receptor for smell lies in the olfactory epithelium in the nasal mucosa ($\rightarrow$ **A**, **B**). In man, there are 10^6 receptor cells in an area of 500 mm². The receptor cells are primitive neurons. They serve both for reception and for neural connection. Resting activity of the receptors is normally low to absent. Volatile substances reach the receptors during inspiration and dissolve in the mucous lining of the nasal cavity before they are recognized. In man, up to 4000 different substances can be distinguished by smell. The receptor cells have a low threshold for **recognizing** the presence of a smell, but to **identify** a particular substance, the concentration must be 50 times greater. **Discrimination** ($\Delta I/I$) ($\rightarrow$ p. 284) is poor; the intensity of an odor must be changed by 30% before the difference can be detected. Thus, there are different thresholds for smell: detection, identification, and discrimination.

When the olfactory epithelium is stimulated, it generates a spike potential. The axons leave the epithelium in small bundles that conduct the signals to synapses in the olfactory bulb where 200 receptor axons converge on one neuron. After two synapses, tracts course to the cerebral cortex, hypothalamus, reticular formation and the limbic system. The olfactory receptors adapt quickly ($\rightarrow$ p. 254) and often completely. After 3 min of smelling octanol, a concentration 300 times above threshold can no longer be recognized. Such adaptation may persist for hours and is expressed by the low to absent frequency of propagated action potentials ($\rightarrow$ **C**). Adaptation is specific for the substance and does not influence the threshold for other substances.

Taste

The primary receptors are in the taste buds, found principally on the tongue and palate. Each taste bud is innervated by approximately five neurons; each neuron receives input from five taste buds. Four basic sensations are recognized: **salt, sour, sweet,** and **bitter**. Conventionally, these tastes are shown to be distributed as in Fig. **E**, but responses to taste stimuli occur in many other areas of the oropharynx. The palate, for example, is the most sensitive area for sour and bitter. There is a circadian variation in taste acuity, the sensitivity being highest in the early afternoon. Taste depends to a large extent on simultaneous smell reception and is influenced by texture of food, temperature, and other variables. The threshold for intensity discrimination ($\rightarrow$ p. 284) is 20% for taste. The concentration of the substance determines in part whether the taste is pleasant or unpleasant ($\rightarrow$ **G**).

Hypogeusia is a reduced taste acuity. Lesions of the tongue reduce acuity for salt and sweet. Dentures that cover the palate reduce acuity for sour and bitter. The endocrine system is also involved in taste sensitivity. Adrenal insufficiency results in a 100-fold increase in acuity for each taste. Treatment with glucocorticoids severely restricts the acuity, raising the taste threshold. A higher than normal threshold for sour and bitter occurs in pseudohypoparathyroidism. There is evidence that copper is involved in regulating taste acuity; depletion of plasma ceruloplasmin, a copper-containing protein, results in elevated taste threshold.

The taste buds are penetrated by dendrites to cranial nerve cells: the chorda tympani of VII to the anterior two thirds of the tongue, IX to the posterior one third of the of the of the tongue, X to the palate and pyarynx. The trigeminal (V) has no connections to taste buds but does register temperature, touch, and pressure in the mouth, which influence interpretation of taste. The primary fibers from taste buds (in chorda tympani VII, glossopharyngeal IX, and to some extent vagus X) converge on the medulla (nucleus of the solitary tract). Second order fibers course to the thalamus; third order fibers from the thalamus reach the postcentral sensory cortex. In man, there is genetic

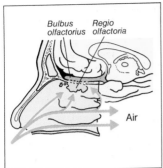

Bulbus olfactorius Regio olfactoria

Air

A. Nasal cavity

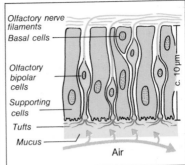

Olfactory nerve filaments

Basal cells

Olfactory bipolar cells

Supporting cells

Tufts

Mucus

c. 10 μm

Air

B. Olfactory epithelium

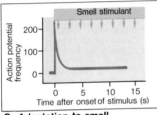

Smell stimulant

Action potential frequency

200

100

0

0   5   10   15

Time after onset of stimulus (s)

C. Adaptation to smell

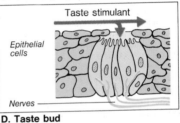

Taste stimulant

Epithelial cells

Nerves

D. Taste bud

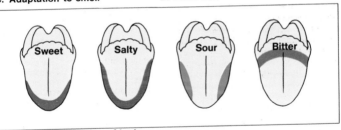

Sweet   Salty   Sour   Bitter

E. Classical localization of taste

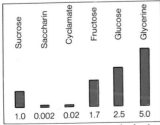

Sucrose Saccharin Cyclamate Fructose Glucose Glycerine

1.0   0.002   0.02   1.7   2.5   5.0

F. Concentration for equivalent sweet taste

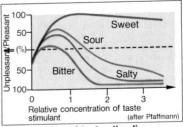

Unpleasant/Pleasant

100

50

(%)

50

100

Sweet

Sour

Bitter

Salty

0   1   2   3

Relative concentration of taste stimulant

(after Pfaffmann)

G. Evaluation of taste stimuli

Vestibular Function

The vestibular system responds to movement and changes in position. The vestibular organ lies near the cochlea (→ p. 299) in the inner ear. Its membranous labyrinth consists of three component parts: the semicircular canals (at one end of each of which is an ampulla) (→ **A, 2**) with a common chamber, the utricle, and the saccule. On each side of the head there are three semicircular canals (→ **A, 1**), each of which is perpendicular to the other two so that they can detect changes in three-dimensional space. The canals are surrounded by perilymph and contain endolymph. The receptor organ is the **crista** (→ **A, 2**), located in the ampulla of each canal. Each crista consists of hair cells (**cilia**) (→ **A, 3**) crowned by a gelatinous **cupula** (→ **A, 4**), which moves like a swinging door toward one or the other side of its canal. Nerve endings (acoustic nerve, VIII) are in contact with the hair cells.

The receptors in the semicircular canals respond to **angular** acceleration. When the head is rotated, the endolymph, because of its inertia, is displaced in a direction opposite to the movement of the canal. This displacement swings the cupula and bends the cilia. Movement of the cilia in one direction depolarizes the sensory cell, in the other hyperpolarizes it. Since the canals on either side of the head are mirror images, they produce opposite responses that amplify the signal. The threshold for detection is an acceleration of c. 0.5°/s. When a constant rotational velocity is reached, fluid and canal move at the same rate and no signal is generated. When rotation slows or stops, the endolymph, because of its inertia, continues to move; the stimulus is the reverse of that at the outset.

The **saccule** (→ **A, 5**) and **utricle** (→ **A, 6**) each contain an otolithic organ, the **macula**; this contains hair cells surmounted by a membrane (→ **A, 7**) in which calcium carbonate crystals (**oto-**liths) (→ **A, 8**) occur. The saccule and utricle respond to **linear** acceleration. Since the otoliths are denser than endolymph, they are more sensitive to changes in position relative to earth's gravity. Inertial displacement of the otoliths stimulates the hair cells.

The nerve endings of the bipolar cells (→ **A, 9**) in the vestibular ganglia contact the sensory cells and conduct their impulses to the vestibular nuclei and to the cerebellum. Ascending fibers from the vestibular nuclei associate with oculomotor III, trochlear IV, and abducens VI nerves in the vestibulo-ocular tract, which mediates reflex movements of the eyes. Descending fibers to the vestibulospinal tract in the spinal cord mediate postural reflexes. Additional associations with the reticular formation occur that are probably responsible for extension-flexion reflexes (→ **B**). When the body axis is tilted, the downhill side extends, and the uphill side flexes in a protective reflex to maintain balance (→ **B, 2**). In vestibular disturbance, the righting reflex is lost and the individual tips over (→ **B, 3**). The close relationship between the vestibular organ and the oculomotor system (→ **C**) accounts for vestibular **nystagmus** (→ p. 294) which is a reflex occurring at the start and finish of a period of rotation. The eyes are fixed on a stationary point; when the limit of vision is reached, the eyes move quickly in the direction of rotation, and the gaze is fixed on a new point. This process contributes to spatial orientation (→ p. 294).

The vestibular system is unable to distinguish whether the entire body has moved or only the head. This important postural distinction is made by neural associations between vestibule and cerebellum with muscle spindles and receptors in the cervical vertebral joint capsules.

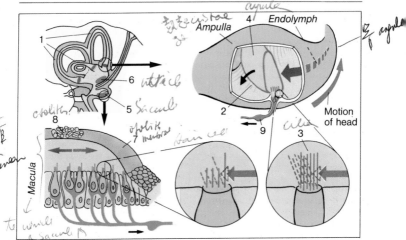

A. Vestibular organ

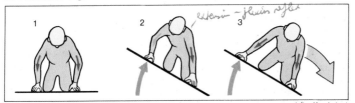

B. Vestibular organ: effect on equilibration

(after Kornhuber)

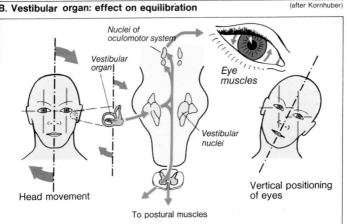

Nuclei of
oculomotor system

Vestibular
organ

Eye
muscles

Vestibular
nuclei

Head movement

To postural muscles

Vertical positioning
of eyes

C. Vestibular organ: effect on eye movement

Eye Structure, Tears, Aqueous humor

The eye consists of a lens system for focusing light, a light-sensitive receptor area, and a system of nerves to conduct information from the eye. Light that reaches the sensory cells of the retina must pass, in sequence, the cornea, aqueous humor, lens, vitreous humor, and several nonreceptor layers of the retina before it reaches the light-sensitive receptors (→ **A**). The overall distance from cornea to retina is c. 24 mm. There are several phase changes before light is focused on the retina where, as in a camera, the image of the external environment is inverted. Transparency, optical surfaces, and shape must be maintained optimally for an undistorted reproduction.

The cornea is protected by **lachrymal fluid** which is continuously secreted by the *lachrymal glands* (→ **B**). A fluid film of tears protects the cornea from dehydration. Stimuli for reflex secretion of tears and for blinking are irritation of the cornea or the nasal mucosa, bright lights, coughing, vomiting, yawning etc. Emotional upsets may result in psychic weeping. The tears exceeding the loss due to evaporation are drained by the lachrymal canaliculi, the lachrymal sac, and the nasolachrymal duct (→ **B**). Tears serve additionally to wash away irritating particles and chemicals, and to correct unevenenness of the cornea surface. Tears also contain lysozyme, a bactericidal enzyme.

The **iris** (→ **C, 3**) functions in the same way as the diaphragm in a camera. It contains radial fibers to dilate (mydriasis) and circular fibers to constrict (miosis) the pupil. Adrenergic stimulation produces mydriasis; cholinergic stimulation produces miosis. The form of the ocular bulb is maintained by its tough **sclera** (→ **A; C, 1**) and by its internal pressure, which is normally 2 to 3 kPa (= 15 to 22 mmHg). The pressure tends to increase with age.

Aqueous humor. This fluid is produced by active transport of electrolytes in the ciliary process (→ **C, 2**) of the posterior chamber (→ **C, 3**). It flows over the iris into the anterior chamber (→ **C, 4**) and exits at the angle of the anterior chamber through the canal of Schlemm (→ **C, 5**), where it is carried away by the venous blood flow; complete replacement of aqueous humor occurs every 60 min. The angle is made smaller when the iris is dilated (mydriasis), and efflux through the canals may be partially impeded. Production of aqueous humor maintains internal ocular pressure and shape. If production (inflow) of fluid exceeds its outflow, the ocular pessure may rise, causing glaucoma, pain, and visual disturbances; ultimately, the increased pressure may cause blindness by damaging the retina. Treatment of glaucoma involves (1) inhibition of the formation of aqueous humor (e.g., by carbonic anhydrase inhibitors) and (2) stimulation of miosis to increase the size of the drainage angle.

The **lens** is suspended on the zonular fibers (→ **C, 6**), which influence tension on the lens and thereby its shape, its dioptric power, and its focal point. The **ciliary muscle** (→ **C, 7**) contracts for near vision (increased curvature) and relaxes for distant vision (→ **D**). The inner surface of the eye is covered by the choroid, a pigmented area that contains the nutrient blood vessels.

The **retina** covers the posterior two thirds of the eye except where the fibers of the optic nerve (II) leave the bulb at the optic disc. Near the optic disc is a depression, the **fovea centralis** (→ **A**). Retinal receptors are the **rods** (120×10^6) and the **cones** (5 to 7×10^6). These connect with *bipolar cells*, which then synapse with the *ganglion cells*, the axons of which form the 10^6 fibers of the *optic nerve*. The rods (→ **F**) and cones are associated with the light-sensitive retinal pigments (→ p. 284). The *horizontal* and *amacrine* cells provide connections among the receptor cells of the retina (→ **E**; p. 292).

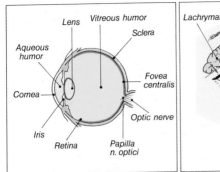

A. Right eye (horizontal section)

Lens
Vitreous humor
Sclera
Aqueous humor
Cornea
Iris
Retina
Fovea centralis
Optic nerve
Papilla n. optici

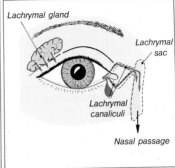

B. Right eye: tear flow

Lachrymal gland
Lachrymal sac
Lachrymal canaliculi
Nasal passage

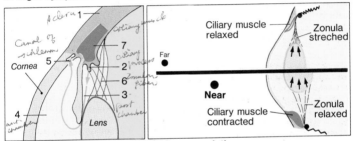

C. Aqueous humor

Sclera 1
Canal of schlemm
Cornea
5
7 ciliary muscle
ciliary process
2
6 zonular fiber
3
post chamber
4 ant chamber
Lens

D. Accommodation

Ciliary muscle relaxed
Zonula streched
Far
Near
Zonula relaxed
Ciliary muscle contracted

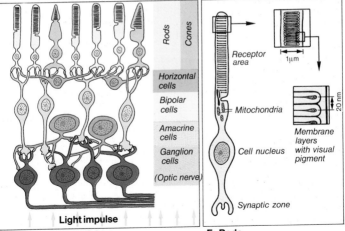

E. Retina (schematic)

Rods
Cones
Horizontal cells
Bipolar cells
Amacrine cells
Ganglion cells
(Optic nerve)
Light impulse

F. Rods

Receptor area
1μm
Mitochondria
Membrane layers with visual pigment
20 nm
Cell nucleus
Synaptic zone

Optics

Light rays that pass through an interface (e.g., air to glass) are refracted (bent) and dispersed (→ **A**). Light rays from a distant source may be considered to be parallel. When they pass through a spherical interface, parallel to the optical axis, they will meet at the **focal point** (F_h) (→ **A**, **1**, red dot). If they pass through the interface at an angle to the optical axis, they will meet at some other point (→ **A**, **1**, lilac dot). All such points define the **focal plane**. If the light rays are generated from a point source closer to the lens (→ **A**, **2**), they are no longer parallel and will meet on a plane farther from the lens.

The eye is a complex optical system with many interfaces, but it may be considered as a simple refractive surface that can change its curvature. The *nodal point* of this simple optic system (reduced eye of Listing) is situated c. 7 mm behind the anterior corneal surface (→ **B**, N) and c. 17 mm in front of the retina. The less the curvature of the lens, the less it refracts light; thus, for distant vision, the lens curvature is minimal. In the eye accomodated for distant vision, parallel rays converge at F_h, the focal plane of the retina (→ **B**, **1**, red dot). The plane for an object closer to the eye is behind the retina (→ **B**, **1**, green dot) and is seen as blurred. To accomodate for near vision, the lens must increase its curvature and its refractive power to bend the rays onto the retinal plane. In this state, a distant object is seen in front of the retina and is seen as blurred (→ **B**, **2**, red circle) because the *posterior focal length* (→ **B 2**, distance H-F'_h) is now shorter if compared to eye accomodation for distant vision (→ **B1**, distance H-F_h). The eye accomodates faster for near than for distant vision.

Refraction in the eye is measured in diopters, the reciprocal of the anterior focal length in meters (→ **B 1**, distance F_v–H). When the eye is accomodated for distant vision this focal length is

0.017 m and the refractive power is 1/0.017 m or c. 60 diopters; for near vision, the lens can increase its refractive power over a range of c. 10 diopters in young people. The reciprocal of this **range of accomodation** defines the **near point** (m) of the eye, which is the nearest point to which the lens can accommodate. In young adults the near point is 1/10 = 0.1 m. The **far point** at which the lens can accommodate to produce a sharp image normally is at ∞.

The lens stiffens with age and the range of its curvature decreases; as one grows older, the near point becomes more distant and the range of accomodation decreases (**presbyopia**) (→ **C**, **1–3**), although distant vision remains unaffected (→ **C**, **1**). Reading becomes difficult because the book must be held farther from the eye to come into focus, but at this greater distance the size of the retinal image is too small to permit resolution of the letters.

Accordingly, old age has been defined as that time when the arms are no longer of sufficient length to permit reading the newspaper. Correction for reading in presbyopia is achieved with a convex lens.

In **myopia**, the ocular bulb is relatively too long, and the light rays focus in front of the retina (→ **C**, **4**). Normally, the far point is at ∞; in myopia it is much closer (→ **C**, **5**). A divergent lens (minus diopters) is needed for correction (→ **C**, **6**, **7**). The refractive power (diopters) of the lens needed can be calculated from the reciprocal of the far point in meters. In **hyperopia** or farsightedness, the bulb is relatively too short and focus occurs behind the retina. Accomodation changes for near vision occur at longer than normal distances (→ **C**, **8**) so that the range of accomodation is insufficient for near distances (→ **C**, **9**). A convergent lens (plus diopters) is needed for correction (→ **C**, **10**, **11**).

A **cataract** is a decrease in transparency of the lens. The lens may be removed surgically and is replaced by glasses having a converging lens of at least 15 diopters.

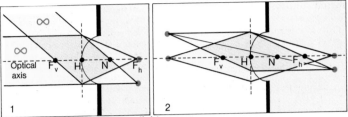

A. Optics: (1) distant, (2) near light source

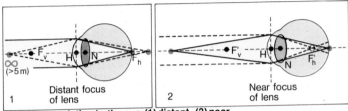

B. Accommodation in the eye (1) distant, (2) near

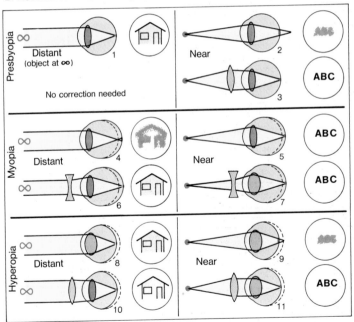

C. Visual defects

Visual Acuity, Retinal Receptors

Visual acuity is a measure of the resolving power of the eye. It can be measured as the minimum visual angle that will permit resolution; it is the shortest distance by which two points can be separated and still be seen as two distinct points. With adequate light, the normal eye should be able to discriminate two points that subtend an angle, α, of 1 min (1/60°) ($\rightarrow$ **A**). The reciprocal, $1/\alpha$, represents the visual acuity (1/1 min = 1). Clinical testing of acuity is based on the minimum separation threshold: symbols of progressively smaller size are presented for recognition. They may be letters of the alphabet or split rings in which the split subtends 1 min of arc ($\rightarrow$ **A**). Acuity is then calculated as a ratio. Example: 20/200 means that the subject discriminates at 20 ft a symbol that an individual with normal vision can discriminate at 200 ft (acuity = 0.1). Visual threshold is distinct from acuity; it is the minimum amount of light that stimulates a sensation of light.

The retinal receptors are the light-sensitive **rods and cones** ($\rightarrow$ p. 281, **E**). They are not uniformly distributed in the retina: the fovea contains exclusively cones; the density of cones declines rapidly as the distance from the fovea increases ($\rightarrow$ **B**, left). The rods are greatest in number 20° from the foveal center ($\rightarrow$ **B**, left). The fovea is capable of the greatest visual resolution because the signals of its cones have the lowest convergence towards the visual centers of the cortex ($\rightarrow$ p. 290). The fovea is therefore the visual center of the eye. The gaze adjusts to the foveal axis for the most precise visual inspection. Visual acuity declines as the image becomes distant from the fovea ($\rightarrow$ **B**, right) because the convergence of the signals of the peripheral receptors is higher ($\rightarrow$ pp. 290, 292). At the optic disc, which subtends an angle of 3° at a distance of 15° from the fovea, there is

no visual capacity at all; this is the blind spot because it contains no receptors. After dark adaptation ($\rightarrow$ p. 286), the distribution of visual acuity changes and corresponds to the distribution and convergence of the rods ($\rightarrow$ **B**, right). Thus, cones are specialized for discrimination of color in bright light (photopic vision) and rods for discrimination of form (black and white) in poor light (scotopic vision). Visual acuity is sacrificed for dark adaptation.

Dark adaptation is a lowering of the visual threshold ($\rightarrow$ p. 286). Full adaptation requires c. 20 to 30 min and depends to a large extent on the regeneration of visual pigment (rhodopsin) in the receptor cells. Since vitamin A is involved in the synthesis of visual pigments, vitamin A deficiency is associated with various visual abnormalities. Chief among these is **nyctalopia** or night blindness; in this state, there is not only a deficiency of visual pigments but also anatomic changes in the rods and cones and progressive degeneration of the neural layers of the retina.

Photochemistry. When quanta of light energy are absorbed by the retinal receptors, a photosensitive pigment is changed chemically and transduces light energy into electric energy. The pigment in rods is **rhodopsin** (visual purple); it is composed of a protein (opsin) and 11-*cis*-retinal ($\rightarrow$ **C**). Light produces a steric change to 11-*trans*-retinal (lumirhodopsin), which initiates changes leading to generation of an action potential. Lumirhodopsin is split to its component parts (bleaching) by hydrolysis. Rhodopsin is then regenerated by an energy-dependent process (photochemical cycle). All of these reactions, except the formation of 11-*trans*-retinal, can occur in the dark as well as the light. Thus, the amount of rhodopsin in the retina varies inversely with the amount of incident light. The greater the amount of light, the more rhodopsin is converted to 11-*trans*-retinal.

Cone (fovea centralis) ⇒ VA ↑↑. 2r 5 color in bright light

Rod(s) — dark adaptation = scotopic vision, VA ↓↓↓, 2r5 ¼ to ½ form in darkness

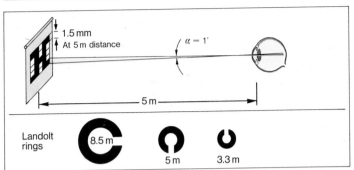

A. Visual acuity

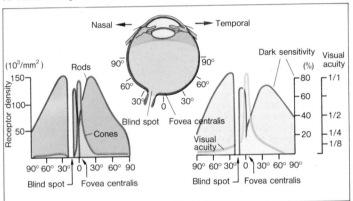

B. Retina: distribution of rods, cones, dark sensitivity and visual acuity

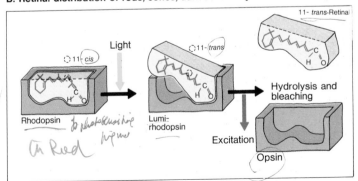

C. Photochemistry of rhodopsin

Adaptation to Light

The eye is capable of responding to a wide range of light intensities – from the low intensity of a distant star to the glare of the summer sea shore or a sunlit snow field (range 10^0 to 10^{11}). Response to such a broad range requires special mechanisms for adaptation. For example, in going from daylight into a darkened theater, the eye is unable to perceive the low intensity light signals because they are temporarily below the retinal threshold. Important changes must occur before it is possible for the crude low luminance (scotopic) vision (→ p. 284) to operate optimally (→ **A**); maximum adaptation may require up to 30 min. The minimum light that can be detected after dark adaptation is the **absolute visual threshold**. For the first few minutes of adaptation (→ **A**), the threshold declines rapidly (10 min) and then more slowly to the absolute threshold (20 to 30 min). At c. 2000 times the absolute threshold, the bend in the curve represents transition from cone receptors to rod receptors (→ **A**, purple curve). The transition level is the threshold for daylight (cone) vision. In night blindness or nyctalopia, the transition does not occur, and it is possible to record the adaptation curve for cones alone (→ **A**, red curve).

The capacity to discriminate between two light stimuli is defined by the luminance threshold or **resolution threshold**: for two light sources of intensity I and I', the absolute threshold, $\Delta I = (I - I')$ and the relative threshold $= \Delta I / I$. A relative threshold ratio of 0.01 represents a high capacity for resolution, allowing recognition of a change in intensity of 1%. Point resolution of this sort is greater for cones than for rods. Variations in light intensity initiate several mechanisms for adaptation (→ **C, 1–4**):

1. The **pupillary** reflex influences the amount of light falling on the retina by a factor as great as 16. The pupil is larger (8 mm) in dim than in bright light (2 mm) (→ p. 290).

2. The concentration of **visual pigment** varies. Since bright light bleaches more pigment, concentration of pigment falls (→ p. 284). In dim light, a greater degree of regeneration of pigment is possible because breakdown of rhodopsin occurs to a lesser degree.

When visual pigment is in low concentration, the probability of its contact with a photon is reduced; in the dark, the concentration of pigment and the probability of contact with a photon is maximal (increased sensitivity).

3. The visual receptor cells can be modified in a form of **spatial** summation, possibly by a feedback mechanism. In effect, the organization of the receptive field (→ p. 292) is changed. It is as though the individual nerve fiber alters the number of cells from which it receives excitatory impulses (change in convergence of excitation on conduction). The "gain" (number of cells) is higher in dim light than in bright light (→ p. 292).

4. **Temporal summation** occurs when a *sub*threshold light stimulus is presented for a long time. The accumulation of subthreshold stimuli eventually can exceed the threshold. This response occurs because the product, light intensity times stimulus duration, is relatively constant.

A **local** kind of **adaptation** is seen in the phenomenon of successive contrast (→ **D**). If one stares at a black and white pattern for 20 s and then shifts his gaze to a plain field, the pattern will reappear, but with a reversal of black and white areas.

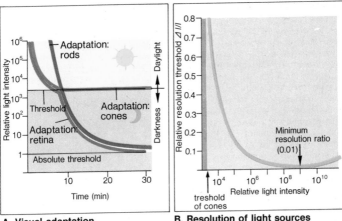

A. Visual adaptation

B. Resolution of light sources

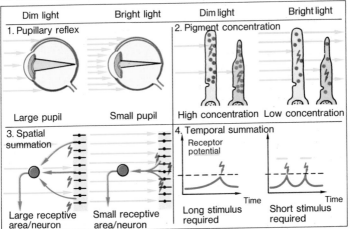

C. Mechanisms of adaptation

D. Successive contrast ("local adaptation") (see text)

Color Vision

The spectrum of light to which the human retina is sensitive ranges from a wavelength (λ) of 650 to 700 nm in the red-infrared to c. 400 nm in the violet-ultraviolet range ($\to$ **A**). In other species, the visual organ has a different spectral range; insects, for example, can see infrared light sources. It has been demonstrated that white light is the sum of its monochromatic components. White light can be produced as well when any two complementary colors, such as blue (490 nm) and orange (612 nm) light, are mixed additively. Primary colors (red, green, blue) can be mixed in varying proportions to produce any color; no two of them produce white light when they are mixed additively, but all three together produce white light. These principles are exemplified in the classic color triangle ($\to$ **B**). The horizontal limb is undetectable light energy, the upper two limbs incorporate the visible spectrum. Central in the diagram is a point to represent white light. A line through the white light point intersects the upper limbs of the triangle at two complementary wavelengths of light (e.g., 490 and 612 nm).

Additive mixture of red and green yields yellow ($\to$ **C**); more red yields orange, less red yields yellow-green. The primary colors for additive mixture are red, green, and blue; for subtractive mixture by reflected light they are magenta, cyan, and yellow.

Since all colors in the visible spectrum and white light can be produced by mixing no more than three primary colors, it may be concluded that color vision requires a minimum of three discrimination channels. These are represented by three different pigments occurring in three different classes of retinal cones. These pigments absorb light signals of the appropriate primary color wavelength. Since rhodopsin in the cones absorbs **all** visible light energies, it cannot serve for color discrimination.

It has a maximum energy absorption at 500 nm, causing blue-green light to be perceived as the brightest at night and red as the darkest ($\to$ **D**). As a result, dark adaptation can be hastened by wearing red glasses. The pigments for color vision have other absorption maxima in the blue-violet, green, and yellow-orange ranges ($\to$ **E**). The Young and Helmholtz theory of color vision is based on this principle. In a broad range of the visible spectrum, the retina can discriminate colors that differ by only 1 to 2 nm (discrimination threshold ($\to$ **F**, p. 286).

In **color blindness**, there is a high discrimination threshold for particular colors ($\to$ **F**); c. 9% of men and 0.5% of women are color blind. Where all three cone color systems are present, the individual is designated a trichromat (normal), where only two are present, a dichromat. Dichromats can match their color spectrum by mixing only two primary colors. The completely color blind are monochromats (or achromats); they match any spectral input with a single monochromatic light by adjusting its luminance. For the most part, they lack normal cones. Their vision is scotopic and acuity is low; their luminance function corresponds to the dark-adapted retina so that high luminance is disturbing. They see photopically in one color and scotopically in gray shades.

A color weakness is designated an -anomaly, a blindness an -anopia. Red and green color blindness are protanopia and deuteranopia, respectively; red and green are confused and are seen as yellows. Tritanopia (rare) confuses yellows and blues. A partial defect is a prot-, deuter-, or tritanomaly.

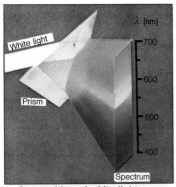

A. Composition of white light

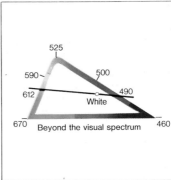

B. Color triangle   (after Kries)

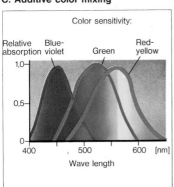

C. Additive color mixing

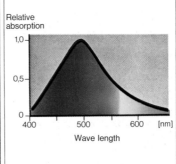

D. Spectral absorbtion curve: rhodopsin

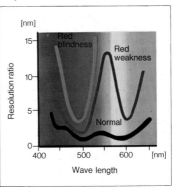

E. Spectral curves of color pigments **F. Discrimination of color**

Visual Fields and Visual Pathways

The visual field of each eye is the segment of the external world seen by the fixed eye. Theoretically, it should be circular, but the nasal and orbital portion is limited (→ **A, 1**). The visual fields are measured by a **perimeter**, a concave hemispheric surface. The eye is positioned at its center, and a light source (white or monochromatic) is moved into the field from the periphery. The points at which the light source are first recognized are marked on the background (→ **A, 1**). Fields for colored light are smaller than for white light; as a result, a colored light is seen as a light source before its color can be identified. The central parts of the visual fields of the two eyes overlap (binocular vision, → p. 295, A). The retina contains c. 130 × 10^6 receptors (rods and cones), but the optic nerve has only c. 10^6 neurons. The receptors are said to **converge** on the neurons. The convergence is greatest in the peripheral retina (1000 rods : 1 neuron) and smallest in the fovea (1 : 1). Low convergence relates to high visual acuity and low light sensitivity, high convergence to lower visual acuity but higher light sensitivity (spatial summation) (→ pp. 284, 286, 292). The axons of the ganglion cells in each eye unite at the optic disc, pass through the sclera, and emerge as the **optic nerve**. The two optic nerves unite in the X-shaped **optic chiasma** where fibers representing the nasal half of the retina cross to the other side (→ **A, 2**). The temporal fibers remain uncrossed. Because the image is inverted on the retina, the nasal part of the visual field registers on the temporal half of the retina and the temporal part on the nasal half. A lesion of the left optic nerve before the chiasma (→ **A, 2a, 3a**) results in loss of the entire left optic field (scotoma). A lesion of the left optic tract behind the chiasma (→ **A, 2b, 3b**) results in loss of the right half of both visual fields. A lesion of the chiasma (→ **A, 2c, 3c**) re-

sults in loss of both temporal fields when the nasal halves of both retinas become nonreceptive.

The optic tract fibers make synaptic connections with several structures in the brain. Most of the third order neurons (→ p. 281) end in the superior colliculus (SCol) and in the lateral geniculate nucleus (LGN), the major visual thalamic relay. Radiation fibers, fourth order neurons from the LGN, the geniculostriate bundle, project to the occipital visual cortex on the same side. Some fibers from the fovea (→ p. 280) cross to the opposite side (→ **A, 2**, broken line); thus, lesions in the radiation bundles (→ **A, 2d, 3d**) often do not affect central vision.

Fibers from the LGN and of the optic nerve also connect with the oculomotor, trochlear and abducens nerves (III, IV and VI), which are involved with eye movements. The **pupillary reflex** (→ p. 286) is initiated by an increase in light intensity. The efferent limb of the reflex follows parasympathetic fibers of nerve III and constricts the pupils. Both pupils react simultaneously even when the stimulus reaches only one eye. The **corneal reflex** is protective: touching the cornea (afferent, nerve V) or approaching the eye (afferent, nerve II) initiates the reflex and closes the eyelid (→ p. 280).

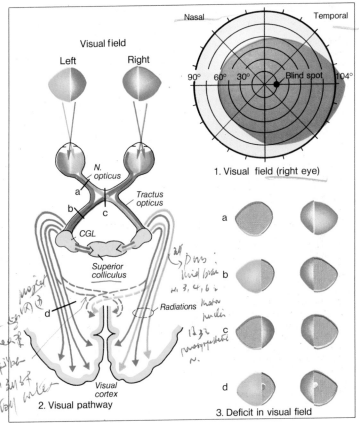

Visual field

Left   Right

Nasal   Temporal

90° 60° 30°   Blind spot   104°

1. Visual field (right eye)

N. opticus

a

b   c

Tractus opticus

CGL

Superior colliculus

Radiations

d

Visual cortex

2. Visual pathway

a
b
c
d

3. Deficit in visual field

A. Visual pathway and field

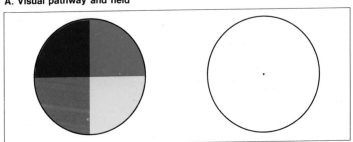

B. Successive contrast, color (see text, p. 292)

Processing of Light Stimuli

The eye responds to light intensity by generating a so-called secondary **receptor potential** (→ **A**, left). Unlike other receptor cells (→ p. 254), the rods and cones hyperpolarize instead of depolarize (potential becomes more negative). Magnitude of the receptor potential varies with the logarithm of light intensity; by this means, the information is considerably compressed when being translated into electric impulses. When the receptor potential reaches threshold, it permits an **action potential** (AP) spike (→ **A**, right) to be propagated in the *ganglion cells*. The basic mechanism of the receptor emphasizes change: initially, there is a rapid firing frequency to signal the onset of illumination, but adaptation is rapid (→ p. 254) and the firing frequency is reduced.

There are many lateral connections in the pathways from receptor cell to ganglion cell. In the retina, *horizontal cells* furnish contacts between receptor cells, and *amacrine cells* between ganglion cells (→ p. 281). Thus, to stimulate a particular ganglion cell, a beam of light need not strike the retina at only one specific point. Each ganglion cell is responsive to light striking a wider area of the retina. This area is the **receptive field** (RF) for that ganglion cell. The RF of a neuron of the optic nerve is concentric, with a *central* and a *surrounding zone* (→ **B**). The zones are antagonistic in their effects. Stimulation of the center increases the spike frequency (→ **B**, **1**), and stimulation of the periphery reduces spike frequency.

Turning off the peripheral light stimulus results in excitation (→ **B**, **2**). This type of RF is called *center on*. In other RFs, the converse situation occurs; light stimulates the periphery and suppresses the center; this type of RF is called *center off* (→ **B**, **3**, **4**). The action within an RF is a form of **lateral inhibition** (→ p.

255, D) in which activation of one unit inactivates surrounding units.

The function of the zones in an RF is to increase the contrast of a stimulus and to improve discrimination. At a border between light and shadow, the dark is perceived as darker, the light as brighter. A gray circle appears darker against a white field than against a dark field (simultaneous contrast) (→ **C**, left). In a field of crossed lines (→ **C**, right), the cross points appear darker if the lines are white, lighter if black. This phenomenon is explained by the balance sheet for stimuli within an RF (→ **C**, center); it demonstrates that there is less contrast of the lines at the cross points.

In dark adaptation, the influence of the central zone increases progressively at the expense of the peripheral zone until, in full adaptation, there is no more antagonism within the RF. Contrast and therefore visual acuity decrease (→ p. 285, B), but spatial summation (→ p. 286) increases as a result of this reorganization of the RF. In summary, lateral inhibition in the retina (mainly by the horizontal and amacrine cells) enhances spatial resolution in bright light but decreases during dark adaptation to allow spatial summation to become optimal.

RFs can be demonstrated also in higher visual centers. In these, the direction as well as the form of the stimulus play an important role. An RF may also allow antagonistic responses to red-green or to yellow-violet light sources to increase *contrast of color vision* centrally. Thus, if one stares at a colored field (→ p. 291, B) for 30 s and then looks at a neutral background, the complementary color will appear (successive contrast).

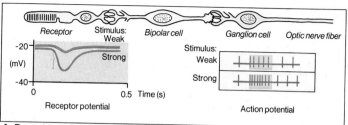

A. Receptor and action potentials of the retina

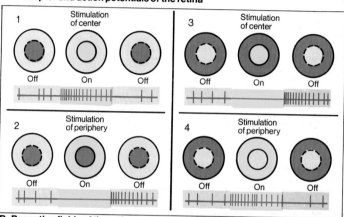

B. Receptive fields of the retina: center ON (1,2), center OFF (3,4)

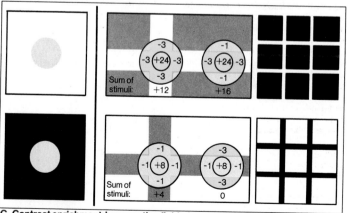

C. Contrast enrichment by receptive fields (center ON)

Eye Movements

The eye has six **extrinsic muscles** which rotate the eye in vertical, horizontal, transverse, and oblique axes. These muscles are innervated by nerve **III**, **IV and VI**. Their activity is controlled by **occulomotor centers** situated in the reticular formation of pons and mesencephalon.

Several types of eye movement may be distinguished:

1. **Conjugate movement**, in which both eyes move in the same direction, as in pursuit movement. Reflectory conjugate movement adjusts the eyes for precise visual inspection by the fovea (→ p. 284) if for instance a moving object appears at the periphery of the visual field (→ p. 290).

2. **Saccadic movement**, as in reading. As a line is being read, the eyes are fixed, then the eyes move quickly to the next fixation point; four to five fixations per line of print take place before the eyes return to the beginning of the next line. In large saccadic movements, the angular velocity may reach 500°/s.

3. **Smooth pursuit movement**, as in tracking a moving target. The speed of this movement depends on the velocity of the target, the eyes being able to follow a rate of 30°/s accurately. If a repetitive pattern is continuously moved horizontally in front of the eyes, the eyes pursue in an **optokinetic reflex**. When telephone poles pass a train window, the eyes follow slowly and return quickly, producing a reflex **nystagmus**.

4. **Vergence movement**: when the gaze moves from a near to a far object, the eyes diverge; from a far to a near object, they converge. Vergence movements are accompanied by pupillary and lenticular accomodation reflexes. The degrees of accomodation and of convergence do not always match. In hyperopia, convergence becomes excessive because of the greater degree of accomodation that the visual defect demands (→ p. 282). Double images are seen in this case. If the hyperopia is not corrected the image of one eye will be suppressed by cortical centers, and blindness of this eye may develop.

5. **Cyclo-rotatory movements** (→ pp. 278, 279)

Three-Dimensional Vision

The visual fields of the eyes overlap in their central portions; an object that falls on this overlapping area is seen with binocular vision. A point (→ **B**, point A, B or C) will stimulate the two retinas on **corresponding points** (→ **B**, A_l and A_r, etc.); if the eye is displaced so that corresponding points are not stimulated, **diplopia** (double vision) results. When the gaze is focused on an object, a circle (three-dimensionally: a spheric surface) can be described by the object and the two nodal points (→ **B**, point K) of the eyes; all objects on this circle (→ **B**, points A, B, C) also have corresponding points.

An object beyond this circle (→ **C**, point D) will produce a double image (→ **C**, D' D''). If the spatial separation, D' D'', is small, the image is integrated in the cortex to give the impression of depth (D' is registered from the left eye and D'' from the right). Likewise, if an object (→ **C**, point E) is within the circle, a double image again results but with opposite orientation (E' is registered from the right eye and E'' from the left).

At great distances and in monocular vision, spatial interpretation is based on other cues: overlapping contours, haze, shadows, size relationships, perspective, etc. (→ **D**, 1–4). Additional three-dimensional information is obtained by moving the head or as a whole, for instance in a train. A near object appears to move relatively faster over the visual field than a distant object (→ **D**; compare sun with station sign).

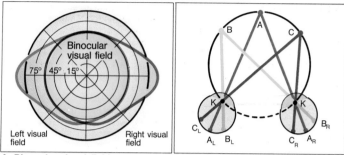

A. Binocular visual field

B. Corresponding points

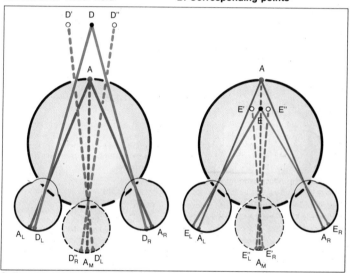

C. Binocular visualization (stereoscopic)

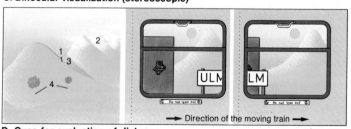

→ Direction of the moving train →

D. Cues for evaluation of distance

Physical Principles: Sound

Hearing is based on detecting vibrations in an elastic medium. The concept of **sound** is applied only to those frequencies that the human ear can detect. An adequate stimulus for hearing is a pressure wave through a medium, producing periodically alternating areas of compression and rarefaction of that medium ($\rightarrow$ **A**). There is no sound in a vacuum. Velocity of sound transmission in air at 0°C is 332 m/s; in water it is c. 1400 m/s (c. 3200 mph). In general, velocity (C) is the product of frequency (f) and wavelength (λ): $C = f \times \lambda$. Since velocity is a constant, f and λ vary inversely. A sound wave is depicted graphically in Fig. **A**. Wavelength (λ) is the distance between two corresponding points of the sound wave form; frequency is the number of periods or cycles/s. Amplitude (**a**) refers to the height of the wave form ($\rightarrow$ **A**). A small λ produces a high tone; a large λ produces a low tone. A small **a** produces a quieter, a large **a** a louder tone. The pitch of a sound is conventionally defined by its frequency although it is just as readily determined from its wavelength. The unit of frequency is the hertz (1 Hz = 1 cycle/s).

A pure tone is a sinus-shaped vibration; the sounds of speech and music are considerably more complex; they are a superimposition of many frequencies and amplitudes. These differences characterize various sound sources (e.g., musical instruments, human voice); an A (f = 440 Hz) played by a flute and a violin does not sound the same; although the fundamental frequency is identical, each instrument superimposes different higher frequencies (overtones) that allow the listener to distinguish the sources. In contrast, a broad band of frequencies having similar amplitudes produces "white" noise ($\rightarrow$ **A**).

The human ear can detect sound frequencies in the range of 16 to 20,000 Hz, although the upper limit declines with age and may be as low as 5000 Hz (presbyacusis). The range of sound pressures is greater. The threshold for detection of a sound stimulus varies with the frequency; at a frequency of 1000 Hz, it is c. 3.2×10^{-5} pascal (1 Pa = 1 N/m² = 10 dyne/cm²) ($\rightarrow$ p. 6). Maximum sensitivity is at a range of 2000 to 5000 Hz. A sound pressure up to c. 20 Pa and more (depending on the frequency, $\rightarrow$ **B**, uppermost blue curve) can be tolerated without pain or damage to the ear.

Because of the wide range of sensitivity, sound pressure level is described by a scale in which the units are the **bel** and the **decibel** (dB = 0.1 bel) (named after A.G. Bell, the inventor of the telephone). The bel scale is logarithmic: 0 dB is not the absence of sound but a standard sound pressure (p_o) at the auditory threshold: $p_o = 2 \times 10^{-5}$ Pa. The **sound pressure level** (unit: dB SPL) at a higher sound pressure (p_x) is calculated from:
$SPL = 20 \times \log p_x/p_o$ (dB SPL);
A 10 fold increase in sound pressure represents an addition of 20 dB, a 1000 fold increase an addition of 60 dB ($\rightarrow$ **B**, left). *Sound intensity* I (W/m²) is proportional to the square of sound pressure p and describes power flux (at $p = 2 \times 10^{-5}$ Pa I = 1 pW/m²; p = 2 Pa, I = 10 mW/m²).

Subjectively, although two sounds of different frequencies have the same sound pressure, they are not perceived as being equally loud. Example: a sound at 40 Hz will not seem as loud as one at 1,000 Hz unless its sound pressure is c. 100 times greater (+40 dB). The unit for subjective sound level is the **phone**. The **isophones** ($\rightarrow$ **B**, blue and red curves) in the dB-Hz diagram show the sound pressures required for all sound frequencies to have the same subjective sound level. At a frequency f of 1,000 Hz, the phone scale and the dB scale are identical by definition. The isophone at a level of 4 phon represents the acoustic threshold, that of 130 phon an intolerable (painful) volume. Phone values are obtained by subjective search for the *same volume* at *different frequencies*. The **sone** scale which is also obtained subjectively, describes how many times louder or less loud (**loudness**) a sound is perceived at the *same frequency* if compared to a standard sound of 1000 Hz and 40phon (= 1 sone). 3 sones means tripled loudness, 0.5 sone halved loudness etc.

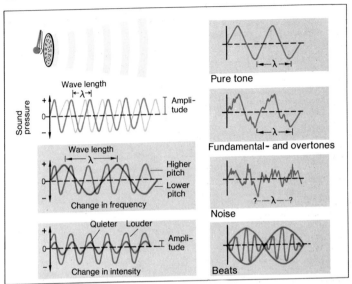

A. Characteristics of sound

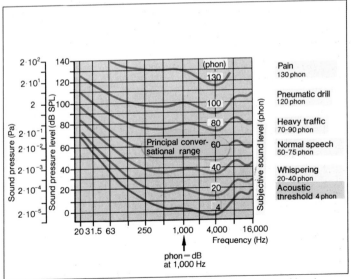

B. Sound pressure, sound pressure level, and subjective sound level

Sound Conduction and Detection

Sound pressure waves are transmitted through the air along the **external auditory canal** to the tympanic membrane. From this point, sound is transmitted by the bones in the **middle ear** to the **oval window**, where the inner ear (labyrinth) begins. In the **inner ear** sound is transmitted through fluid.

Ossicular conduction. Vibratory motion in the **tympanic membrane** is transmitted to the small bones of the middle ear, the **malleus, incus,** and **stapes** (→ A) in sequence. The shape and linkages of these bones are such that they produce an advantage in transmitting sound from the tympanic membrane to the **oval membrane** of the inner ear. The main function of the small bones is to match the resistance (impedance) of air for sound waves to that of the inner ear. This matching is brought about mainly by transformation of the pressure (force/area) because area of the oval membrane is more than 20 times smaller than that of ear drum, and force does not change. Ossicular conduction improves hearing acuity by 10–20 dB. Without the small bones there would be greater reflection of the sound waves at the oval membrane. Therefore, *direct* conduction of sound from air to the inner ear is of low efficiency. The linkages between the small bones can be influenced by two muscles, the tensor tympani and the stapedius muscle. **Bone conduction** allows transmission of vibrations through the skull to the inner ear. It is not of physiological importance, but serves for testing hearing acuity.

The **cochlea** is a coiled tubular labyrinth in the inner ear. In man, it is 35 mm long and takes $2^3/_4$ turns. Two membranes divide it along its length into three chambers: in descending sequence, these are the **scala vestibuli** (SV), separated by Reissner's membrane from the **scala media** (SM), which is separated by the basilar membrane (BM) from the **scala tympani** (ST). The SV is continuous with the

oval window, which contacts the stapes. The ST ends in the *round window*. SV and ST contain perilymph and connect with each other through the *helicotrema*.

Sound vibrations are transmitted through the oval window. The oval and round windows equalize perilymph pressure; when the oval window bulges in, the round window bulges out. The SM contains endolymph and does not connect with the SV and ST. The auditory receptor cells, located in the **organ of Corti** on the BM, are hair cells in two rows; they are surmounted by the **tectorial membrane,** in which the cilia are embedded. In man, there are c.224.000 receptors and 28.000 fibers in the auditory nerve. (For function, → pp. 300–303).

Hearing acuity is tested by the **audiometer**: through earphones, pure sounds are presented in varying frequency and intensity. The threshold intensity for each frequency is compared with normal threshold. Hearing loss is defined for each frequency separately. Hearing loss is measured in dB (*not* in dB SPL, because in audiometry 0 dB is the normal threshold at *all* frequences (compare with p. 297, B). The audiogram provides a measure of the degree of deafness and the frequency range that is affected. Hearing loss may occur from inflammatory disease of the middle or inner ear, from pharmaceutical agents, and from increasing age (presbyacusis). Hearing loss may develop because of a defect in the transmission sequence from outer to inner ear (conduction deafness), from a defect of the organ of Corti (deafness of the inner ear), or from a neural defect in the acoustic pathway (nerve deafness). The first type of deafness can be distinguished from the latter two by a number of tests such as Weber's test: a vibrating tuning fork is placed in contact with the vertex of the skull in the midline. Since bone conducts sound and bypasses the middle ear bones, the tuning fork will sound louder in the diseased ear in conduction deafness because the sound is not masked by environmental noise. In inner ear and nerve deafness, the fork will sound louder in the normal ear. When both ears are normal, there is no difference in perception by the two ears.

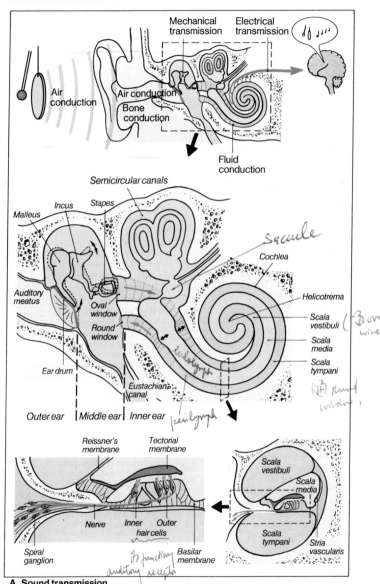

Mechanical transmission

Electrical transmission

Air conduction

Air conduction

Bone conduction

Fluid conduction

Semicircular canals

Incus

Stapes

Malleus

Saccule

Cochlea

Helicotrema

Auditory meatus

Oval window

Round window

Scala vestibuli (oval window)

Scala media

Scala tympani

(round window)

Ear drum

Eustachian canal

Endolymph

Outer ear | Middle ear | Inner ear

perilymph

Reissner's membrane

Tectorial membrane

Scala vestibuli

Scala media

Nerve

Inner Outer hair cells

Stria vascularis

Spiral ganglion

Basilar membrane

Scala tympani

auditory receptor

inf. colliculi

med geniculate body

auditory cortex

A. Sound transmission

Sound Processing in the Inner Ear

Sound waves are conducted to the inner ear via vibration of the foot of the stapes, which is in contact with the oval window (→ p. 299). Vibration of the oval window is transmitted to the perilymph of the scala vestibuli (SV) and of the scala tympani (ST). These vibrations set up a volume displacement in the perilymph, which is transmitted to the cochlear duct or scala media (SM) and its adjacent membranes. Thus, the vibrations are short-circuited from SV through SM to ST. The *basilar membrane* (BM) is relatively stiff at the base of the cochlea but becomes broader and more flexible as it approaches the apex at the helicotrema. Vibrations of the BM are transmitted along its length, but they decrease as they approach the apex, becoming more and more *damped* as the stiffness of the BM decreases. The higher the frequency, the greater the overall attenuation. Up to 200 Hz, the entire SM is in motion. Above 200 Hz, less than the entire SM will be in motion; the higher the frequency, the smaller the segment in motion. These movements of the SM correspond to a wave that travels along its length. As the **traveling wave (→ A)** moves up the cochlea, its amplitude increases to a *maximum* and then quickly falls off. The distance from the oval window to the point along the BM where the maximum amplitude appears is characteristic for the frequency of the sound that initiates the traveling wave (*place discrimination*). The form of this standing wave is identical and reproducible for identical stimuli. Vibrations of high frequency produce their maxima near the oval window; lower frequencies reach their maxima at greater distances from the oval window (→ B). In this way, each sound frequency generates a traveling wave with a maximum amplitude at a characteristic specific distance from the oval window. The nature of the stimulus thus determines the characteristics of the traveling wave, which produces a characteristic pattern of excitation of receptors along the BM. The system analyzes sound both by **spatial** (vibrational) and **temporal** sound pressure changes patterns. Low frequencies are discriminated mainly by sound pressure changes. Above 120 Hz spatial discrimination becomes more and more important.

The traveling wave sets the scala media including the *tectorial membrane* (TM) in vibration (→ C). The TM is in contact with the outer and inner hair cells of the organ of Corti and encloses a space that is separated from the endolymph. When the BM is depressed, the TM produces a shearing motion across the ends of the **hair cells (→ C)**, the actual receptor cells of the inner ear. The hair cells are so sensitive that they can detect motion of their hairs as small as 10^{-10} to 10^{-12} m, a distance smaller than the diameter of a hydrogen atom. An adequate stimulus in the hair cells generates a *receptor potential* which in turn starts an *action potential* for transmission along fibers of the acoustic nerves, which are in contact with the receptors.

The hair cells produce a receptor potential. The **cochlear microphonic potential** is generated in linear proportion to the displacement of the BM when a hair is bent. The cochlear microphonic potential is the sum of the single receptor potentials (→ p. 254). The endolymph has a much higher K^+ concentration and a lower Na^+ concentration than the perilymph, the latter being similar in composition to the ECF. These gradients are brought about by active transport mechanisms in the *stria vascularis* (→ p. 299, right, bottom) and create a *positive* potential in the endolymph of c. +80 mV relative to ECF (→ p. 303, C). Like other cells the hair cells have a resting potential of some 70 mV (inside *negative*). The sum of these two potentials (150 mV) is an extraordinarily high electrical driving force for ions. Even very small changes in the conductivity of the hair cell membrane, probably initiated by a shearing action on the hairs, create a receptor potential.

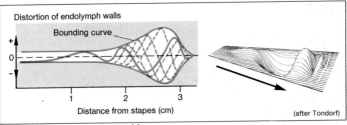

Distortion of endolymph walls

Bounding curve

Distance from stapes (cm)

(after Tondorf)

A. Traveling waves in the cochlea

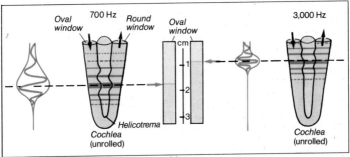

Oval window   700 Hz   Round window   Oval window

Helicotrema

Cochlea (unrolled)

3,000 Hz

Cochlea (unrolled)

B. Frequency identification in the cochlea

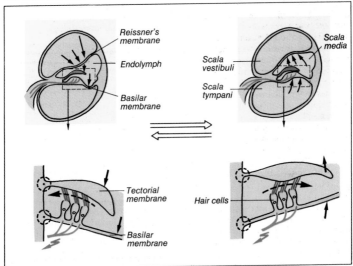

Reissner's membrane

Endolymph

Basilar membrane

Scala media

Scala vestibuli

Scala tympani

Tectorial membrane

Hair cells

Basilar membrane

C. Excitation of hair cells by a shearing action on their hairs

Central Processing of Acoustic Information

Central processing analyzes and interprets encoded acoustic information for (1) frequency, (2) sound pressure, (3) direction, and (4) distance.

1. **Frequencies** are detected in the organ of Corti in the cochlea ($\to$ pp. 298–301) and are conducted in individual fibers of the acoustic pathway. Central decoding of the fundamental frequency with its overtones leads to characterization of the sound generator and permits identification of individual voices. The capacity to distinguish different frequencies is determined by the discrimination threshold ($\to$ p. 286). A normal ear can distinguish readily between 1000 and 1003 Hz, a difference of 0.3%.

If two tones of nearly the same frequency (and a similar loudness) are heard at the same time, a variation in loudness of a frequency equivalent to the *difference* of the frequence of the two tones is heard: *beats* ($\to$ p. 297, A). This is often used by musicians in tuning their instruments.

2. **Intensity** or loudness discrimination is much cruder. A difference in loudness can be detected only after a 10% change in sound pressure. This value corresponds to a change of c. 1 dB. Increased sound intensity increases the number of action potentials or spikes produced in acoustic fibers and stimulates adjacent fibers to contribute to transmission of information: *recruitment* ($\to$ **A**).

In inner ear deafness the acoustic threshold is high and frequence discrimination is impaired, whereas the recruitment of lateral neurons is not disturbed. Therefore, at high sound intensities, perception of loudness is normal in these patients. This fact is used for discriminating inner ear deafness on the one hand from neural and conduction deafness ($\to$ p. 298) on the other hand. In the latter patients recruitment is much less because the sound does not reach the inner ear at adequate intensities; in neural deafness recruitment also is poor because the defect of central processing also includes neurons necessary for recruitment.

3. **Direction** of a sound can be detected by several cues: (a) Sound originating to one side of the midline reaches one ear later than the other. A 4° difference produces a time lag of 10^{-5} s between the ears ($\to$ **B**). (b) For continuous tones, there will be a phase difference between the ears. (c) The head will cast a shadow and sound intensity will be attenuated in the farther ear. A less intense sound takes longer to generate an action potential. This increased latency time allows central interpretation. Differences in time or phase are more important for low frequencies (under 1400 Hz), while differences in intensity are important for high frequencies. When these cues are insufficient, the head will be turned to change the orientation angle and to increase the contrast of the binaural information input.

4. **Distance** of a sound source is detected by the content of high frequencies. As sound travels, it loses intensity; the loss is more pronounced in the higher frequencies. Thus, the greater the distance traveled, the less the content of high frequency components.

The synapses of the acoustic pathway ($\to$ **D**) are: organ of Corti, spiral ganglions ($\to$ **D, 1**), anterior ($\to$ **D, 2**) and posterior cochlear nucleus olive ($\to$ **D, 4**), accessory nucleus ($\to$ **D, 5**), lateral lemniscus ($\to$ **D, 6**), inferior colliculus ($\to$ **D, 7**), medial geniculate nucleus ($\to$ **D, 8**), and primary acoustic cortex ($\to$ **D, 9**). In the cortex, sound frequencies are separately represented, as if the cochlea had been unrolled on the cortex. Discrimination and identification of timber (overtone components) take place in the acoustic centers. In the auditory cortex spezialized neurons can be found which are stimulated only at the onset or at the end of an acoustic event; others react only to long-lasting or repeated sounds, to noise, to binaural or monoaural stimuli. Efferent nerves also influence the acoustic pathway. Their significance, however, is not well understood. Destruction of the temporal cortex does not cause deafness; its function is not the recognition of sound but the identification of sound patterns and sound localization.

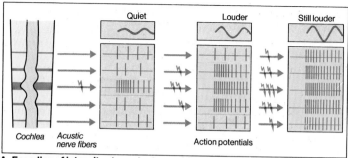

A. Encoding of intensity at constant sound frequency

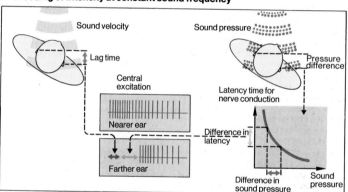

B. Stereophonic hearing

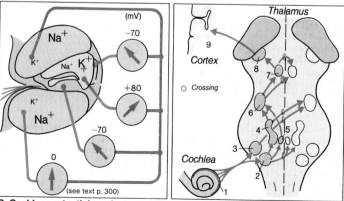

C. Cochlear potentials and electrolyte distributions

D. Afferent acustic pathways

Voice and Speech

The human voice is primarily an organ of communication that augments the information received by the acoustic organ, the ear. The voice is a wind instrument (reed) (→ **A**). Air from the lungs is driven past a narrow slit in a double reed (vocal cords); vibration of the reed produces sound, the character of which is affected by a resonance chamber (chest and oronasal cavity). The sounds of speech can be influenced, as in any wind instrument, by velocity of the air column, tension of the vocal cords, shape of the slit, and shape of the resonance chamber.

The joints and muscles of the larynx adjust the position and tension of the vocal cords. Normally, the gap between the resting cords during respiration is c. 8 mm posteriorly, halfway between full adduction and abduction. The voluntary efferent signals arise from the precentral motor cortex (→ p. 264) and reach the nuclei of the vagus (X), which is directly responsible for motor and sensory control of the larynx. The superior laryngeal branch supplies the upper larynx (sensory) and the cricothyroid muscle. All other muscles are supplied by the recurrent laryngeal, which also sends sensory fibers to the lower larynx. The sensory stimuli are important in protective reflexes (coughing) and for voice production. Afferent fibers in the mucus epithelium and in the muscle spindles (→ p. 258) of the larynx continuously send signals centrally describing the position and tension of the vocal cords. There are also close associations via the acoustic pathways with bulbar and cortical speech centers, which are essential for fine control of the voice. These primary speech centers are influenced by secondary centers (e.g., Broca's center), which affect the content of speech. Failure of the secondary speech centers leads to *motor aphasia* even though all of the motor pathways remain intact. Failure of secondary *acoustic* centers (e.g.,

Wernicke's center) leads to a loss of understanding of speech (*sensory aphasia*). Normally, a person hears his voice c. 1 ms after he speaks a word. If this acoustic feedback is delayed as much as 0.2 s, stuttering may occur. In deafness, speech is defective because the feedback is interrupted. Feedback is operative in singing: when a note is held, the voice approaches the desired frequency in a series of progressively smaller oscillations above and below the desired pitch.

Voice production. When the vocal cords are set to vibrating by a column of air, the cords are not simply brought together (adduction) but move vertically in a "rolling" movement in the same direction as the air stream (→ **B**). In the production of low-frequency sounds, the cords are closed longer than they are open (5 : 1 at 100 Hz); for higher frequencies, they are open longer (1 : 4 at 400 Hz); in singing with head voice (falsetto) or whispering, they remain open.

The frequency range of the human voice is from c. 40 to 2000 Hz. Sibilants (s, z) are high-frequency sounds that are poorly reproduced in radio and telephone. The tonal range of speech is about one octave (→ **C**), of singing two octaves, although some singers can span three octaves.

Vowels, even though they have a similar basic frequency (100 to 130 Hz) (→ **D**), can be differentiated by their overtones (*formants*). The three "primary" vowels ɑ:–i:–u:make up the vowel triangle;œ: ɔ:, ø:, y:,æ: and e: are the intermediates within this framework.

The phonetic notation used here is that of the *International Phonetic Society*. The symbols mentioned here are as follows: ɑ:as in glass; i:as in beat; u:as in food; œ: as in French peur; ɔ: as in bought; ø:as in French peu or in German hören; y:as in French menu or in German trüb; æ:as in bad; e:as in head.

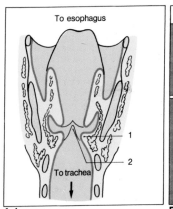

A. Larynx

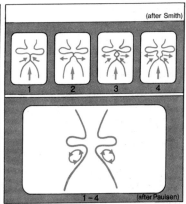

(after Smith)

(after Paulsen)

B. Motion of vocal cords

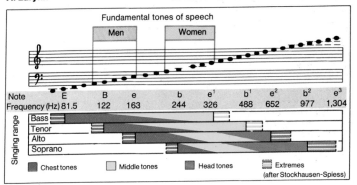

Fundamental tones of speech

Men

Women

Note	E	B	e	b	e¹	b¹	e²	b²	e³
Frequency (Hz)	81.5	122	163	244	326	488	652	977	1,304

Singing range

Bass
Tenor
Alto
Soprano

Chest tones   Middle tones   Head tones   Extremes

(after Stockhausen-Spiess)

C. Vocal ranges

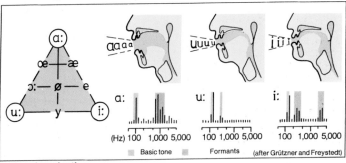

(Hz) 100 1,000 5,000   100 1,000 5,000   100 1,000 5,000

Basic tone   Formants   (after Grützner and Freystedt)

D. Vowel production